The Winning of The Carbon War

THE WINNING OF THE CARBON WAR

POWER AND POLITICS ON THE FRONT LINES OF CLIMATE AND CLEAN ENERGY

JEREMY LEGGETT

This first edition published 2016 by Jeremy Leggett.

Colophon

Set in Minion Pro, 11pt on 14pt leading.

For Aki

Humanity is in a race, a kind of civil war.

On the light side the believers in a sustainable future based on clean energy fight to save us from climate change.

The dark side defends the continuing use of fossil fuels, often careless of the impact it has on the world.

Jeremy Leggett fought for the light side for a quarter of a century as it lost battle after battle. Then, in 2013, the tide began to turn. By 2015, it was clear the the war could be won.

Leggett's front-line chronicle tells one person's story of those turnaround years, culminating in dramatic scenes at the Paris climate summit, and what they can mean for the world.

"Given how vital developments in energy and climate will be for the future global economy, a front-line chronicle of events as they unfold in the make-or-break year ahead promises to be fascinating. With a past in oil and gas, a present in solar and finance, and two decades of climate campaigning along the way, we can rely on Jeremy Leggett to provide an interesting and insightful perspective."

Richard Branson

Founder, Virgin Group

"Jeremy's clear and penetrating eye-witness account of the Paris climate summit captures beautifully the story both of the historic agreement reached and of the understandings on which it is based. It is great fun to read; kind, tough, and often very funny."

Professor Lord Nicholas Stern

Chair of the Grantham Research Institute, London School of Economics

"This is one of the first books on climate change that I truly could not put down. A brilliant page turner, alternately daringly funny, coldly sobering, starkly terrifying. A deeply rewarding look into the most significant problem we face today – poetic, personal, beautiful and intensely urgent. If you want to understand why fossil fuels may not just tank the environment but might also bring down the world economy, read this book, and – for all of our sakes – read it fast."

Josh Fox

Emmy-winning film director

"Leggett has a unique vantage point that he shares frankly, openly and with candid humor and humility. I don't tend to gush about books on climate politics, but I can't recommend this resource highly enough. I've been eagerly reading what actually comes across as a thriller ....and can't even begin to list all the insights gleaned so far."

Sami Grover

Writer and Creative Director, *The Change Creation*, on *treehugger.com*

"If you are interested in @COP21 and are not reading Winning the Carbon War by @JeremyLeggett you should be"

Tom Carnac
Senior Advisor to the Executive Secretary of the UNFCCC

"In all my years as an energy analyst in investment banking, I have rarely encountered an entrepreneur with the factual grounding Jeremy Leggett has across the full spectrum of the energy and climate space. From his beginnings in the oil industry to his leadership in campaigning for new clean, green energy, he has a vital message that deserves as wide an audience as possible. We see the signs of what he has predicted beginning to unfurl around us with severe ramifications for the future of the oil industry and fossil fuels."

Mark Lewis
Head of European Utilities, Barclays

"Jeremy Leggett has been right about climate change longer than just about anyone, and he's consistently thought of innovative ways to get the message across–here's one more powerful medium for helping spread the most important message ever."

Bill McKibben
Founder, 350.org

"One of the crucial themes in Jeremy Leggett's narrative is shale, on both sides of the Atlantic. As a long-time opponent of fracking, it is fascinating for me to read Jeremy[1]s perspectives, particularly given his past life in the oil industry where he has researched shale funded by both BP and Shell – and his current generous sharing of his expertise with British parliamentarians of all political parties."

Caroline Lucas MP
Green Party of England and Wales

"An unusual book by an author with an unusual professional life. I do not know any person who brings together intimate knowledge of relevant fields the way Leggett does. He has not only studied the climate problem scientifically but seen the innermost workings of multilateral climate policy. He also has inside knowledge of fossil energy, especially the oil industry and the prevailing corporate culture. Purely academic scholars lack this expertise."

Prof. Jochen Luhman

Wuppertal Institute for Climate, Environment and Energy, Germany

"I found it entertaining, informative and almost impossible to put down, like a James Bond adventure it puts you at the heart of the action in the main theatres of the carbon war—US, China, Russia, Germany, Saudi Arabia, and Africa—in solar factories, the Bank of England, World Economics Forum, Occupy camp in London... It's a real eye opener and a colourful read that makes a complex subject easy to digest."

Nicci Talbot

Huffington Post

Contents

Note on the project

The book is my personal account observing the events of the carbon war from May 2013 to the United Nations Conference on Climate Change in Paris in December 2015. It was published online in monthly instalments from 1st March 2015. Each episode republished an edited full version of the book, incorporating suggested changes and corrections received since the previous version. The final edition, published as both an eBook and paperback, has benefited from more than a year of crowd-sourced editing.

Sources can be located on my website, www.jeremyleggett.net using the word search facility in the Chronicle of the Carbon War.

Aiming as I am for as wide a non-expert readership as possible, I have tried not to write a technical book. There are plenty of those in the climate and energy field. As an ex-academic I was often tempted to go into further detail to help make my points. But as a writer I knew that would quickly make the book dense and easy to put down. In the more detailed chronicle on my website are links to vital work by others that expands on the relevant technical arguments.

The diary extracts recount real events and conversations, but I made no tape recordings at them, so dialogue is from memory, usually written up immediately after the scenes described. I vouch completely for the sense of the dialogue, but cannot obviously vouch for the exact words. I also use extracts from conversations, not the whole. When making these unverifiable quotations, I do not use quotation marks although I take nothing out of context. The same applies to speeches, press conferences and the like. If I have to rely on my slow note taking, I do not use quotation marks. If there are transcripts or quotes available from reliable media outlets, or I am 100% certain of short sentences, I use quotation marks.

Thanks

Many people have read sections of this book to help me iron out glitches. You are too many to mention, but you know who you are, and you know you have my sincere appreciation. I acted on many but not all of your suggestions. Your feedback convinces me that crowd-sourced editing is an immensely powerful literary tool. That said, nobody but me has responsibility for any mistakes that might have slipped through.

Chapter 1

We have good quality assets

Shanghai, 2nd May 2013

I throw open the curtains in my room at the Peace Hotel and look across the Huangpu River at the Financial District. The skyscrapers leap into a clear blue sky through the faintest yellow haze.

Good. The air pollution will be low today. A wind must have blown across the Yangtze delta overnight. Maybe I'll only be forced to smoke the equivalent of a few cigarettes on this trip, rather than whole cartons.

Air pollution has broken all records in China recently. In January, the PM2.5 levels – particulate matter of diameter 2.5 micrometres or less, the deadly stuff that lodges deeply in the tissues of the lungs – reached almost 1,000 micrograms per cubic metre, turning the whole city into the equivalent of an airport smokers' lounge.

Coal burning in power plants and oil burning in vehicles does most of the damage. I first looked out of the windows of this hotel in the early 1980s. The view was single-storey then, and bicycles cruised the Bund – the boulevard on this side of the river – in their thousands. Now, after 30 years of an economy growing at an average rate of over 10% a year, the skyscrapers soar to more than a hundred floors, and cars clog the Bund. Coal barges plough the Huangpu, their cargoes – the primary fuel of China's spectacular economic growth – black humps on their decks.

But concern is rising amid the tangible damage being created by China's emulation of the 20th-century model of development. That is one reason I am here in the city today.

I shower, slap my jet-lagged face in the mirror, slug a coffee, go downstairs to meet my hosts. I wait for a while in the marbled lobby, squinting to imagine how this palatial hotel must have felt in its heyday, back before the Second World War. It doesn't take much mental conjuring. The hostesses still dress in full-length coloured silk, the suits on the business travellers wouldn't have looked much different. Broader lapels, maybe.

The hosts arrive, apologise for their delay: the traffic is terrible today. Words I hear in most every city I travel to these days.

We drive in a people carrier, stop-start for an hour to the district of Fengxian. Here, in an industrial park, built on scrubland, we pull into the parking lot of one of JA Solar's six factories.

I am taken for tea and introductions before touring the facility.

Accompanied by a small group of executives – Chinese bosses and expatriate European managers – I don a lab coat, a hairnet, and plastic covers for my shoes. I look ridiculous, but at least my dandruff won't be adversely affecting product quality.

The doors to the plant swing open, and I see before me a "gigawatt fab", meaning a factory capable of manufacturing a thousand megawatts of solar panels per year.

I have been visiting solar factories around the world since I set up my company, Solarcentury, in 1999. This is the first time I have seen one capable of a gigawatt of annual production. The machines stretch in ordered rows many football pitches into the distance. Hundreds of workers, dressed just like the touring party, attend them. A thousand people work under this roof, in alternating shifts, 24 hours a day, 7 days a week.

I have been interested in solar energy since 1990. Back then, the total annual global market for installed solar photovoltaic panels was less than 50 megawatts a year: some 5% of the capacity of this single Chinese plant today. The market was mostly limited to off-grid electricity generation, and consumer applications. In the decade of the 1990s, the global market grew at an average of rate of 20% per year, pulled along by a market-building programme in Germany involving subsidies for 100,000 solar roofs, and a 70,000-roof equivalent in Japan. By 2000, when Solarcentury was installing its first solar panels on British barns, the global market was still only around 300 megawatts a year. Then, Japan dominated the manufacturing.

Now? China joined the game only in the noughties. And what a difference it made. The average annual growth of the global market leapt to 52%. By 2010 the global market was about 20 gigawatts – 20,000 megawatts – a year.

Sometime in 2012, the total installed capacity of solar photovoltaic panels in the world crossed 100 gigawatts of peak power. This is equivalent to 85 gigawatts of power, or 65 full-size 1.3 gigawatt nuclear reactors.

Having completed the factory-floor tour, I sit with the Chinese executives in a sterile conference room, talking the language of our business: prices of modules, state of the national markets in which we operate, scope for partnership, current market problems, and so on.

I wonder what they think of my motivation in all this: a conviction that humankind must stop burning fossil fuels if we are to survive climate change, and that solar energy is a vital tool in the mixed arsenal we will need if we are to make it happen in the years to come. It is widely understood by my peers that I wouldn't be playing the businessman otherwise.

I'm pretty sure I wouldn't like to know what my Chinese business partners think. But then, I reflect, they don't have to be as worried as me about climate change to be allies in the cause. They just have to be worried about the air their children breathe.

London, 19th June 2013

The thing about global warming, and the climate change it causes – the thing that really gets me about it – is the insidious scope for a point of no return. As greenhouse gases pour into the atmosphere, mostly from the burning of fossil fuels: coal, oil and gas – the global average temperature rises decade by decade. As the global average temperature rises more greenhouse gases are released from the warming planet: from drying soils and forests, melting permafrost, and so on. At some point, potentially, we won't be able to stop the warming by cutting the burning of the stuff that started the process. "Warming" becomes a singularly inappropriate term then.

That is why governments have been trying to negotiate an effective treaty on climate change since 1990. That is why they have pledged to keep the level of global warming below 2°C: a ceiling that many climate scientists profess gives us an acceptable chance of avoiding the more extreme dangers of a "warming" planet, such as a runaway greenhouse effect.

In 2011, two old friends, financial analysts Mark Campanale and Nick Robins, resolved to try and bring the financing of fossil fuel burning into line with the professed aspirations of international policymakers. They called themselves Carbon Tracker, and invited me to be their chair. I accepted with delight.

The first Carbon Tracker report, in 2011, set a very stark target. To have a reasonable chance of keeping global warming below the intended two-degree ceiling, fully 80% of existing fossil fuel reserves would have to stay in the ground unburned. The second Carbon Tracker report, published in May 2013, looked at the capital expenditures fossil fuel companies were making each year to develop existing reserves and find new reserves to add to the "carbon bubble", as we called it, of already-unburnable carbon. That sum, $670 billion, was in danger of being wasted, we argued. Currently all reserves are accounted on the books of fossil fuel companies as though they are assets at precisely zero risk of being stranded. That can't be correct in a world where around 200 governments are engaged in negotiations to try and find a way of slowing the burning. Or busy trying to cut air pollution, for that matter.

These arguments, couched in the language of the capital markets and written by financial experts as they are, have begun to hit targets. Investors, politicians and journalists are seeking briefings. So too are the fossil fuel companies themselves.

BHP Billiton, one of the largest global mining companies, is one such. The Carbon Tracker team knows these sessions are a kind of "weigh your enemy" exercise on the part of energy-incumbency companies. But they are also a two-way learning experience, so we generally agree to do them, notwithstanding the bandwidth challenges they present us with.

It is clear why BHP wants this meeting. On June 1st, a Bernstein Research report spoke of "the beginning of the end of coal". It concluded that the "once unthinkable" was now in sight: declining Chinese demand by 2016. Air quality concerns were a part of this, but so were other factors, including the rise of renewable energy.

Three days later, mining giant Rio Tinto, BHP's major rival, announced it was seeking to get rid of $3 billion worth of Australian coal assets. It no longer fancied its chances of selling the product in China, its main market.

Then last week the International Energy Agency released a report suggesting that the need for a new global climate change agreement is now urgent. Waiting until 2020 is not an option, because doing so would mean an end to hopes of limiting global warming to less than two degrees. To have a good chance of staying below two degrees, the IEA says, 60 to 80% of coal will need to stay in the ground.

Today Carbon Tracker's core team is on the other side of the Atlantic. Founder Mark Campanale is in Rio de Janeiro, talking to the Brazilian stock exchange. Research Director James Leaton is on Wall Street, talking to US investors. The Chairman is standing in for them.

In the BHP Billiton boardroom in a tower block near Victoria I deliver Carbon Tracker's stump 15-minute position statement to four senior executives of the company.

The body language among the mining execs is not good. They shift in their seats as I set out the case. They are Australians, forced to listen to a Pom. Experience tells me that this contact session will be less polite than most.

When I have finished, the senior exec allows hardly a space after my final full stop.

Well mate, he says, his accent thick. We have good quality ("qualidee") coal assets, close to market. And we also have gas resources, unlike most of our competitors. We're well diversified. So where is our problem exactly, d'ya reckon?

I look at him, attempting a poker face, processing this.

His response is consistent with what we are finding with other companies. Their first line of defence is not to question the carbon arithmetic of 80% unburnable reserves, 20% burnable. It is to argue that their reserves are in the 20%.

You may well be right, I say. But we would need to do a deep-dive into data, company-by-company, to know for sure, would we not?

And that's what we intend to do at Carbon Tracker, in the year ahead, I add. We will be providing data to help your investors look company by company. Fossil fuel by fossil fuel. Fossil fuel species by fossil fuel species. Fossil fuel project by fossil fuel project. Fossil fuel province by fossil fuel province.

Meanwhile, I continue, we have noticed an interesting thing, in presentations like this. The first line of argument tends not to be a defence on behalf of your entire industry, but one specific to the company – that your particular set of assets can be burned safely.

You can't all be right, can you?

London, 2nd August 2013

In an appropriate office in the citadel on Threadneedle Street, Mark Campanale and I sit drinking tea with Andy Haldane, the man responsible for the stability of the capital markets. He is Deputy Governor of the Bank of England, and chair of the Financial Policy Committee, the body set up by the Bank, in the blinding hindsight of the financial crisis, to scan the horizon for potential future shocks to the financial system. Mark and I are trying to persuade him that potential over-valuation of the fossil fuel companies that are the backbone of the stock markets might be a candidate for scrutiny by his committee. We are to be considerably helped by a late addition to our team, Saker Nusseibeh, Chief Executive of Hermes Fund Managers. He is a man who knows about asset valuation: he leads the management of £26 billion worth of them not far from here.

We know enough of Andy Haldane from a previous meeting, by reputation, and from reading his speeches, to suspect that he has an open mind. We are also confident he will take us seriously, and not just because of our eminent draftee. On July 2nd Norwegian pensions and insurance giant Storebrand announced it is excluding nineteen fossil fuel companies from its investment portfolio, specifically because of carbon asset-stranding risk as described by Carbon Tracker. The exclusions, of coal and tar sands companies, cover all of its funds and investment vehicles. This is not an ethical move, the company made clear. It is a search for longer-term, more stable, returns on investment.

Meanwhile, a growing movement is arguing on campuses and in city halls around America that ethics are a perfectly valid imperative to stop investing in fossil fuels. The campaign group 350.org, set up by Bill McKibben and others in 2008, began a divestment campaign after a landmark article written by McKibben in Rolling Stone magazine, based on Carbon Tracker's 2011 report. Given the way fossil fuels stoke climate-change, 350 argues, you should divest whatever the returns on fossil fuel investments happen to be.

Saker Nusseibeh leads off the discussion. He leads an informal club of senior City of London figures who worry that the financial-services industry is not learning enough from the crisis of 2008. We made such a bad job, collectively, of misvaluing assets in the sub-prime mortgage saga, Saker says. And now I share Carbon Tracker's view that the industry might be in danger of repeating the same mistakes in energy.

The problem I have, Andy Haldane tells us, is that I need hard evidence that there might be systemic risk to capital markets. The Bank is seeking that by adding a question on fossil fuel asset stranding in the questionnaire we send regularly to financial institutions seeking their perceptions of risk in the markets. The responses show that financial institutions do not perceive much or any risk.

I let Mark do the talking. He grows more impressive with every meeting I accompany him to. He is of Italian ancestry, and I know him to be a man of passion and mischievous humour. Yet he radiates analytical cool in these meetings.

The problem you have, Mark tells Haldane, is that they would say that, wouldn't they? If you had asked the same question about mortgage-backed securities in the run up to the credit crunch they would have said the same. No-one was collecting any data on Collateralised Debt Obligations, and all the other instruments subsequently discovered to be toxic, and as a result the risk was only material in hard evidence once Lehman Brothers collapsed.

Carbon Tracker has some ideas for data the Bank could collect, in the case of carbon-and-climate risk. You could for example look at all the loans made through the UK banking system for fossil fuel projects. You could easily measure the amount of embedded carbon dioxide held in the reserves of the largest publically traded companies on the London Stock Exchange. You could ask whether this amount breaks the budget of carbon implicit in the collective climate-policy goals of governments at the ongoing international climate talks. You could check whether it is growing, and if so by how much. You could ask if the risk of fossil fuel asset devaluation is being transferred on to UK pension schemes. All this data collection you could think of as a 'no regrets' thing to do. If you just ask companies to provide this kind of information, you could be sure they would start monitoring it themselves, in parallel with any collation your own team would do here in the bank. Then they might all have a better chance of taking a view on whether a carbon bubble is emerging, and if so what might be done to deflate it sustainably.

Haldane begins his response to our case by reminding us that concerns about the aftermath of the financial crisis are far from over. The Bank of England remains troubled by many aspects of the episode, he says. Risk-taking by banks remains troublesome and it is far from unimaginable that there could be a recurrence of some kind of crisis.

I watch him as he carefully picks his words, and can't help but feel empathy. He bears a huge responsibility. The British Banking Commission, set up by UK Chancellor George Osborne, has recently concluded that bankers have escaped meaningful constraint to date, and that the culture and antics that spawned the financial crisis of 2008 continue. Its recommendations include the jailing of any bankers found guilty in the future of "reckless misconduct". They also favour forcing bankers to wait up to ten years for their bonuses, so that their short-term gambling might be reined in.

As for collecting the data you suggest, Haldane continues, given all the other priorities of the Financial Policy Committee, it is difficult to decide whether such actions would be helpful or unhelpful. Perhaps we should direct our suggestions at others in the first instance, he suggests. The ratings agencies might be top of the list. If one of them were to downgrade an investment or two based on particularly marked overexposure to the kind of fossil fuel asset-stranding risk we describe, that could really trigger scrutiny of carbon-asset risk in the markets.

Given how pitifully the ratings agency performed during the financial crisis, I think to myself, what chance is there of that? They gave triple-A ratings to whatever piece of toxic junk the investment banks floated past them.

But this, I suggest to Haldane, is the kind of thing everyone is saying, right across the financial sector. The accountants say it is not for them to move first on insisting fossil fuel asset-stranding risk be recognised in company accounts. The brokers say it is not for them to move first by insisting IPO prospectuses discuss the risk. And so on. Everyone is looking at everyone else to make the first move. And nobody does. With respect, you are the regulator. A little action from you would go such a long way in lifting the risk blindness. What we fear here, in essence, is that the energy industry is in the process of doing much the same thing that the banking industry did in the run-up to the financial crisis. They are overstating the value of their assets, and clocking up huge debts to bankroll a growing portion of their operations that is uneconomic. Given the importance of that asset value on stock exchanges, on bank loan books and all the rest of it, there must at least be some case that they might be posing a threat to the capital markets.

Andy Haldane does not look convinced.

Chapter 2

The men behind the wire

Balcombe, 18th August 2013

In America, the shale boom is in full swing. Gas and oil are being sucked from rocks that were once thought unproductive, by the drilling technique known as fracking. Drillers have found a way to drive horizontal wells through shale horizons, force open the natural fissures in the rock and free hydrocarbons long thought irrecoverably trapped. As well as a high degree of drilling wizardry, large amounts of water and toxic chemicals are involved. For the oil and gas industry, and many people besides, this is an example of how innovative people find new ways to create wealth. Cheerleaders talk of an America en route to oil-and-gas self sufficiency because of the boom. "Saudi America" here we come, they are saying.

For a wearying minority, this is another example of one of those bubbles humans seem so good at creating: a sudden eruption of apparent asset value that spreads fortunes around for a short while and then deflates, once it becomes clear that the value is delusional. The doubters believe the delusion begins with the simple economics. The industry is losing money, we point out, even with the oil price as high as it is today. The shale boom might have created benefits for American industry, with the cheap gas and reduced oil imports that it has provided, but how can it be sustainable, when so many drillers are producing oil and gas at a total cost – including the cost of money borrowed to finance drilling – in excess of the price they can sell it for?

The oil and gas industry insists there is a way around this: some combination of the gas price going up and the cost of drilling going down. In any event, new prosperity can be created in multiple countries if they would but import the shale narrative.

In the UK, the British Geological Survey has announced that the country is sitting on shale gas deposits that could supply the nation for a quarter of a century. Chancellor George Osborne has duly unveiled the most generous tax breaks in the world for fracking. Big energy suppliers are pressing the government to support gas-fired power and shale exploration, warning that without growing gas the UK could face a shortfall in generating capacity. The Institute of Directors speaks of "a new North Sea".

Prime Minister David Cameron insists fracking will bring jobs and cheap gas to the UK. He intends to encourage that process. Local communities in fracking areas will receive lump sum cash handouts and a 1% share of revenues. He has hired an election strategist, Lynton Crosby, well known for promoting shale-gas companies as a lobbyist. His government has instructed planners to ignore protests at fracking sites when considering applications to drill. It even bans the planners from considering proposals from communities for renewables projects as alternatives to fracking.

The government minister keenest on shale is Chancellor George Osborne. His father-in-law, former Conservative government minister Lord Howell, was a gas lobbyist. Howell elicited gasps of astonishment in the House of Lords in July by recommending the industry frack in what he calls "the desolate north" of Britain. There, he reasons, any environmental problems won't matter so much.

Osborne, it seems, favours fracking north, south, east and west. On August 5th he went so far as to say that he will fight any backlash against fracking, even if it is in the Conservative party's heartland in the south.

I live in that Conservative heartland, in an area of southern England called The Weald. So it is today that I stand among thousands of people moving at shuffle pace down a leafy road in Sussex towards the drilling site that the oil industry hopes will be the first fracked oil well in the south of England. Many of the people protesting are clearly residents of this rich and rural county: conservatively dressed, devoid of the rucksacks, banners or other paraphernalia of the experienced green protestors.

The road, closed to traffic by police roadblocks, is lined with tents where protestors intent on a long stay are camping. Some tents serve as improvised market stalls, offering tea and food to the marchers. Much of the food has been donated, some from as far away as Manchester, in the desolate north.

From one of these tents, a man reclining in a camping chair observes me from behind the smiling face mask favoured by those who seek anonymity from the roving police cameramen. Next to me, an elderly couple march with their labrador on a lead, the lady with a bemused expression on her face. She has never been on a demonstration before in her life. Ahead of us, a young family strolls with a child in a pushchair. There are lots of first-time marchers, lots of dogs, lots of pushchairs.

There are also lots of police. A few days before this mass weekend march, the papers had carried extraordinary images of massed ranks of policemen marching ahead of a lorry carrying drilling equipment to the site. Had they not done so, it could never have passed the hundreds of early arrivals at the protest.

Is this what it comes to, in the fossil fuel endgame, I ask myself. The storm-troopers of what looks suspiciously like an embryonic police state effectively employed in the service of energy companies to force into place the apparatus needed to squeeze oil from the ground in ever more extreme ways?

I am nervous. There are reports of police snatch squads diving into the crowds, removing protestors, bundling them into vans and driving them off, ignoring questions from bystanders and journalists as to why. I also hear tales of police seemingly trying to provoke assaults by protestors, who are resolutely peaceful as far as I can see up and down the road, and as far as I have read in any newspaper.

All this is part of a pattern of policing in modern Britain that unnerves me more with each incident I see or read about. Campaigners are becoming convinced that the police are using extreme tactics in order actively to discourage ordinary citizens from protesting against the plans of the frackers to turn the garden of England into the morass of burning gas flares, toxic waste-water handling sites, and intense lorry movements that a typical shale-drilling operation requires in America. They are using random excess, as much as they think they can get away with, to persuade people to stay at home.

Because I believe this, I have had to force myself to go on this protest. I do not do so lightly. By placing myself in harm's way – a strange and sad thing to have to say about a police force – I am taking a non-trivial vocational risk. If I have the misfortune of being arrested, and convicted of some trumped-up offence, I will no longer legally be able to be a director of my own company.

I suspect that most people who profess to support fracking in the UK have no appreciation of what a North Dakota-type shale-drilling operation looks like, or the social impacts imposing it on the English countryside would have, if commercial quantities of gas or oil are ever found in British shale. When the drillers target oil, which can fetch a high price, they care not about gas, they simply flare it. These days the burning gas flares from the oil drilling in North Dakota shale light the area up so much at night that from space it looks like a huge metropolis. Even at low US gas prices, one calculation has put the value of the gas going up in smoke at more than a billion dollars.

Aerial photos of this and other hydrocarbon-rich shale regions, whether producing gas or oil, look like road maps of towns where the houses have been stripped away. The drill "pads" – the flattened areas where the rigs are sited for the actual drilling – are often within a few hundred feet of each other. The lorries carrying water and toxic drilling chemicals in, and gas and oil out, can extend nose to tail down roads never built for such intense use.

The oil and gas industry in America does a good job at suppressing bad news. Bloomberg journalists have reviewed hundreds of regulatory and legal filings and discovered a shocking if predictable thing: American gas frackers are buying out owners of homes polluted by past drilling and trying to gag them by making them sign agreements not to talk to anyone about their experiences. For example, the Hallowich family in Pennsylvania complained that nearby shale-gas drilling by three companies caused headaches, burning eyes and sore throats. The companies offered a $750,000 settlement provided the Hallowiches committed not to tell anyone. This kind of thing is going on from Wyoming to Arkansas and Pennsylvania to Texas.

The problems for the shale drillers involve more than human health. When the era of mass shale exploitation began in America in 2005, drillers succeeded in exempting the fracking process from the rules of the US Clean Drinking Water Act: an act of regulatory piracy that became known as the "Halliburton Loophole," after its architect, the drilling-services company once led by Dick Cheney, US Vice President under George Bush Jr. It became inevitable, the day that happened, that chickens would one day come home to roost. Now Pennsylvania regulators are linking gas and oil drilling with some 120 cases of water contamination between 2009 and 2012.

In the UK, aspiring frackers do their best to mask all this from the British public. Cuadrilla, the company doing the drilling at Balcombe, professes that their rig is merely drilling a conventional oil exploration well. They say they are doing what has already been done on numerous occasions in the south of England. Indeed, they say, they are on a site of conventional oil drilling abandoned in the 1980s.

But they are targeting shale samples 3,000 feet or so below the surface, to check if this site is suitable for fracking. There is no reason they would drill other than with an intent to frack.

The column of marchers reaches the drill site, which is hidden from the road. I see people pushing their way through bushes to get a view of the drill pad. Seeing no police in the immediate vicinity, I do the same.

The pad is surrounded by a tall wire fence. Cuadrilla has stopped drilling today. A few dozen site workers and security men stand around on the other side of the wire, regarding the protestors with blank faces. They look like yellow-jacketed inmates in some kind of prison camp.

I stand for a while staring back, imagining what the mindset must be, inside the wire. I know enough of oil-industry cultural machismo to have a good idea. It isn't a pretty picture.

Next to me, a man stands holding the hand of a small boy, his grandson it looks like. The boy stares wordlessly at the men in yellow. His guardian offers him no explanation of what he is looking at.

All things being equal, the boy can reasonably expect to live for more than a hundred years, the medical profession tell us, proud of their advances. But what will his century look like, if the men behind the wire have their way?

Geneva, 28th August 2013

Behind security gates worthy of an American embassy, the headquarters of the World Economic Forum look north from terraced gardens across Lake Geneva. On the far side, ordered Swiss countryside rolls into the distance. Inside, business leaders mingle with United Nations officials, eating a stand-up buffet lunch.

This place has long been regarded by many an environmentalist as the heart of darkness: the citadel where the business elite meets to polish its addiction to endless growth in a globalised economy on a finite planet, and to plot how to shield its consequent despoliation of the global environment from public and regulatory scrutiny.

Today, those environmentalists would be surprised. The subject of the plotting is cutting carbon emissions on a global basis, and seriousness of intent is in the air.

We file into a large meeting room, and sit around a square of tables, behind nameplates, in the style of a UN negotiation.

I look around the room and tell myself to feel encouraged. The World Economic Forum, the United Nations, and the Secretariat of the UN Framework Convention on Climate Change have launched a collaboration that is the first of its kind: a project to entrain the business community in a co-operative effort to shift climate policymaking worldwide onto a path to success. At no point in the quarter century of my concern about climate change has this kind of thing been feasible before. Much less my own presence: a known maverick, a trouble maker in the eyes of the energy incumbency, a dreamer about unattainable clean-energy alternatives.

Times really might be changing, I dare to hope.

The specific aim of the High-Level Workshop on Climate Change is to boost the prospect that meaningful targets and timetables for greenhouse-gas emissions reductions will be adopted at the Paris climate summit in December 2015, where governments have agreed in principle to take such a long-delayed step. The intended outcome is for a draft agreement to be reached at the 2014 climate summit – to be held in Lima, Peru. A further key element of the plan is for world leaders to convene at the UN headquarters in New York in September 2014 to agree that it should all happen as the UN plans.

This was what was missing in Copenhagen, the 2009 climate summit widely deemed a catastrophe, it is now widely acknowledged: a prior agreement by world leaders that would mandate negotiators to deliver a treaty with teeth.

Christiana Figueres, the executive secretary of the UN Framework Convention on Climate Change, is the woman charged with the awesome responsibility of delivering that treaty in Paris. She sits across the room from me now, with senior UN and WEF officials beside her.

I have heard so much about this woman. She comes from a Costa Rican dynasty. Her father was President. When he came to power he did something amazing: he went to all his neighbouring heads of state, persuaded them to agree to come to his aid if anyone ever invaded Costa Rica, and when they said yes, he got rid of Costa Rica's armed forces. He spent much of the money he saved on social services.

Today, in surveys of which country in the world has the happiest citizens, Costa Rica tops the poll on a regular basis.

Christiana's brother, Jose-Maria, was elected President after his father. He kept the tradition going. When he left office, he headed up the World Economic Forum. Now he runs Richard Branson's climate-campaigning organisation, the Carbon War Room. I have spent quality time with Jose-Maria in Costa Rica and in Spanish mountains. I count him as a friend.

Now, I sit and listen to his sister outline her plan for winning a war to save civilisation in a decisive battle in Paris little more than two years from now.

Tunbridge Wells, 27th September 2013

The UN's Intergovernmental Panel on Climate Change, a panel of thousands of climate scientists from multiple countries, has written four reports since 1990. All were warnings about global warming so severe that governments have felt compelled to keep negotiating at annual climate summits, and preparatory sessions in between, for nearly a quarter of a century. The fifth IPCC scientific assessment is published today.

The UN has warned that elements of big business are intent on undermining the report. "Vested interests are paying for the discrediting of scientists all the time," a senior UN official has told the press. It is clear that this form of sabotage has been a long-running effort. An Oxford University report has showed that 8 out of 10 news reports published between 2007 and 2012 on climate change have focussed on uncertainty, in a manner guaranteed to mask the strength of the IPCC consensus.

Today, it is the duty of people like me to try and redress that, as best we can when trading soundbites with the media-trained professional sceptics often pitted against us.

I am up at the crack of dawn to check the news websites before heading off to the nearest BBC studio to talk live to Radio 4's Today Programme. Representatives of the hundreds of climate scientists from around the world who compiled the IPCC report are giving a press conference later today in Stockholm. But there have already been enough leaks of content for the news media to be carrying their basic message. A key new conclusion is that if the world wants to stay below two degrees of global warming, half the available carbon budget has been used up in fossil fuels burned to date. And the report makes it very clear that there is far too much carbon in existing fossil fuel reserves to be burned safely.

The Today Programme wants to talk about what all this supposedly unburnable carbon means for the business world. It helps that my book *The Energy of Nations* was published yesterday. I recount its basic message on carbon-bubble risk: in essence, I say, business risk is high for the energy incumbency and those overly dependent on it, business opportunity is substantial for those moving early to do something about the risk, like abate it with clean energy. This conclusion is amply borne out by the IPCC's deliberations, I contend.

To my amazement, there is no default denier to confront me in this interview. I have fully eight minutes, one-on-one with the Business Correspondent, live over many a breakfast nationwide. This is a rare privilege. Most of these media opportunities are short debates with sceptics: a distortion of manifest scientific consensus that the BBC is frequently criticised for these days.

I have time to spell out the similarities I see between the risk blindness the banks were clearly guilty of in the run up to the financial crisis, and the risk blindness most of big energy is in the grip of today. I am able in the time available to build what I hope will sound like a credible concluding bottom line: what the IPCC is saying, in essence, is that if we don't collectively dismantle climate-change risk with deep cuts in emissions, there will ultimately be no viable economies left to do business in.

I drive home, and spend the day monitoring news emerging from Stockholm.

The 2,000-plus page report has been written by 209 lead authors from many countries, finalised at a meeting lasting four days.

"30 years to climate calamity if we carry on blowing carbon budget," reads the headline of the main Guardian report on their deliberations.

But the sceptics are busy spinning their disinformation elsewhere. In the live webcast of the Stockholm press conference, I see the Daily Mail's man note that global average temperatures have been flat for 15 years. Surely this absence of global warming since 1998 is a problem for the IPCC's story?

On the webcast, I watch Thomas Stocker, co-chair of the report working group, deal with this. Stocker looks exhausted. I know from reports inside the meeting that the lead authors are by now hugely sleep deprived. But he betrays no hint of impatience. Measuring recent years in comparison to 1998, an exceptionally hot year, is misleading, he observes. Temperature trends can only be observed over longer periods, of about 30 years. The eighties were hotter than the seventies. The nineties were hotter than the eighties. The noughties were hotter than the nineties. The trend of the last century is clear: a rise of 0.9°C, the IPCC report concludes. The last decade has clearly been the warmest on record, even if the trend within it is flat. Heatwaves, sea-level rise, melting ice and extreme weather are the outcome, with much worse to come, unless there are deep cuts in global greenhouse-gas emissions. If there aren't, average temperature between 2080 and 2100 will be 2.6–4.8°C higher than today.

The IPCC's projections for the average sea level rise by then, in the absence of deep emissions-cuts, range from 45–82 centimetres higher than now. Shanghai and New York would be among many coastal cities under threat. And these estimates assume no collapse in ice sheets in Greenland and Antarctica, which would greatly add to the rise. Reaching between 1 and 4°C of global warming locks in a complete melt of the vast Greenland ice sheet over the course of a millennium, ultimately adding 7 metres to sea level. The equivalent figures from Antarctica are equally scary. Our descendants would lose the coastal plains. Most of the current global economy is on the coastal plains.

As the day goes on, Twitter feeds show sceptics featuring more and more in media reports and interviews. Thousands of scientists worked for many months on distilling their consensus report on climate change for the world. Yet a diehard few contrarians – almost all of them non-scientific ideologues, many openly funded by fossil fuel interests – are able to insert into much of the mainstream media, in multiple countries, the mantras of their concrete-encased belief systems, as though these are opinions rooted in science, and of equal standing to the IPCC's assessment of risk. The situation is completely dysfunctional.

Why does the media allow such distortion? Some outlets, like those of the Murdoch empire, are owned by deniers who insist on a sceptical editorial line on climate. Others, like the BBC it seems, simply take the view that a debate is better than a documentary, even if the grounds for debate are the stuff of black-arts propaganda, often funded by vested interests.

I feel a deep sympathy for the climate scientists who have to deal with this, day-in day out. I used to be an academic myself. I can well imagine their frustration.

Today, I have to soak up my own small share of it. The BBC invites me back to their studio in Tunbridge Wells, this time to do a live TV debate. I have an opponent this time, a denier who calmly recites the mantra about a complete lack of global warming being evident over the last 15 years. He will have been told, many times, by qualified people, that this is a misrepresentation. He also knows that if he keeps calm and looks sensible, he will confuse hundreds of thousands of people around the world, maybe more. His label will help him: he represents an organisation with the kind of neutral name favoured by most sceptic organisations and their backers: The Global Warming Policy Foundation.

For my part, I know I have at best two minutes in total both to counter the toxicity of his argument, and to paint a credible picture of what thousands of climate scientists are saying, and its implications. And all the media coaches say the same thing: this must be done calmly, in a manner that will make a listener generally only half aware of the actual words being spoken simply like you more than him.

This is not an experience conducive to low blood pressure.

London, 5th October 2013

On the Arctic oil and gas frontier, Greenpeace has had a run-in with the Russian state. On September 18th in the Pechora Sea activists left their ship the Arctic Sunrise in four inflatables to try and hang banners protesting against drilling by Gazprom's Prizalomnaya rig. Had they been targeting BP or Shell in the North Sea, they would probably have been offered tea and biscuits. European companies have long since learned that failure to use water cannons is the best way to starve Greenpeace protests of media coverage. The Russians, by contrast, fired warning shots with AK-47 rifles and a cannon not of the watery variety, and the next day armed commandos descended from helicopters onto the Arctic Sunrise and seized the ship, though it was in international waters. 30 activists were arrested, and taken to Murmansk. The news media now relay the ridiculous and ghastly news that they have been charged with piracy, a crime which carries a sentence of up to 15 years' imprisonment.

I stand now on the Bayswater Road in a crowd of several hundred protesting outside the Russian embassy. There are demonstrations like this at other embassies around the world. International outrage is running high that a state with a seat at G8 summits can treat environmental protestors undertaking a non-violent direct action little differently from terrorists.

I am here alone, and can see nobody I know in the crowd. I feel at home though. Around me is a type of person that has that effect. The banners show a seriousness of purpose but the demeanours show a relaxation of spirit, a calm confidence that non-violent protest is a fundamental right in a democracy. A couple next to me share a laugh with a policeman. Actor Jude Law and pop star Damon Albarn stand in the crowd without there being a clue that they are megastars. I could talk to anybody here and be met with an open face, and most likely an open heart.

There won't be too many scenes like this one today in modern Russia.

How to persuade a petrostate like Vladimir Putin's to leave hydrocarbons underground unburned, and in significant quantities, in the years ahead? It doesn't look likely that non-violent protest and people-power are very likely to do it. The democracies are struggling badly enough on that front at the moment. The scientific advice is so clear, the emerging impacts of global warming are so worrying. Yet carbon emissions go right on rising, in most countries, despite huge concern in public polls, whatever the flavour of the governments.

I console myself with the thought that the nations of the world have negotiated treaties that make a start. The Framework Convention of Climate

Change, agreed at the Rio Earth Summit in 1992, did commit to keeping global warming from harming economies and ecosystems, though it had no teeth in terms of targets-and-timetables for emissions reductions with which to do that. The 1997 Kyoto Protocol did provide some first-step targets and timetables for minimal emissions reductions by the developed nations. Russia did ratify that protocol, even though more than half the government's tax income derives from oil and gas.

But can the world, Russia and other petrostates included, go the long distance that nations will need to go beyond Kyoto, in Paris? Can they commit to leaving much of their hydrocarbon endowment underground, unburned?

Whatever America does will clearly be key. Here the news is very mixed. On the one hand the USA is poised to overtake Russia as the number one combined producer of oil and gas. To speak against that is to invite accusations of treason. On the other hand, President Obama seems to be contemplating ways of making climate change a key theme of his legacy with an attack on coal. In June, in a speech at Georgetown University, he went further than any previous US President in signalling intent to act on greenhouse-gas emissions. He pledged to bypass Congress, so great is the opposition to action on climate change there. Much of the Republican party seems to be in total denial that there is even a problem. Some of the most outspoken contrarians see the issue as a leftist scam. On this, as so much else, American politics seems to have tribalised itself.

But Obama was clear where he is going to stand in his second term. "I refuse to condemn your generation and future generations to a planet that's beyond fixing," he told the students. He pledged to use his presidential powers by issuing an executive memo to the Environmental Protection Agency calling for new rules curbing greenhouse gas emissions from power plants, aiming to put the US on track to meet its commitment to cut carbon emissions 17% from 2005 levels by the end of the decade.

"Power plants can still dump limitless carbon pollution into the air for free," Obama said. "That's not right, that's not safe and it needs to stop.... We don't have time for a meeting of the flat earth society."

The President also turned his attention to investors in fossil fuels. "Convince those in power to reduce our carbon pollution," he urged the students. "Push your own communities to adopt smarter practices. Invest. Divest. Remind folks there's no contradiction between a sound environment and strong economic growth."

On September 20th, the Environmental Protection Agency announced, for the first time, rules to limit carbon emissions from future US power plants. They would, in effect, block the construction of any new coal plants not fitted with carbon capture and storage technology. New rules are to be announced for limiting emissions from existing power plants in June 2014.

This promises to be the opening salvo in what will become a long running political struggle in the United States, one that will intersect with developments in China to become a key determinant of success or failure at the Paris climate summit in December 2015.

Carbon Tracker continues to live in hope that we can help forge a win in Paris. Our dream is that by then there will be so much pressure on fossil fuel companies not to waste capital on piling up ever-more unburnable reserves that climate negotiators might have the impression that "the markets are moving", even ahead of any decisions they might take to commit to carbon-emissions cuts. In this way, we reason, policymaking might prove easier by dint of being conducted with a sense of playing catch-up.

As for the Russians, they will need foreign capital if they are to keep the oil-and-gas taps on in Siberia and the Arctic. Suppose the capital taps begin to turn off?

I stare at the shuttered windows of the Russian embassy, and imagine the Greenpeace protestors shivering in dark Arctic jail cells. My mind wanders to memories of the Soviet Union, as it was known when last I visited.

If there is a return to the Cold War, or anything like it, then the task of achieving a multilateral agreement on climate change will become appreciably more difficult.

Chapter 3

Not responsible

Daegu, 14th October 2013

Five thousand delegates from more than a hundred countries mill around in a cavernous convention centre in South Korea for the opening reception of the World Energy Congress. Walking into the vast exhibition hall, the first stands that loom at me are those of the Russian gas and oil giants. Gazprom's is populated by a troop of blonde Russian fashion models. Rosneft has opted for a large model of a polar bear and her cub. Nearby, the Saudi Aramco stand bears a slogan reading "Bringing Petroleum to Life."

I drift slowly, people-watching, glass of champagne in hand. A bigwig – a minister or a CEO – strides by with a media circus in train.

I note that the stands of the nuclear companies – France's Areva and Russia's Rosatom – are thinly populated. Carbon capture and storage seems to be faring even worse. The Global CCS Association has a tiny stand, with nobody there.

The State Grid Company of China stand is dominated by a model of a sizeable solar farm with a wind park on low hills around it. That's more like it, I think.

The sessions begin the morning after the reception. A giant LCD screen looms above the delegates in the foyer of the Congress centre. It is filled with a spinning globe, that familiar blue pearl in space, and some writing: "Nature has provided our energy needs for thousands of years. As we make the choices to meet our future energy needs" – the camera dives in to pan across sweeping rainforests – "Nature is relying on us."

The very first speaker, Saudi Aramco CEO Khalid al-Falih, is typical of many who will follow him over the week.

The earth is blessed with a colossal endowment of fossil fuels, he intones. Fossil fuels are the crown jewels of the world's energy mix. We have 50 years of oil supply and 250 years of gas supply. And we must let market forces decide how much of it we use.

He does not mention climate change, or the message on the screen outside.

GDF Suez CEO Gerard Mestrallet sits relaxed on a stage being interviewed by a journalist quite unprepared to ask any hard questions. He explains why he wants the shale gas boom in America exported across Europe. He and the CEOs of nine other European utilities want to see subsidies for renewables ended. They are touring European capitals, telling governments to put the lid on renewables.

Why do they want this?

We are for the security of Europe, he says. We are for the climate. Gas can provide all the heat, electricity, security, and stable climate that we need.

The journalist does not ask him how he thinks such proliferation of gas can lead to the end of fossil fuel burning less than four decades from now. He does not ask how the cutting of subsidies for renewables – along with energy efficiency, the main route to replacing fossil fuels – can be "for the climate". Instead, he wants to know the top three places in the world GDF Suez would go to first in order to frack for shale gas.

This is going to be interesting. Mestrallet cannot list his own country. The French government banned fracking back in 2011. Just three days ago the Constitutional Court upheld the ban under challenge from a Texan oil company. French environment minister, Philippe Martin made things very clear. "With this decision the ban on hydraulic fracturing is absolute". He went on to air a consideration I hardly ever hear in coverage of the American shale boom. "Beyond the question of fracking, shale gas is a carbon emitter," he said. "We must set our priorities on renewable energies."

Deprived of his homeland, the top of Mestrallet's list shocks me. We are looking at the UK, he says, since the government there wants to favour unconventional gas.

Had the uninquisitive journalist asked about the climate implications of all this gas, Mestrallet would doubtless have responded that burning gas creates less emissions than burning coal, which would have been true. But he would have omitted the worries about fugitive emissions of gas all the way from the wellhead to the home cancelling out the advantage of gas over coal. Big Energy bosses usually do. He would have dodged the issue of all the investment flowing to gas depressing the development of renewables and efficiency. Big Energy bosses are good at that.

The World Energy Council's latest energy scenarios, published during the Congress, offer a window on the dysfunctional group-think at work in this industry. One, the so-called Jazz scenario, envisages total world primary energy increasing 61% to 2050, amid little multilateral effort to co-ordinate fossil fuel reductions. The other, the Symphony scenario, envisages an increase of 27%, amid a degree of policy co-ordination. Fossil fuels in 2010 provided 79% of world primary energy. Their share by 2050, by which time climate scientists tell us they must be phased out in energy use, or nearly so, would be 77% in the Jazz scenario, and still 59% in the Symphony scenario. In both scenarios gas expands significantly from its current share.

And the climate implications? The target at the climate negotiations is a ceiling of 2°C increase in global average temperature. A reasonable chance of staying below that requires returning carbon dioxide-equivalent atmospheric greenhouse-gas concentrations below 450 parts per million. At present the figure for the greenhouse gases included in the Kyoto Protocol is more than 420 parts per million. The Jazz scenario would take us to between 590 and 710 parts per million of carbon dioxide equivalent. The Symphony Scenario would take us to 490–535.

GDF Suez's Gerard Mestrallet was not asked to explain how his pro-fracking anti-renewables vision would get us to 450 parts per million.

The top 20 European utilities, including GDF Suez, were worth a trillion dollars in 2008. Today they are worth half a trillion. The growing success of renewables, plus their own mistakes and oversights, have done this to them.

"How to lose half a trillion dollars", the Economist scoffed a few days ago. "Europe's electricity providers face an existential threat."

On current course, indeed they do. These are companies with unworkable business plans. They should be embracing renewables with open arms, but instead seem set on a last ferocious and open assault on them. Their culturally-ingrained and institutionalised belief system evidently compels them to do this.

For the time being, so I hope.

One of them has to break from the pack, surely?

Oslo, 31st October–5th November 2013

The Norwegian Government Pension Fund is the biggest sovereign wealth fund in the world. It has around $850 billion under management and owns one percent of all shares in global stock markets. Many observers view the "oil fund", as it is often called, as too skewed towards fossil fuel investments, especially considering that the money it invests largely derives from Norwegian oil and gas in the first place. Any recognition by this fund of fossil fuel asset-stranding risk would obviously send the mother of all signals to the capital markets, to the general detriment of fossil fuel investments, and default betterment of renewables investments.

The Norwegian Labour Party, when in government, opposed any change to rules allowing fossil fuel investments by the fund. In an election last month, Labour was replaced by a minority conservative coalition. The environmental organisation WWF has now arranged a day of Carbon Tracker briefings for politicians and business leaders, including the former Labour minister of foreign affairs, hoping to change their minds about fossil fuels and the fund. This is one of those occasions where the responsibility falls not to one of the star Carbon Tracker analysts, who are off around the world doing what they do, but to the chairman.

What the analysts do is increasingly jaw dropping. Citywire, a City of London news outlet, wrote recently that Carbon Tracker has "caused a sensation" in the capital markets. Mainstream coal and oil analysts at institutions including Goldman Sachs, Citi, Morgan Stanley and Deutsche Bank have started engaging with carbon-bubble risk in reports to their clients. Terms from Carbon Tracker reports like "unburnable carbon" and "stranded assets" are appearing in investment-bank research reports with increasing regularity. Questions about the wisdom of capital expenditure of fossil fuel companies – capex, for short – are flooding into fund managers' inboxes. Carbon Tracker's Director of Research, James Leaton, has been voted by peers into 6th place internationally in the Socially Responsible Investment Research Analyst ranking, from a very large field.

And we are fresh off another milestone, one that the Norwegians must be mulling carefully. One of the five Swedish state pension funds has just retreated from fossil fuel investing, saying it is seeking relief from carbon-bubble risk. The $38-billion AP4 fund plans to invest in a tailored emerging markets fund consisting of companies that have both low-carbon emissions and low fossil fuel reserves. Chief executive Mats Andersson tells the financial press that "if it works, we will increase our exposure so that hopefully it will be a much bigger part of our portfolio. We want to do this on a global basis." He explains his rationale purely in business terms. "In 10 years' time, carbon will be priced and valued in a different way so that companies with a high carbon footprint will perform worse. This sustainable approach isn't about charity, but about enhancing returns."

I increasingly wonder whether it is good tactics for Carbon Tracker to have a known long-term climate campaigner as chairman, especially one so regularly accused of hyping the climate problem simply because he wants to sell more solar panels. Perhaps a City of London grandee, one with a closet desire to see action on climate change, might provide better optics. But the analysts seem happy enough with me chairing them.

I do my duty for a long day in Oslo. It would be far more time efficient if I could do a single two-hour Q&A briefing for all the politicians, rather than a repetitive series of one hour briefings one-on-one. But, my hosts tell me that if they had gone that route, none of the politicians would have turned up. I suppose that would be the same in every country.

Three days later, the former foreign affairs minister announces that the Norwegian Labour party will support the national pension fund's complete withdrawal from coal, and put oil and gas under watch too. Combined with minority-party support, this gives a voting majority in the new Norwegian parliament in favour of extracting the oil fund from coal.

WWF are over the moon, amazed at the success of their briefings.

A day later, the new Norwegian prime minister Erna Solberg speaks at a climate conference for the first time. The Zero Emission Conference, in a historic Oslo theatre, is packed with the youth of Norway: stalls and balconies of fair hair as far as the eye can see from the stage off into the gloom.

Chelsea Clinton is the keynote speaker. She does a good job trying to walk in her parents' footsteps, pressing all the hot buttons of climate change: the roles of cities, communities, solar, youth, and so on.

Erna Solberg does not mention the oil fund in her pedestrian climate debut speech. But afterwards, she is surrounded in the foyer of the theatre by a media scrum of TV, radio and print journalists. I watch from the fringes of the scrum as impossibly young journalists fire questions at her about the Labour move on fossil fuel investments in the national pension fund.

She says that ahead of any support for withdrawal by the sovereign wealth fund she will look at coal companies to check that they aren't investing in renewables.

I am asked for a comment by Aftenposten, a national paper.

She won't have to spend too long on that exercise, I say.

It had seemed impossible to dare hope this, but it looks as though the world's biggest pension fund may well withdraw from coal next year. The entire coal industry would look to be in danger of investor withdrawal as a result. This is beginning to look like a very gratifying few days' work.

Dagens Naeingsliv, the financial daily, wants to know what I think of Norwegian investors putting their money into yet more oil.

That's a big bet now, I venture. Investors should take a hard look at whether their money might be wasted on high-cost frontier projects such as those in the Arctic.

I fly home, telling myself not to be over-encouraged. I know how easily, in the world of politics, today's apparent advance can be reversed tomorrow.

Amsterdam, 13th November 2013

Every year, the Dutch energy industry holds a retreat where executives review the state of the energy markets, national and global. A mixed group gathers. The oil industry is well represented, Shell in particular, with the Netherlands being its home country. A good deal of the modern Netherlands has been built on taxes from the vast gas field under Groningen, the largest gas field in Europe, jointly owned by Shell and ExxonMobil. But thirteen years into the twenty-first century there are many Dutch clean-energy aspirants, and they are well represented too. One of them, my friend and Solarcentury Chief Executive Frans van den Heuvel, has invited me.

The presentation and conversations are in Dutch. At this year's gathering, in a mansion near Amsterdam, there is one exception. The opening presentation will be in English, and I have been asked to give it. I will have fully 45 minutes to offer participants my view of systemic risk in modern energy markets.

In the front row listening to me is Jeroen van der Veer, former CEO of Shell. He is a regular speaker at these events, and will address the gathering immediately after me.

We are old sparring partners. In March 2007, we debated on BBC TV for more than an hour, in a programme on climate and energy in which I was pitted against four executives from the World Business Council for Sustainable Development. Then our debate was a verbal tennis match. Now we have two long set pieces for the participants to compare. Before he can deliver his, van der Veer has to sit and listen to me for three quarters of an hour. It cannot be a comfortable experience for him.

I elect to major on carbon-bubble risk, and the vast amounts of capital the fossil fuel industries will have to raise if they are to keep the world where they want, mostly reliant on fossil fuels for decades to come. In our 2007 debate, I majored on the moral case: burn all the fossil fuel you are set on burning, I argued, and you will lead the way in the slow cooking of our planet. At the climax of the debate then, I managed to tease from him a revealing oil-industry argument that I heard a lot in private, but rarely in public. Shell is not ultimately responsible for the energy used in the world, van der Veer said. We just meet the demand.

Drug pushers deploy the same case.

Drug pushers also fight hard and dirty to keep their demand in place.

Today I ignore ethics and morality and focus on dollars. I argue that Carbon Tracker and other organisations working on pressuring oil-and-gas capex, or arguing for divestment, or both, are together posing a danger of removing financial licence from the fossil fuel companies. Without access to the trillions held in pension funds and suchlike, I tell the oil-and-gas executives, you are not going to be able to explore, drill, and add to your reserves. You are not going to be able to grow. You are going to lose your business model.

I have plenty of time to lay out my case, as logically as I can. I couch it all in terms of risk. The march of events seems to be handing me increasingly potent ammunition. Many companies are struggling to make money in shale oil, including Shell, even with the oil price as high as it is today. For two and a half years now the price has been in the $90 to 120 per barrel range, and mostly well over $100. Yet Shell announced in August that it is writing down $2.1 billion of assets in its quarterly results, mostly as a result of poor results in shale-oil drilling.

Peter Voser, the man who took over from Jeroen van der Veer as CEO, is retiring. He has told the Financial Times that he regrets the company's huge bet on US shale. "Unconventionals did not exactly play out as planned," he said. "We expected higher flow rates and therefore more scalability." He ventures that projections of the US shale boom being exported to other countries are "hyped". The rest of the world should steel itself for "negative surprises", he warns.

I remind the audience of this, as diplomatically as I can. I'm not saying you are sure to lose your business model, I say, I'm saying there is risk that you might – much of it wilfully unrecognised by the energy industry, shielded from investors.

I can see smiles of the faces of some of the renewable-energy practitioners as I speak. They like this new way of making a default case for their technologies, and business models, it seems.

Following me, van der Veer begins in English, before switching to Dutch. He tells the audience how he and I are working together at the World Economic Forum on one of their Global Agenda Councils. Together with a dozen or so other people, under his chairmanship, we are weighing the possibility of "black swan" events in modern energy markets.

It is true, counterintuitive as it may seem.

Black swans involve supposedly low-probability or barely conceivable risks: "surprise" events that, if they happened in the real world, would have huge negative consequences.

Van der Veer agrees with me that there is a lot of risk around in energy markets today. There are plenty of potential black swans to worry about he says.

Yes, I think to myself, but some events that might be a "surprise" to some are to others high-probability, or even near certainties. I am arguing precisely this case, as best I can, at the moment in our World Economic Forum discussions. I am in a minority, of course.

I wonder if Peter Voser would count the risk of a shale bubble as a black swan. He fears "negative surprises" similar to his own humiliating asset write off, and has not invested in British shale, the second most attractive national shale target for at least one of his peers. It doesn't sound, on that basis, as if Shell would put shale on the black swan list.

Van der Veer is in a conciliatory mood. Jeremy & I may disagree if the glass is half empty or half full, he volunteers, but we both agree the glass is too big.

He means we agree that too much fossil fuel is being burned, I suppose: that there is a climate problem as a consequence.

He switches to Dutch, and I cannot follow further. But his slides show me that he is talking about Shell's two latest scenarios for the future. They are called "Oceans" and "Mountains", and they are well known to me. Like the World Energy Congress scenarios for the future that I heard about in Korea, one is awful (Oceans, a version of business as usual), and the other merely very bad (Mountains, wherein gas replaces a lot of coal, and carbon capture and storage is used at industrial scale for both coal and gas burning). Neither comes close to delivering a world where global warming can stay anywhere near a two degrees ceiling.

Chapter 4

Comrades in arms

Istanbul, 25th & 26th November 2013

Turkey has become a strategic front line in the carbon war. It is more perfectly suited for a low-carbon energy future than most, yet its government has embarked on a high-carbon energy plan that is the worst of its kind in the world. Campaigners at the European Climate Foundation have invited me to go to Istanbul to help them pluck what they hope will be some low hanging fruit.

Turkey's plan is remarkable. The sun shines in this nation more than 7 hours per day on average, yet electricity generation from solar power is currently almost non-existent. Installed solar photovoltaic capacity per inhabitant is higher in freezing Finland. The national target is a mere 3 gigawatts by 2023, about as much as the cloudy Britain's current capacity. Yet with current low solar costs, solar electricity would be the cheapest option in many settings around the country, and in a few years in all of it. To meet 100% of Turkey's projected electricity in 2050 would need only 0.21% of Turkey's land area, much of which is not fit for other purposes anyway.

Yet in a nation dependent on foreign imports for three-quarters of its energy, Turkey has declared 2012 the "Year of Coal".

The plan, if that is the correct word for it, is to utilise the nation's lignite and hard coal resources by 2023 in more than 50 new coal plants. If all those are built, Turkey's greenhouse gas emissions would grow by 75%.

I spend a frantic couple of days with the European Climate Foundation's campaigners talking to banks about financing this madness, and plotting with the Turkish solar industry and NGOs. This is one of the last half dozen or so bolt holes around the world for the coal industry's aspirations of survival.

Based on what I see and hear from worried bankers and an impressive Turkish NGO movement, the Year of Coal plan won't work. The bankers who would have to bankroll the new coal plants fear stranding their loans. The NGOs, almost exclusively run by powerful women, are confident of confronting the power plants one by one, location by location, with irresistible people-power on the ground.

If only this could prove to be typical of what fossil fuels can expect more widely on the global stage.

In Poland, the annual climate summit has just finished. Governments have decided that countries must table national commitments to reductions of greenhouse gas emissions from 2020, and do so well ahead of Paris, by March 2015. (In UN treaty language, these are referred to as Intended Nationally Determined Contributions, or INDCs. I will simply call them Christiana Figueres explains the significance of this in a closing press conference. We have seen essential progress, she concludes. Now governments, and especially developed nations, must go back to do their homework so they can put their plans on the table ahead of the Paris conference. A groundswell of action is happening at all levels of society. All major players came to Warsaw to show not only what they have done but to think what more they can do. Next year is also the time for them to turn ideas into further concrete action.

Turkey's plan will be something of a litmus test. Campaigners will have 16 months to turn it from a coal plan into a solar plan.

Berlin, 29th & 30th November 2013

The European Association for Renewable Energy, better known as Eurosolar, gathers its members for a reception in the headquarters of KFW, one of the biggest banks investing in renewables globally, the bank that has financed much of the German renewables revolution. A video welcomes the crowd. Hermann Scheer, legendary architect of the subsidy scheme that has grown the German renewables industries, the so-called feed-in tariff, says something triumphant. I fail to put my earphones on in time for the translation, but I can imagine what he is saying.

They won't be able to stop this revolution.

Hermann has a twinkle in his eye, as ever he did. He is speaking from beyond the grave: a video filmed before his untimely death three years ago.

This is the twentieth awards ceremony of the organisation Hermann founded. The roll call begins, and as usual it is a rich encyclopaedia of what renewable energy is capable of. A wartime air-raid bunker in Hamburg converted into a clean-energy showcase providing solar electricity and heating to hundreds of nearby homes. A Swiss apartment block wherein energy consumption has been reduced 70% with off-the-shelf technology, including solar. A German mini-utility selling renewable electricity to 250,000 customers, in direct competition to E.ON, RWE and the other giant utilities. And so on.

I accept a prize on behalf of SolarAid's African field team, and am required to say a few words. I try to describe the sense I have, in this historic city, of yet more history in the making. A recent poll shows that 84% of Germans support efforts to shift the national economy to 100% renewables "as quickly as possible". And the projects recognised tonight are living proof of that national dream.

What a great shame it is that Hermann can't be with us to celebrate our growing success, I say. How old solar advocates like me miss him, his friendship, his leadership.

The first person to thank me for my thought afterwards is his widow, Irme. There are a lot more Germans queuing behind her. Hermann was much loved.

I learn that the next day there is to be a demonstration in support of the Energiewende – the German energy transition to renewables. The new coalition government has cut a deal with the big utilities that will slow what the people want. German renewables supporters are descending on Berlin from all over the nation.

I walk alone among the thousands filling the streets in a long loop east from the Hauptbahnhof. Yellow and green balloons float about the noisy column and its many banners.

Kohle killt. Energiewende jetzt. Atomkraft nein danke.

A disco plays off the back of a lorry. They seem to be picking their songs for relevance.

What a day for a daydream.

I am dressed inappropriately, in suit and overcoat. I didn't know I would be at a demonstration. A young woman, dressed in best Berlin-chic informals, looks me up and down and decides I need a pamphlet informing me of the dangers of climate change and the wonders of renewables.

We file past the Reichstag. Here Hermann Scheer served for many years as a parliamentarian. I imagine for a moment that he is walking beside me. He would be smoking a cigar, a big beam on his face. We would be joking, no doubt, gently pulling each other's leg. Maybe not so gently: a German and a Brit, both with big egos, walking together in this city of so much shared dubious history. But now? Allies, in blood, dead or alive, in a global civil war. He saw it all the same way as me. At some point on the march he would he raging at the ten EU utilities, led by France's GDF Suez but including E.ON, who are now calling openly for an end to all renewables subsidies.

As though all the fossil fuels and nuclear they use need no subsidies of their own!

He would be checking that I know necessary detail about the latest outrageous developments in Germany, to carry back to London: especially how the power grid operators hiked the surcharge consumers pay for renewable electricity fully 18% even as wholesale prices fell, to give the impression renewables were to blame for higher bills, not the failed big-energy model for a power grid.

Maybe we would be working a laugh or two from a contrast between the rainbow coalition of people in this crowd and an imagined demonstration of fracking advocates protesting in Dallas, or Arctic oil advocates marching in Moscow. Not that they would need to protest of course. Governments, whether in petrostates or oil-importing nations, mostly give them no particular need to.

We would certainly be celebrating our steadily growing successes, maybe with the help of a hip flask. How annual solar installations around the world are set to beat wind for the first time in 2013. How solar bonds are being raised by American superstar company SolarCity, backed by cash flow from the rooftop solar panels leased to US home owners. How Solarcentury, the solar company I founded in 1999, is now a service partner in IKEA's selling of rooftop solar in their stores.

As though solar roofs were wardrobes!

I can see him laughing at that thought so clearly.

The final destination appears, and I stop daydreaming. Angela Merkel's Chancellery is already completely encircled by many thousands of demonstrators.

I wonder if she is hearing the message from her streets.

I communicate with my tiny fraction of the world, including some German marchers, on Twitter, that ever interesting instrument that I once thought a waste of space, until younger minds like those on this march educated me otherwise.

It seems that many in Germany are rather keen on renewable energy, I tweet, with a link to the spectacular pictures of the demonstration by now appearing on the web.

I march with you wishing my countrymen could see this and be inspired.

The sad truth is that it is impossible to imagine a spectacle like this in the UK. Hermann Scheer and the other German pioneers have done a so much better job in Germany than I and my fellow renewables advocates have so far been able to do in Britain.

London and Washington, 20th November 2013

I have a new ally, in an unlikely place. For a long time now I have been e-mailing and Skyping with a US Army Officer. Lt. Col. Daniel Davis first wrote to me from a tour of active duty in Afghanistan. He has done four tours there and in Iraq, and been decorated more than once for valour along the way. He is a war hero, yet has grave doubts about his government's general support for the energy incumbency's narratives of shale- and oil plenty. He and I have elected to try and ventilate our concerns together in an unlikely transatlantic alliance.

An example. In mid November every year the International Energy Agency brings out its World Energy Outlook. This year Lt. Col. Davis and I critiqued the messaging in these reports since 2005, including this year's, in a co-written article for the Huffington Post.

Here is what worries us, and others like us. The IEA's reports are schizophrenic. On the one hand they are replete with useful data on global energy markets, much of it unavailable elsewhere. On the other hand, they are written by officials of an agency created by rich-nation governments essentially to promote fossil fuels, officials that are required – like all civil servants – to walk a political line between statistics and service to the perceived interests of the governments in power in those nations. To simplify the challenge for readers of IEA World Energy Outlooks, it is often to read between the lines: to filter the facts from the politics.

After years of being very relaxed about global oil supply, in 2005 the agency reversed, sounding a warning that collective oil production in non-OPEC countries, including the USA, would peak "right after 2010", leaving huge pressure on OPEC to keep supply in pace with demand. In 2006, they repeated the warning. Global energy demand would surge 50% by 2030, within less than 25 years, requiring 116 million barrels a day of oil, with most of the increased supply having to come from Saudi Arabia, Iraq and Iran. "This energy scenario is not only unsustainable but doomed to failure", said the then Secretary-General of the IEA.

In 2007, the IEA chief economist warned that the world had only ten years to turn round its energy policy. To meet projected oil demand by 2030, OPEC would have to double supply. This looked infeasible.

The oil price was approaching $100 at the time. On June 10th, 2008, with the oil price near $140, the IEA announced that the world officially faced an oil crisis.

In the 2008 World Energy Outlook, the IEA published an oilfield-by-oilfield study of the world's existing oil reserves for the first time. It showed current crude oil fields running out alarmingly fast. The average depletion rate of 580 of the world's largest fields, all past their peak of production, proved to be fully 6.7% per annum.

As for the IEA forecast that year, crude production began a steep descent, falling steadily all the way below 30 million barrels a day by 2030. The depletion factor, so I wrote in *The Energy of Nations*, could better be called a fast-emptying factor.

Even with demand for oil being destroyed fast by recession in the west, driven in large part by the then high oil prices, the rate of demand growth – led by China, and India – was such that the world would need to be producing at least 106 million barrels a day by 2030. Could this be done?

In December 2008, I watched senior IEA officials wrestle with that question in public.

Yes, the party line went, but only if massive investment was thrown at the challenge, especially by the OPEC nations. Global oil production at the time totalled 82.3 million barrels a day if we subtracted biofuels and added to existing crude production the 1.6 million barrels a day of 'unconventional' oil squeezed from the tar sands and 10.5 million barrels a day of oil produced during gas-field operations. To reach production of 106 million barrels a day by 2030, in the face of this, would require oil-from-gas to expand almost to 20 million barrels a day, unconventional production to expand almost 9 million barrels a day, and on top of that more than 45 million barrels a day of crude oil capacity yet to be developed and yet to be found. All this added up to 64 million barrels a day of totally new production capacity needed on-stream within 22 years.

To put that another way, the sum of new oil production needed, according to the International Energy Agency, would be fully six times the production of Saudi Arabia at the time.

It was clear to me in reading between the lines of the IEA report, in watching the body language of officials as they presented it, and in comments made to me and some journalists in confidence that the then IEA leadership didn't think there was a snowball's chance in hell of pulling this off.

How quickly things changed in the next three years. Shale gas and shale oil production had not appeared in the IEA reports until 2010. But now they leapt to the fore. Throughout 2010 and 2011, oil-industry executives everywhere repeated that shale was what they called a "game changer".

Entering 2012, the oil price in euro equivalent was nearing a return to its July 2008 peak, bouncing slowly back from a big fall during the recession after the financial crash of 2008. In 2011, the average oil price had been a record $107, up 14% on the previous record year, 2008. In 2012 it would be $111.

On March 15th, 2012, with warnings of renewed recession filling the financial pages of newspapers, the US and UK governments actually floated the prospect of emergency releases of oil stocks from reserve. The margin for error was perceived as being that tight.

There was by now a surreal element to the oil-depletion risk debate, I wrote in *The Energy of Nations*. It was polarising into two greatly contrasting belief systems.

The 2012 World Energy Outlook knew which belief system met payroll. It projected an oil price in 2035 not much higher than the then price, with production at nearly 100 million barrels a day. The shale gas boom would spill over into oil production from shale to such an extent that the USA could reasonably expect to become almost self sufficient in oil and gas by 2035, the Agency enthused. The USA would be the new "Saudi America". Much of the world would be not far behind, as the good news about shale spread.

This irrational exuberance would have been funny, were it not so deadly serious. The contrast with 2008 was breathtaking.

As usual, many seasoned IEA watchers responded with dissenting views. But these appeared only on specialist websites. The mainstream media seemed to be almost universally reserved for the oil industry's cheerleaders, people like Daniel Yergin and Leonard Maugeri. Those who could be relied on to repeat the vital industry self-defence mantra that "peak oil is dead".

Peak oil can never be thought of as dead. Oil is a finite resource, and so there will have to be a time that its production reaches a record level and falls never to return to that level. That peak can be defined by both geological and geopolitical factors: below-ground, and above-ground factors, as analysts tend to call them.

The 2013 IEA World Outlook continues the politicised optimism of the 2012 volume. The net North American requirement for crude imports will all but disappear by 2035, the report concludes, and the region will become an exporter of oil products. There is a caveat, that "this does not mean the world is on the verge of an era of oil abundance", but still the underlying reasons for concern about fast crude depletion, on a global basis, are well shielded from direct view.

Lt. Col. Davis and I are keen to make clear our deep reservations. Man seems psychologically hardwired to prefer good news to bad, we write for the Huffington Post. But the reality of oil supply and demand fundamentals has not substantively changed since the IEA began issuing its warnings in 2005. Global conventional crude oil production in fact stopped growing and flattened out at around 74-75 million barrels a day in 2005 and despite herculean efforts of the most cutting edge technology, it has remained essentially flat since. Additional unconventional liquids – which are not equivalent to crude in energy terms, on several important counts – bring the global total to around 90 mbd.

How much of the decline in crude production has been counterbalanced by the new US shale oil production to date? The answer is about 2 million barrels per day, currently. That has required more than half the world's oil rigs outside Russia and China to produce.

Many governments are pinning their hopes – indeed, some their national economic health – on this feel-good narrative of perpetual abundance. Yet it is based on thin evidence and a great deal of industry bluster. We believe that considerable evidence suggests the decline trend in global oil production will resume before the end of this decade.

We recall how those few analysts in 2006 who warned of a looming banking crisis were ignored prior to the collapse, and the considerable price paid by so many when the truth finally imposed itself. In the years following the resulting global recession, Western governments simply printed trillions in bank notes in an attempt to mitigate the disaster. However, once the truth – as we see it – of oil supply and demand imposes itself on the world in the future, we won't be able to "print" oil: we'll have to find much more painful ways to adapt to the new reality.

Lt. Col. Davis and I think hard about our closing sentence. We decide to speak of mission.

We know, as tellers of a story almost no one wants to hear, how difficult our mission is. But the consequences to a world caught unprepared for such an oil crisis are too great to allow us to remain silent.

London and Washington, 10th December 2013

Lt. Col. Daniel Davis and I elect to expand our coalition of the worried. We convene video-linked gatherings in Washington and London of people who share our concerns about the risk of a global oil crisis. Those joining us are a list we have chosen carefully. It includes retired military officers, security experts, senior executives from a wide spectrum of industry, and politicians of all the main parties, including two former UK ministers. We call our event the Transatlantic Energy Security Dialogue.

The two roomfuls of people sit facing each other on massive TV screens, as though simply separated by a glass wall, not three thousand miles of ocean. The Americans are in a room on Capitol Hill, the Europeans are in a room in the London headquarters of engineering firm Arup.

We begin with a presentation by Mark Lewis, a former head of energy research at Deutsche Bank. With this background, you might expect Lewis to be a disciple of the conventional narrative-of-plenty in oil markets. Most of his peers are. But he is now a member of Carbon Tracker's Advisory Board.

Lewis suggests that three big warning signs in the oil industry point to a counter-narrative of impending problems for supply: high decline rates, soaring capital expenditure and falling exports. The decline rates of all conventional crude-oil fields producing today are spectacular, he shows. The International Energy Agency projects output falling from 69 million barrels per day (bpd) today to just 28 million bpd in 2035. Current total global production of all types of oil is some 91 million bpd.

Consider the spending needed to try to fill that gap, Lewis reflects. Capital expenditure for oilfield development and exploration has nearly trebled in real terms since 2000: from $250bn to $700bn in 2012. The industry is spending ever more to prop up production, and its profitability is reflecting this trend, notwithstanding an enduringly high oil price.

Meanwhile, consumption is soaring in OPEC nations. As a result, global crude-oil exports have been declining since 2005. It is difficult to conflate this data and not see an oil crunch ahead, he concludes.

What of the recent addition of two million bpd of new oil production from American shale: the boom that has even been cast as a "game-changer" and a route to "Saudi America" by industry cheerleaders?

We have asked Geological Survey of Canada veteran David Hughes to address this question, straight after Mark Lewis. Hughes has conducted the most detailed analysis of North American shale of anyone outside the oil and gas companies. He now offers some sobering views that fall well outside the incumbency narrative. His data show that spectacularly high early decline rates in existing shale gas and shale oil wells mean high levels of drilling are needed just to maintain production, and that this problem is compounded because "sweet spots" – zones of rich hydrocarbon accumulation – become exhausted early in field development. As a result, Hughes says, shale-gas production is already dropping in several key drilling regions, and production of shale oil in the top two regions is likely to peak as early as 2016 or 2017. These two regions, in Texas and North Dakota, comprise 74% of total US tight-oil production.

Like Lewis, Hughes believes that the oil and gas industry is leading the world by the nose towards an energy crisis.

In *The Energy of Nations*, I described how military think-tanks have tended to side with those who distrust the cornucopian narrative of the oil-and-gas incumbency, like Lewis and Hughes. One 2008 study, by the German army, puts it thus: "Psychological barriers cause indisputable facts to be blanked out and lead to almost instinctively refusing to look into this difficult subject in detail. Peak oil, however, is unavoidable."

It is interesting to me that the militaries of the world are tending to show so much more interest in systemic energy risk issues than politicians are. That interest is extending beyond analysis to action. The US Army now has a $7 billion renewable procurement programme, one of the biggest in the world, and the US Navy and Marine Corps have started training veterans for civilian jobs in clean energy. You can almost persuade yourself that they see writing on the wall.

One of the attendees at the transatlantic dialogue is Rear-Admiral Neil Morisetti, who was Commander of UK Maritime Forces between 2005 and 2007. He is so concerned about climate risk that he spent 2013 as Special Representative for Climate Change at the UK Foreign Office. At the transatlantic dialogue, he shows that he harbours grave misgivings about oil security too.

I am encouraged by all this, but very cautiously so. Edward Snowden's revelations about illegal intelligence gathering by the spy agencies on both sides of the Atlantic, the NSA and GCHQ, are still echoing around the world since he leaked them back in June. Both agencies have been spying on the OPEC oil cartel, leaked documents show. I suppose that is to be expected, given the primacy of energy supply in national security. What is not expected is the extent to which these agencies have been spying on their own citizens.

As a result of his leaks, Snowden is variously viewed as a hero or a traitor. Whichever he may be, he faces the prospect of 30 years in jail if ever he returns to the USA.

On the day of the transatlantic dialogue, 500 of the world's leading authors make their views about Ed Snowden's status very clear. They issue a statement professing that state surveillance of personal data is theft. They demand a digital bill of rights to curb the abuses. I'm with them on this.

Which means I can barely imagine who is listening when I have my Skype calls with my collaborator in the Pentagon.

I raised the issue with Daniel once.

Yep, he said. Morning fellas!

Chapter 5

If only he had the money

New York, 13th January 2014

Eight people sit today in the home of the richest man who ever lived, John D. Rockefeller. Rockefeller founded Standard Oil, the most powerful oil company ever to have drilled the planet, parent of ExxonMobil and other modern American oil giants.

The eight are talking about climate change and the oil industry. What do you think the drift of their discussion might logically be?

If the latter-day antics of ExxonMobil and its peers are anything to go on, then surely the conversation would be about how to take the air out of the tyres of anyone suggesting climate change might be a reason for cutting oil dependency.

But no, the eight people are talking about the reverse: how to take the financial air out of big energy's tyres. How to turn off the lifeblood of the oil, gas and coal industries, and force them to change beyond recognition.

Let me explain this counter-intuitive state of affairs.

Rockefeller co-founded the Standard Oil Company in 1870. With kerosene and gasoline growing increasingly vital in the industrialising world, he accumulated a staggering fortune by the time of his death in 1937. By 1880, Standard Oil was refining over 90% of all oil in the USA. By the time of his retirement in 1897, Rockefeller's company was part of a tight cartel with a handful of other oil majors including the forerunners of BP and Shell, controlling all aspects of oil production, refining and distribution, everywhere in the commercialised world.

The business methods Rockefeller used to gain such control of the exploding oil market were controversial, to say the least. A New York city newspaper of the day stated the case succinctly. Standard Oil, as the New York World saw it, was "the most cruel, impudent, pitiless, and grasping monopoly that ever fastened upon a country."

Along the way, Rockefeller bought a 3,000-acre estate, Pocantico Hills, and built a 40-room mansion on it, Kykuit, with views sweeping across the Hudson Valley and off to the New York skyline twenty five miles to the south. It was and is a beautiful place. Someone once said of it: "It's what God would have built, if only he had the money".

Standard Oil spent a lot of time in court, facing much ire from actors ranging from small drillers, claiming they were being cheated, to the US government, feeling their democratically-bestowed power was being usurped. But even Rockefeller's biggest legal setback turned into an eventual triumph. In 1911, the Supreme Court of the United States found the company to be in violation of a piece of legislation called the Sherman Antitrust Act. The court ruled that Standard Oil operated illegal monopoly practices and ordered it to be broken up into 34 smaller companies. Among these were Standard of New Jersey, which became Esso (and later, Exxon), now part of ExxonMobil; Standard of New York, which became Mobil, now part of ExxonMobil; Standard of Indiana, which became Amoco, now part of BP; Standard of Ohio, which became Sohio, now part of BP; and Standard of California, which became Chevron.

You get the picture. The US Supreme Court succeeded in chopping the head off one giant only to create a veritable hydra of giants.

But like many a human being, John D. Rockefeller was complex. One of his biographers wrote of him: "his good side was every bit as good as his bad side was bad. Seldom has history produced such a contradictory figure." The good side turned him into one of the most prolific philanthropists the world has ever seen.

Among his many substantial acts of benefaction, he set up the Rockefeller Foundation in 1913, an organisation with a simple mission: "to promote the wellbeing of humanity throughout the world". And so it has, on the whole, perhaps especially in the field of medicine. In 1940, members of the Rockefeller family continued the tradition by setting up the Rockefeller Brothers Fund. This foundation became concerned about climate change before governments even began negotiating a climate treaty in 1990.

The Rockefeller Brothers Fund is an early funder of Carbon Tracker. And it runs a conference centre for the use of its grantees on the Pocantico Hills estate. That is why the Carbon Tracker team is here today, plotting. Key foundation advisors, including Tom Kruse of the Rockefeller Brothers Fund and Richard Mott of the Wallace Global Fund, are here in the room with us. Another advisor cannot be with us, and she too is Rockefeller-connected. Joanna Messing, principal advisor to the Growald Family Fund, essentially kick-started Carbon Tracker. She called me after the Copenhagen climate summit in 2009 asking if I intended just to write about the disaster of all the capital that would now flow to coal, or whether I intended to do anything about it. I said I knew a man who was intent on doing something about it. That is how Mark Campanale found his first major funding for Carbon Tracker: from Paul and Eileen Growald, who run the Growald Family Fund, via Joanna Messing. Eileen Growald is a fourth-generation member of the Rockefeller family.

I find the irony of it all somewhat disorientating. I wonder what the man who essentially spawned the global oil industry would make of the fact that a tiny but impactful part of his fortune is now being used, with the licence of his descendants, to try and confound the industry he so loved.

What would he have made of climate change itself, I wonder, had he been alive half a century after his death, when the concerns about greenhouse-gas emissions began to emerge? One thing he would surely have been was well informed. He ran a formidable spy network.

To my great relief, Mark and I have found a Chief Executive to lead the tiny team of analysts at Carbon Tracker. Anthony Hobley, a lawyer who specialises in climate change, is bravely taking a significant drop in salary in order to do the job. But he now leads an encouragingly influential climate-change think tank, one with perhaps an outside chance of influencing the outcome in Paris and the course of global carbon emission reductions beyond.

As I watch the team in action today, Anthony is clearly in his element.

Also with us is Mark Fulton, former head of research at Citi. He has set up his own think-tank, Energy Transition Analytics (ETA), and networked a small group of old and new star analysts to work on the Carbon Tracker mission. His involvement with us dates back to a retreat for friends and advisors back in June 2013, in the Scottish castle home of my friend and legend of responsible investment, Tessa Tennant. At that event, pushed by Mark to aggressively engage markets on their own terms of supply-demand and commodity prices, we decided that Carbon Tracker should have a major focus on capex, and that a twinned think-tank approach would prove to be an advantage. ETA would stick closely to the pure analytics of carbon in the capital markets. Carbon Tracker would do that too, led by James Leaton, but will also act as an advocacy organisation.

We are not the only people plotting. One of Anthony Hobley's first actions has been to send an e-mail to the team saying that an insider has informed him that the oil industry's environmental front group, IPIECA, has set up a working group solely focussed on the carbon bubble idea, with a brief to shoot it and Carbon Tracker down. All major oil and gas companies are involved. They are gearing up quickly, no doubt in strategy deliberations not unlike ours today. They will be retaining their own experts to go through Carbon Tracker's work with a fine tooth comb, looking for bricks to pull out.

It seems we are destined to be engaging in a public battle of words with them. On the tab of their main founder.

New York, 15th January 2014

A blue shiver of a day. Ice chunks float in the East River. The North Polar Vortex has shifted south and sits over the northeastern states, pushing temperatures well below zero, way below normal. Scientists talk of a sudden warming in the upper atmosphere having been responsible for all this inconvenience in the lower atmosphere.

They call it climate *change*, after all.

The extremes seem increasingly clear around the world. Mostly they involve heat, not cold. In Australia, in a record-breaking heatwave, bats have been falling dead from the sky. Tennis matches have been postponed at the Australian Open. Bushfires threaten the exhausted suburbs once again.

I could fill paragraphs like that, for every continent, over this last year. Thirteen of the hottest-ever years have been in the fourteen since 2000.

Here in America, the US State Department said on the second day of the year that it will be making a priority of climate change. Secretary of State John Kerry has told all his embassies that the USA intends to become the lead broker of a climate deal in Paris, and that diplomacy on the ground should reflect that goal from now on.

If governments are to be brokered in this way, much must surely change in the interim in the way the capital markets view climate change. And here too America is shaping as though to lead. Today in the UN's historic Economic and Social Council chamber, a remarkable event is taking place. Five hundred investors have gathered for a summit to discuss investment and climate risk. These are people with fully $22 thousand billion of funds under management, much of it in fossil fuel companies, inevitably.

Scott Stringer, newly elected Comptroller of New York City, the man with responsibility for the pensions of New York city workers, is the first speaker of the afternoon. He bats for a city hit by devastating recent reminders of climate change and its power to harm economies, he reminds everyone: hurricanes Sandy & Katrina. We must take strong proactive steps without delay, he says. Climate change is the defining issue of our times. As fiduciaries it is our responsibility to shape our portfolios for the next generation. The word 'pension' usually makes people think about seniors, but pensions are also for the young.

He offers an example. A 25-year-old firefighter.

This is an emotionally symbolic understatement that will be lost on nobody here.

New York is one of the most vulnerable cities in the world, Stringer continues. The symbolism of sandbags piled up against the New York Stock Exchange during Sandy should be a message to all. We in New York are ready to act. The city has joined a campaign to ask the world's leading fossil fuel companies how they are going to plan for climate change. These companies must begin to plan for the long-term transition to renewable energy.

Stringer refers to a letter circulated by Ceres, an American organisation dedicated to mobilising business leaders to advocate a sustainable future, in partnership with Carbon Tracker. It is Ceres that has convened this summit at the UN today. Our letter has been signed by a group of 70 global investors managing more than $3 trillion of collective assets, from the $22 trillion represented at the summit. It has gone to the world's 45 top oil-and-gas, coal, and electric-power companies. It draws their attention to Carbon Tracker's April 2013 Unburnable Carbon report. It reminds them that Carbon Tracker has found that in 2012, the 200 largest publicly traded fossil fuel companies collectively spent an estimated $674 billion on finding and developing new fossil fuel reserves some of which may never be utilized. The letter suggests there is an opportunity today to redirect this capital, rather than it being wasted on high carbon assets that could very well become stranded. The financial institutions request detailed responses from the 45 companies ahead of their annual shareholder meetings in early 2014. Investors signing the letters include California's two largest public pension funds, the New York State Comptroller, F&C Management and the Scottish Widows Investment Partnership.

On October 25th last year, the Washington Post ran a story about this letter. "Climate regulations could cost fossil fuel firms trillions", the headline read. Jack Ehnes, the head of California's State Teachers' Retirement System, one of the pension funds, explained why his organisation signed. "As long-term investors, we see the world moving toward a low-carbon future", he says, "in which fossil fuel reserves that companies continue to develop may actually become a liability."

A liability.

The American Petroleum Institute was dismissive. "This is either delusion or wishful thinking on the part of some folks who just don't like fossil fuels," its chief economist told the press.

Really? $3 trillion worth of wishful thinking?

We will see. Events here today in the UN are very much the beginning of the seeing.

Ceres's chief executive, Mindy Lubber, formerly a senior Environmental Protection Agency official during the Clinton years, architect of the summit and driving force behind the letter, now takes the microphone. She charges me with chairing a small panel to discuss the carbon bubble. On that panel is Anne Simpson, a senior executive at CalPERS, the pension fund of California's public employees, the largest pension fund in the United States. They have $270 billion under management.

Speaking into a microphone from which government ministers will have discussed matters of life and death many times, looking out over so many people controlling so much money – people with such power to move societal change in the life-or-death direction I and so many like me long for – I tell myself I have to be an actor. I must not for a moment show the edges of emotion and desperation that I feel inside.

There is a curiosity in my head that helps me through, on big occasions like this. I was an academic in the 1980s, albeit one on the oil industry's payroll, mostly. There is a part of me that is capable of holding back, of analysing, of chronicling, like a heartless old professor. That part of me can hardly wait to hear what Anne Simpson has to say today.

We have a responsibility to 1.6 million members, she opens, and our liabilities extend the best part of a century into the future. Making sure the capital markets work is absolutely fundamental to paying pensions. We approach this responsibility through a sense of fiduciary duty aware that we are owners of these fossil fuel corporations. The risk we face is systemic. We need the system, not just the companies, to work.

CalPERS is among the pension funds that have signed the Ceres and Carbon Tracker letter. Anne Simpson explains their motivation in so doing. There is no safe investment place for CalPERS, she says, if the world is en route to four degrees of global warming. So we are saying to the fossil fuel companies that if this risk is on your balance sheet, then we want you to go back and rethink.

She moves to the endgame, as her fund sees it. We are going to build demand for renewables, and to help do that we are going to ensure that energy efficiency improves across our asset classes. We see this work with the fossil fuel companies – she means the letter – as one piece of an overall climate strategy for our portfolio.

My job as chair is to lead the questioning of panellists. I ask the obvious one first. How are the companies responding to the letter?

It is early days, Anne Simpson replies. For some of the companies our approach is welcome. Chevron has been calling in its advertising for a debate on energy for a while now. Others have been used to dealing with the issue through their public relations department. Then there is the more difficult group of companies. Here we will need to be talking at length. It's a bit like the conversations that have been going on with the tobacco companies. Except this is different. Fossil fuel companies are about systemic risk.

Another mention of systemic risk, I think. I wonder if the Bank of England appreciates that huge American pension funds view the climate issue through such a prism: that fossil fuel companies might represent a risk to the very viability of the global financial system. I wonder how John D. Rockefeller would view it, if he were alive today. His descendants are fighting hard to turn the tide, with his money. Would he have helped them, or stood in the way, reaching deep into his pocket to bankroll a rearguard action, like the bosses of the modern oil industry mostly do?

Davos, 22nd–26th January 2014

For this one week every year, the sedate Swiss mountain town of Davos becomes the place to be seen for many people in the global elites. They come to the World Economic Forum annual meeting both to network, and to party. Many focus on trying to improve the world while they are here with their peers. In some of these efforts, they do make progress. But in one measure of societal improvement they fail monotonously, each and every year. Global inequality worsens. Stated another way, they and the rest of the one-per-cent get richer and the poor get poorer.

This year, Oxfam has made the remarkable calculation that the 85 richest people on Earth now control wealth equivalent to that owned by fully half the people on the planet, some 3.5 billion souls. I read the press coverage of Oxfam's calculations, amazed. I know that this imbalance exists. But I am constantly surprised at just how extreme it is, and how it just keeps on growing, despite everything we know – or ought to know – about how it threatens social cohesion.

From Zurich I take the train up into the snow clad mountains. Swiss tourism propaganda panoramas unfold around me under a blue sky. Helicopters throb by as we near Davos. I wonder how many other participants have travelled here, like me, via EasyJet and rail.

I arrive, check in through the intense security cordon, make my way through the immaculately dressed crowds at the Hotel Belvedere to my first event. So many very fine suits. So few skirts. Here is evidence of another form of inequality. Only 15% of the attendees are female. Until the evening parties, that is.

The event I am attending is about the problem of short-termism. McKinsey and the Canadian Pension Plan Investment Board have collaborated on a study of how investors might look beyond their ingrained practice of focussing on the next two quarters. Forty business leaders gather over a very fine dinner to hear their conclusions, and discuss them.

Someone has a sense of humour among the organisers. I find myself seated next to the CEO of Suncor, Steve Williams. Suncor, a long-standing operator in the Canadian tar sands, is the world's largest producer of bitumen. Its operations are one of the biggest single corporate point sources of greenhouse-gas emissions on the planet.

We hear that an opinion survey of hundreds of CEOs shows that most of them blame boards more than investors as the primary drivers of short termism. Boards are increasingly populated by white men averaging 57 years of age, we hear.

What about compensation of asset managers, someone asks. Doesn't that skew market behaviour towards short termism?

Yes, says the man from McKinsey. That too is short term. Compensation metrics for asset managers span six months on average.

There's your core problem then, I mutter.

Steve Williams hears me.

Yep, he murmurs.

Lord Nick Stern addresses the gathering on the economic threats of climate change. A former chief economist at the World Bank, Stern is one of the strongest advocates of climate action among the stars in the field of economics.

Williams quietly snorts disagreement this time. I stay quiet.

Later, in the discussion, Williams has a point to make. He has just got a fifty-year investment project away in the tar sands, he says. It was extremely difficult to do. Politicians on 5 year electoral cycles are too prone to go with the latest opinion poll, he grumbles. He is in absolute favour of long-term investment that generates value.

The event ends without a chance for me to offer a counter view to his. So, as we leave our seats, I give it to Williams direct.

If you follow through on that fifty-year investment cycle, I say, and Lord Stern and all the rest of us are half way right, aren't you in danger of people not saying thanks for managing my pension savings so well, but see you in court for wasting them?

People who think that don't understand, Williams replies. We'll still need oil 50 years from now. Nothing has the energy density of oil. Clean energy can't do the job oil does.

I politely explain where I come from, and how very bullish my alternative view is.

Clean energy can't be economic, he scoffs. You won't get the investors.

But we are already in the process of doing that, I say. Investment in renewables is around a quarter of a trillion dollars a year. Doesn't that make you worry just a little about your fifty-year plan?

The parties are something to behold. They begin at six and finish in the early hours the next day. I begin my tour with Unilever's, and rapidly discover that such an occasion is a name-dropper's paradise. The first three people I talk to are the Kingfisher CEO, Ian Cheshire, the Unilever CEO, Paul Polman, and Richard Branson. With the first two, I do my Carbon Tracker duty. They need no persuasion: both are in the vanguard of progressive corporate response to climate change. With Richard, I take that hat off and put on the solar one. He knows all about SolarAid from previous meetings, and wants to hear how it is going.

I resolve to pace myself for the long evening: to alternate water and wine. By the time the Coca-Cola party finishes, that seems to be working. I can see a good few people obviously more animated than myself. By the end of the Time Fortune party, I am not so sure. I shuffle my way through the snow to the Financial Times party, and find myself an early arrival. That wins me a solo chat with Gillian Tett, the FT journalist who made her name by spotting the financial crash brewing, and warning about it when the investment banking industry was at the height of its bullying exhortation of mortgage-backed securities as a route to fabulous new wealth. This is my opportunity, I figure, to try and persuade Gillian that another bubble is brewing in the markets. In the shale boom, I suggest, the energy sector is in the process of repeating much the same catastrophically deluded asset-evaluation that the financial sector did in the sub-prime mortgage fiasco.

I do my best. On my day, I can line arguments up in something resembling a logical train of thought. This is not my day. My arguments seem to be swimming in a dilute river of Time Fortune's very fine wine and my own testosterone.

I recognise the "desperate to be away" look appearing on Ms Tett's face, and I give up, as a gentleman should. She slides away, as decorously as such people do, and I look for my next victim in the by-now VIP-packed room.

Right next to me is Jamie Dimon, CEO of JP Morgan. He has just paid £2.6bn to US federal authorities to head off a criminal prosecution over the Bernard Madoff fraud. JP Morgan bankers had concerns about Madoff for more than ten years as he spun his vast Ponzi scheme, but failed to inform US authorities, according to the US attorney for Manhattan.

Now this will surely be an interesting conversation.

But a small invisible creature now seems to be perched on my right shoulder, whispering urgently in my ear. I have fairly regular visits from this creature, I am ashamed to say. Most spectacularly on the occasion when I took a Harley Davidson out for a test drive.

Go to bed right now, my son, it is saying. Or words to that effect.

On this occasion, I obey.

I'll have to experience the notorious McKinsey nightcap party some other year. Maybe.

I am in Davos with Carbon Tracker colleagues Anthony Hobley and Mark Campanale. We split up each day to try and maximise our chances of interacting with key influencers in the corridors of power. We fancy we have an increasingly strong story to tell them. Thus far in January Storebrand has excluded another ten coal companies from its investment portfolio. HSBC has produced a report concluding that carbon risk to long-term coal demand is growing: that a future wherein governments are intent on low-carbon economies could cut valuations of UK coal mining stocks by 44%. European governments have bolstered the credibility of that scenario for the future by agreeing that the EU will cut its greenhouse gas emissions by 40% by 2030, compared with 1990 levels. This is the toughest climate-change target of any region in the world.

We organise a reception for delegates in the bar of a hotel on the fringes of town. Our budget does not extend to the inner sanctum. A gratifying number of important players trudge through the snow to drink our champagne.

Christiana Figueres opens the evening for us.

I would like to show you what finally showed me exactly what we have to do about climate change, she says. This.

She holds up Carbon Tracker's 2013 report, open at the page where a graphic summarises our recommendations.

This is absolutely the best graph I have ever seen, says the executive secretary of the climate negotiations. You get a prize for communications, for strategic thinking, for developing the message and the task for every one of the constituencies.

Anthony, Mark and I stand among the assembled government officials and executives, trying not to look too pleased with ourselves. James Leaton, designer of the chart, is not even here to listen to this highest of praise for his work.

I do my best to sound alarms on the shale bubble and oil depletion, but it is like pushing snow uphill. "Davos delegates warned of imminent energy crisis," a website story headline reads. "A British businessman will tell world leaders meeting in Switzerland today that it is dangerous to argue that fracking for shale oil and gas can help to avert a global energy crisis." They will be among only a few listening. The World Economic Forum produces a Global Risk Report each year. The 2014 version, compiled by 700 experts, analyses 31 risks in the global economy. An energy crisis is not among them.

My World Economic Forum Global Agenda Council, the expert group chaired by former Shell CEO Jeroen van der Veer, did consider my arguments. They also weighed the happier narratives of shale cheerleaders in the group. I failed to make the impression I had hoped.

David Cameron is very clear where he stands on all this. Before coming to Davos, he had announced that the UK would be going "all out for shale", and that those opposing are "irrational". To demonstrate his commitment, he said councils will be entitled to keep 100% of business rates raised from fracking sites. This would generate millions of pounds for local authorities.

This is the kind of support that renewable energy can so far only dream of in Cameron's Britain.

Now in his speech to the Davos plenary the UK prime minister says he aspires to frack cheap shale gas from below UK soils on a scale sufficient to lure big manufacturers back to the country.

It is, I have to admit, a very alluring narrative for a political leader. How inconvenient he is going to find it when he discovers that it is based on illusion.

My hope is that he will discover this in 2015, in time to change course before Paris. Deprived of cheap gas, he will have only one place to go, so I reckon. Nuclear is being exposed as an unaffordable white elephant. He will have to go for a green industrial revolution. Exactly what a meaningful Paris treaty will mandate.

Chapter 6

He who brings light

Kenya, 3rd February 2014

A crack-of-dawn flight from Nairobi, west to Eldoret. A field team of three from SolarAid's retail arm, SunnyMoney, meets me at the airport, and we drive in a van nearly two hours further west, to Bungoma, a market town near the border with Uganda. Along the way, greenery; fields of maize, cropped; sugarcane, uncropped. This is the breadbasket of Kenya.

A street in that teeming town. The Wells Fargo depot: one of the booths in one of the concrete multi-shop facades with hand-painted signs, next to a wood yard. Every second shop seems to bear a green M-Pesa logo. This pay-by-mobile brand has gone from nowhere to everywhere in the last ten years. A little of this kind of market penetration is what SunnyMoney will need, I reflect, if SolarAid is to sell enough lights across Africa to achieve our mission of playing a lead role in ridding the continent of kerosene lanterns by the end of the decade.

We fill the van with freshly-delivered boxes of solar lights. Thirty boxes, ten lights apiece. I do a rough calculation. We have sold 500,000 in the last six months, across our four countries of operation. That's 1,600 loads like this.

In the centre of town sits a Total filling station. We have seen several such along the way. Total is the second biggest seller of solar lanterns in Africa, after SunnyMoney, retailing from forecourts such as these. They also sell kerosene, so you could say they are hedging their bets. En route we passed the oil refinery that serves this region, dozens of tankers in a queue stretching away from it, sprawling along the roadsides, waiting to be filled with the derivatives of crude oil delivered by pipeline from Mombasa.

Further along the road to Uganda we turn right, and drive five bouncing miles down a red earth road, spewing dust behind us. We pass a school, seemingly miles from the nearest village, yet with several hundred schoolchildren running around in identical green uniforms in a spacious dirt schoolyard.

How do they get to school, I ask Victor, the team leader.

They mostly walk, he says, most of them many kilometres, starting off in the dark.

There are more than 120 schools like this just in the district we are in, Bumula.

We come to the small town that gives the region its name. We park in a yard outside the Education Commissioner's office. Eighty headmasters and headmistresses are gathered in a meeting room.

This is our main route to market: via the education authorities, through the schools. We call it the SunnyMoney Way.

The Commissioner introduces us.

Our guests have a very useful story to tell, he says. I myself have already heard it from Victor. But before you hear it, let us all pray.

A headmaster chosen at random delivers a simple prayer, a set of thoughts about God and wisdom that both Christians and Muslims could easily sign on to.

Victor follows. He has clearly done this many times before. He walks the headteachers through three types of solar light: their benefits, their prices, how they can be ordered. He is a born salesman, working his audience like a revivalist preacher. The head-teachers respond good naturedly. He jokes that once they taught him, and now look, he is teaching them. I hear a lot of laughter this morning.

Jamie Arbib, a funder of SolarAid and Carbon Tracker, is with me on the trip. I catch his eye and he beams, enjoying this as much as I am.

And what is it called, Victor asks.

A Sun King Eco, they chant.

And what does it cost?

A thousand shillings.

And what does kerosene cost on average each year?

Five thousand shillings.

Are we *together* on this?

Yes, they shout.

I am asked to say a few words. I talk of the reasons I am here today: how Solarcentury, a solar company I founded in the UK, gives 5% of its profits each year to a charity of its own design, SolarAid. How that charity has a mission to help bring solar lights in Africa. I recount my own experience of the benefits of solar lights. How the statistics on cost savings of solar versus kerosene inspire donors to SolarAid, the charity that wholly owns SunnyMoney, the retail brand whose profits – once we reach them – we are pledged endlessly to recycle, for social good. How the health impacts of kerosene – the fire deaths, the poisonings, the air quality illnesses – shocked me when I first learned of them. How proud and encouraged I felt when the first stories of rising grades in solar schools started coming through.

Finally, I say, there is a fourth and very important impact of solar lights. It is a connection to a bigger picture: the 25-year-long struggle of governments to deliver a treaty that can stop dangerous climate change by phasing out the burning of fossil fuels, like oil.

Burning kerosene for lighting is fully 3% of global oil use, I say. By working together in our different countries to phase out the kerosene lantern, people like us can create a microcosm to inspire others to phase out all fossil fuels. Please tell your students that together we can light a great big candle for hope in the world.

I stop short of Vincent's tactic of grilling his audience like a headteacher grilling pupils. But I see a satisfying number of heads nodding.

The Commissioner sums up and thanks us. I will never forget his last sentence.

He who brings light, he enthuses, brings…

He holds out his arms in silent invitation.

Life! they explode.

Oxfordshire, 25th February 2014

From African sun to British rain, the worst in more than two hundred years. The winter downfall has been so strong, so persistent, so abnormal that the Chief Scientist at the Meteorological Office has invoked climate change as an explanation. That was this month. UK Prime Minister David Cameron blamed the rain on climate change back in January.

Today I am visiting Cameron's constituency, inspecting a Solarcentury solar farm under construction. Amazingly, as the rain falls once again, work is actually still underway.

With a handful of colleagues and a trio of contractors, I wade through calf-high mud the colour and consistency of parsnip soup, inspecting rows of panels as they rise from the gloop and march off across the terrain. All over the vast site, a platoon of workers carries solar panels from caterpillar trucks and affixes them to frames. They are German, working on contract.

A hundred years ago, I reflect, the Germans and British were engaged in altogether less constructive activities under the rainy skies of Northern Europe. I am prone to such thoughts, by way of celebration that I have been alive in a half century without world war.

The modern work is infinitely less dangerous, but not absent of danger. Everyone in my party has just had an exhaustive health and safety briefing. I have seen the prominent notice in trailers telling the troops where the nearest hospital is. We know the dangers of working in such conditions: health and safety is the first item of business at every board meeting. We only have to face the dangers for a day. The field workers face them every day for weeks per site. Then they move on the next one. Energy is a difficult business, however you come by it.

This 64-acre site will generate 13 megawatts of peak power when the Anglo-German team has done its job. That is enough electricity to power the equivalent of around 4,000 homes while saving 6,000 tonnes of carbon dioxide emissions per year. The site will be planted with wildflower meadow seeds designed to attract butterflies, bumblebees, and birds. Come next summer, this solar farm will be a biodiversity oasis.

Installations like this are going up in many places around the world now. In 2013, China set a new world record for annual solar installation: twelve gigawatts. In the USA, where renewables provided 37% of all new generating capacity, jobs in the solar industry are growing at ten times the national average employment growth. There are more than 140,000 solar workers in America now.

At Stanford University, a professor has envisioned what it would take to provide all the energy of each individual American state just with renewables, and to do it by 2050, 36 years from now. Mark Jacobsen's renewable resource roadmaps cover all electricity, heating and transportation demand. He calculates the mix for each state, without using biofuels. His California scenario, for example, is 35% wind, onshore and offshore, and 55% solar: as solar PV and solar thermal – mirrors used to heat water with the power of the sun – both big solar farms and distributed on rooftops. The rest comes from geothermal and hydropower, plus extensive use of energy efficiency to reduce demand. By blending wind with solar, and combining that well-matched duo with hydro and solar thermal with storage, intermittency problems can be smoothed out, Jacobsen concludes. Society wouldn't have to invent a new technology to get this to work, he insists.

The news is better, for those prepared to believe energy storage enthusiasts. Big advances can be expected in the means to store electricity in 2014, analysts are saying. Important research is underway in Japan and Germany. But it is in America where the most exciting prospects lie, as things stand. In California, the Public Utilities Commission has set a goal of achieving 1.3 gigawatts of storage by 2020. SolarCity, a fast-growing solar company mostly installing leased solar roofs on homes, plans to use the battery technology used by Tesla, California's premier electric-vehicle manufacturer, first to offset electricity costs for businesses, and later to power homes. Tesla founder and CEO Elon Musk, a fabulously successful serial entrepreneur, is also chairman of SolarCity.

If this happens, it will add very tasty icing to Jacobsen's cake. He or she who brings light will be able to put it in a box and use it at night, when the sun isn't shining. He or she will be able to mix and match renewable-energy supply without having to balance the system using different types of renewable. He or she will be able to bring altogether more light than before, and, as electric vehicles are increasingly charged using solar, he or she will be able to bring emissions-free motive power to society.

This, perhaps more than anything else, is what is going to change the face of the energy industry, threatening the business models of both energy utilities and the oil-and-gas industry.

Chapter 7

Duty, bubbles and neuroscience

Brighton, 25th March 2014

A British Member of Parliament is in court today, charged by the police with endangering the public at the Balcombe anti-fracking protests. On August 19th last year Caroline Lucas sat down in the road outside the main gate to the Cuadrilla drill site with several fellow protestors. The police told her to move on, she and the others refused, saying they were exercising their right to protest a dire threat. They were arrested, charged with obstructing a public highway. The highway had been closed by the police earlier, for the duration of the protests. I saw this myself. I passed the roadblocks on the Balcombe road at both ends. Activity had also stopped at the drill site. No vehicles were coming or going.

Caroline is the only Green MP in the British parliament. She has represented the Brighton Pavilion constituency since 2010, and done so with skill and verve. In her first year, she was voted Newcomer of the Year at the Spectator's Annual Parliamentarian of the Year Awards. The Spectator is a conservative organ not known for its appreciation of matters green.

Caroline has asked me to come to the court house today and make myself available for interviews about fracking. She is not able to speak to the media herself, under the conditions of the court case. I am happy to comply. I stand in my suit and tie among less traditionally dressed and much noisier supporters of the Green heroine. A Dutch TV film crew, following me around on a supposedly typical day in my life for a film about my views and their implications for Shell, adds another dimension to the colourful and diverse scene.

ITV News decide they want an opinion on fracking from the man in the suit. First, the protestor with the mobile disco has to be asked to turn it off.

Where do I start, I wonder, standing before a very young journalist keen to emphasise that she has no knowledge of the issues whatsoever. I could talk about the miserable implications of fracked gas and oil for climate change,

Caroline's main motivation for protesting at Balcombe. Or I could major on the increasingly clear downsides of fracking for drinking water quality. Or I could talk about the ruinous economics of the US shale. Or maybe try to squeeze them all in.

It is the economics that is the fastest-emerging story. Bloomberg has reported that independent oil and gas companies will this year be spending $1.50 drilling for every dollar of income from oil and gas sales. This is the case despite the high-price oil ameliorating the low-price gas in the mix of hydrocarbons that the companies sell. Even the Oil and Gas Journal, a stalwart industry defender, has noticed that there might be a problem. They report $35 billion of write-offs in shale investments by 15 of the main drillers, observing in an editorial that this raises "financial questions".

How very astute of them. People worried about their pensions being wasted would have been quite as justified in protesting at Balcombe as Greens worried about climate change. Maybe that is the point I should make. After all, much of the rich south of England votes Conservative, and many Conservatives do not seem able to worry about climate change.

Journalists writing the Financial Times's Lex Column are now among those scanning ahead for victims of the potential shale train wreck in the States. "Are US LNG export terminals the next expensive flop in energy?", they ask. LNG stands for liquefied natural gas. Gas has to be liquefied for export by tanker. Members of Congress are clamouring for faster approval of permits for LNG terminals so that US natural gas can be sold into global markets. Yet the first such terminal will not become operational until 2015, at a cost of around $10 billion. Builders of later terminals, still in planning, will have plenty of time in which to risk stranding their multi-billion dollar investments.

Art Berman, an insider veteran of the US shale saga, a member of the Transatlantic Energy Security Dialogue, is perhaps the most forthright of the critics in America. Production from shale is not a revolution, he says, it's a retirement party. The truth of the matter is that the industry has to make such a big deal out of shale because it's all that it has left. Where does it go once the shale story is done?

But the industry hype just seems to forge on, seemingly oblivious to all this evidence, and increasingly damning commentary on it by elements of the financial and trade press. It is helped in this by geopolitical context. Russia has seized the Crimea, and Russian-backed rebels are making moves on other regions of Ukraine. The Republicans are desperate, as a consequence, to deploy something they call the "oil-and-gas weapon". US Congress experts have touted

LNG exports as a means to punish Vladimir Putin for his expansionism. Hearings of the Senate energy committee are told that the US needs to export US gas to Europe in order to end Putin's "energy blackmail".

This mind-bending wishful thinking is not limited to the Americans. EU leaders have asked Obama for access to US shale gas exports.

Lt. Col. Daniel Davis and I have written an article in *The National Interest*, an American national security journal. Before contemplating the use of US oil and gas as a strategic weapon, it might be useful to review a few key fundamentals, we suggest. First, consider the oil production, consumption and import/export numbers reported by BP for 2012. Russia produced 10.6 million barrels per day (mbd), consumed 3.2 mbd, leaving 7.4 mbd available for export. The United States produced 8.9 mbd, consumed 18.5 mbd, and imported 10.5 mbd.

All the talk of America soon overtaking Russia as the world's largest oil producer comes with a rather sizeable asterisk: even if that eventually occurs, the US will still be required to import an additional five to six million barrels of oil per day, while Russia would have an additional 7 to 8 million barrels of oil per day to export. The truth is that this fact places the Russian Federation in a considerably stronger energy-security position than the United States.

In the UK, an article in the Financial Times concludes that our version of the shale-gas revolution may never reach the stage of being able to morph into a retirement party. The British Geological Survey might have doubled estimates for the shale gas reserves in the north of England, raising expectations of a US-style shale gas revolution, but just because the supposedly abundant gas is theoretically recoverable doesn't mean it makes economic sense to recover it. Explorers must drill actual wells to discover natural gas flow rates, the article observes. Few have been drilled. Few look likely to be drilled. The UK's shale gas potential remains largely unknown.

And then the killer point, inadequately emphasised for my liking: "exploration is expensive and it is easy to spend more on drilling a well than the value of gas that comes out of the ground. Drilling costs are significantly higher in the UK than the US."

The British Prime Minister remains doggedly and didactically unfazed by all this economic bad news. Fracking is good for the UK, he insists. To seek energy independence is our duty. Those who oppose shale gas have a lack of understanding.

Outside the county court, I face the young journalist from ITV, with all this swimming in my head, wondering how I can ever come close to summarising it in a coherent television soundbite or two.

I try.

My effort doesn't make the news bulletin.

Inside the court, Caroline Lucas is making her case direct to the judge.

I'm haunted by the idea that my children and my children's children will turn round to me and say 'what did you do about this overwhelming threat?', she says. And I want to do all I can do, peacefully, to address that before it's too late.

The Chief Constable and the Crown Prosecution Service don't agree. They argue that there is enough evidence to convict Ms Lucas of the criminal offence of sitting down to register her protest on an unused road, and that it is in the public interest to have her convicted.

City of London, 31st March 2014

A meeting with a legendary investor and philanthropist. Jeremy Grantham co-founded one of the world's largest asset management companies in 1977. Today Grantham Mayo van Otterloo, or GMO as it is universally known, has well over $100 billion under management. Jeremy has based much of his success as an investment manager on spotting bubbles and protecting his clients' assets from them. He saw the Japanese economic crash coming in the 1980s, and avoided Japanese equities and real estate. He spotted the internet bubble building in the late 1990s, and avoided tech stocks. Today, we are to discuss another bubble, the carbon one. Jeremy agrees with Carbon Tracker's analysis. The climate research school he funds at the London School of Economics co-authored our 2013 report on capex deployments by the fossil fuel industries and the scope for hundreds of billions of dollars to be wasted thereby.

We sit in a plain room in GMO's London offices, high in Number One, London Bridge, with a sweeping view east down the Thames to Tower Bridge and beyond. Jeremy's chef de cabinet, Ramsay Ravanel, sits with us. I have a sense that time is about to evaporate. There is so very much to discuss.

We start with climate. The Intergovernmental Panel on Climate Change has just published its latest report, this time on the impacts of elevated greenhouse-gas emissions. It is the harshest yet, taking the scientists' warning to a new level of consequence for the world. Climate change is already happening, they conclude. The Arctic is melting. Coral reefs are dying. "New normal" heat waves, deluges, and "megastorms" are hitting most continents. Crucial global food supplies are already declining, especially wheat. And the worst is yet to come, especially in our ability to feed ourselves, provide clean water, and avoid scope for military conflicts.

Jeremy Grantham tells me he views the human predicament as the greatest race of all time. On the one hand we have accelerating climate disaster, and the obduracy of the players rushing us ever faster towards that cliff. On the other hand, we have a global clean-energy shift, and the financing that needs to be mobilised to accelerate it enough to offer a chance of saving humankind.

We consider Carbon Tracker's potential role in this race, and our progress to date. In recent weeks Norway's oil fund has announced that it will debate stopping investments in all fossil fuels. It has been given a mandate by the government, in parallel, to invest in renewable energy. ExxonMobil has agreed to report to investors on fossil fuel asset-stranding risk. BP has confessed in its annual report that it agrees not all fossil fuel reserves can be burned.

Jeremy Grantham tells me that he has no doubt that Carbon Tracker has had a big hand in developments like this.

I ask him where he stands personally on the question of how much fossil fuel can be burned to give a reasonable change of a two degrees ceiling on global warming.

We could burn all the cheaper-production oil and gas, he replies. But we dare not burn all the coal, much of the tar sands and the energy-intensive oil and gas, including from shale. If we do that, we are cooked, we are done for.

We discuss tactical options for Carbon Tracker. Horses for courses, he recommends. Sometimes divestment is the right way to go, sometimes engagement. He agrees with the Carbon Tracker approach of focussing on engagement. He also supports 350.org's divestment campaign.

We move on to renewables. The potential for solar and wind to lead the way in completely replacing coal and gas for utility generation globally is pretty much certain, he argues. The question is only whether it takes 30 years or 70 years. Solar is getting cheaper by the minute, whereas petroleum is getting more expensive. It is only a matter of time before these trajectories cross. Solar and wind are competitive already in many locations. What is needed is a continuing steady drop in the cost-down trend of renewables.

We compare notes on companies seeing the writing on the wall. Software giant SAP is the latest aiming to power its operations worldwide with 100% renewable electricity. They aim to buy renewable energy credits to achieve the goal. Other companies are intent on the more difficult but ultimately resilient pathways to 100%: generating their own renewable electricity, or signing power purchase agreements with renewable-energy suppliers. Corporations on this road include Google, Apple, and Walmart.

We move to oil depletion. I tell Jeremy about the Transatlantic Energy Security Dialogue and the concerns of its members. Here too we find common ground. Frackers are the key marginal suppliers, he says. But the shale phase is going to prove to be another bubble. A great global oil squeeze would be upon us already were it not for fracked US shale oil. We have only put the squeeze off by five years or so.

I feel predictably encouraged that a man who has made a fortune backing his opinions in the capital markets, over more than a quarter of a century, can hold such clear views, today, so congruent with my own.

Which rather raises the question of why I have an overdraft, I suppose.

Inevitably, our conversation turns to the conundrum of why, if things are so blindingly obvious to us, so much of what we believe is disbelieved, ignored, or even ridiculed by so many others. Jeremy has an explanation. Most of the people running companies, federal agencies, and international institutions have both characters and job descriptions prone to focus on the short term. They tend to spend little time looking beyond the quarter's results or the annual budget. They appoint successors who have the same tendencies. So society ends up being run by an army of immediate doers. We need more people with a historical perspective, able to learn from history, who can map forward into the future with a degree of wisdom. Such people don't tend to get the top jobs. They are left to issue warnings – if they choose so to do – on the sidelines of the main game.

Before I know it we have clocked up two hours. Jeremy Grantham has not looked at his watch once in all this time. I am used to twenty minutes at most with people as famous and successful as he, with many glances at watches along the way. It is left to Ramsay Ravanel to remind us that there are other meetings in the diary.

I make my way through the packed mass of commuters walking to London Bridge station outside the office block, thinking about Jeremy Grantham's view of how important neuroscience is institutionally. I first learned something of what the neuroscientists had to teach us about our current dilemmas in 2010, at Oxford University. It was, as I described in *The Energy of Nations*, a revelatory experience. I listened engrossed as some of the most eminent neuroscientists in the world talked about what they had discovered of the "endowment effect": how humans tend to favour what they have to hand, over what they might have, even if the alternative is available and – on rational grounds – an improvement. They described experiments showing how irrational we tend to be in our individual and collective thinking: "predictably irrational", is how they talked of the typical human being. They also described how experiments reveal our desperate preference for good-news narratives over bad-news narratives.

Viewed in that context, it becomes clear why a politician like David Cameron can believe in a narrative of shale-plenty, and be in a majority compared to a politician like Caroline Lucas, with her belief in shale as a road to catastrophe. It explains why investors like those who backed Steve Williams's 50-year capital-expenditure plans for Suncor in the tar sands can be in a majority compared to the relatively few Jeremy Granthams of this world.

The mystery is why this state of affairs should persist, when Jeremy has made so much money, and investors in shale have to date lost so much. Here might lie one vital route to winnability in the carbon war, I reflect. If, as Jeremy Grantham points out, two emerging trends continue – the cost-down surprise in renewables, and the cost-up tendency in fossil fuel production – history might at some point be perceived by a critical mass of key influencers not to be destiny. If these trends turn into megatrends during the rest of 2014, and 2015, much might become possible.

There is something else. I scan the people flooding past me. So few smiles. So many frowns. So many people walking solo through this great city. But in the rest of their lives? Maybe not, in many cases. The neuroscientists have discovered in people a marked tendency to favour community over individualism. You can see why that might be, I reflect, people-watching on the streets of London. Here may lie another strand in the renaissance narrative.

Of course, all this cautious optimism has to be tempered with the obvious thought that we need the speed of physical climate change not to spin out of control, and the speed of policymaking, nationally and at the climate talks, to accelerate.

In her relentless pursuit of the latter, Christiana Figueres was in London earlier this month. Her target was big businesses. She spoke to leaders and laggards both. Her strategy is clear: she needs more laggards to become leaders – or at least not be saboteurs – if she is to deliver in Paris.

Paris, she told them, has to reach a meaningful agreement because, frankly, we are running out of time.

The Guardian shot a three-minute video of her while she was in town. We need to understand we are in a transformation, she says in it, the likes of which we have never seen before. It's a transformation to a completely new economy. And we have a ticking clock in front of us.

The interviewer asks her how she sees her own job.

This is the most important challenge that humanity has ever faced, she replies. Many people say that's the most difficult job in the world. I say it's the most sacred job in the world.

Chapter 8

I am here in learning mode

London, 24th April 2014

In the private dining room of a good Soho restaurant, I sing for my supper. Accenture's energy team have invited me to discuss the future of energy. There are twenty of them, and it soon becomes clear that they have a collective IQ that would win a lot of pub quizzes. The deal is that we eat haute cuisine, I outline my view of the future over the dessert, and then they grill me over the fine wine. They have all read *The Energy of Nations*, they tell me.

Accenture and the other major business consultancy companies hire the brightest and best from the top universities and business schools, and mix them with seasoned veterans of business to create a pool of intellectual brilliance, teams from which they hire out to companies in need of advice. In this world, there are always many companies in need of advice.

I figure it is a good idea to start by asking who they are working with at the moment.

Oil and gas companies? Most of the hands go up. Utility companies? A few hands go up. Renewable-energy companies? Not one.

I smile, shrug. I really don't need to elaborate.

I talk about the very latest news, and what it suggests about where we are going, at least to people like me. The cost reductions in the solar industry are shaping up to be utterly transformative, I say. Financial analysts are beginning to compete with each other in the use of rhetoric. Those at Alliance Bernstein have coined a particularly evocative term for the solar pricing trend, and how it threatens fossil fuel business models. They have produced a diagram, which I am certain will be much reproduced. It shows a range of average fossil fuel projects over the last 30-40 years. This is the lifetime of many such projects – the period they are judged over in terms of economics – be they coal mines, oil fields, central power plants, tar sands projects, or whatever. In 2006, the average solar price appears on the chart far above the lines wiggling along the bottom of the chart that depict the different kinds of fossil fuel project from 1970 to 2012. Between 2006 and 2012 the solar price plunges by 90% – almost vertically, given the multi-decade scale of the chart – down among the fossil fuels.

The analysts call the plunging solar line the "terrordome". The term evokes both a scary fairground ride and the sentiment such a price trajectory might create among owners of the technologies whose markets it threatens.

Evocative phraseology is not limited to the financial analysts. "Global solar dominance in sight as science trumps fossil fuels", reads the headline of a Telegraph article. It is written not by some overenthusiastic cub reporter, but by Senior International Business Editor Ambrose Evans-Pritchard. "Solar power has won the global argument", he concludes. "Photovoltaic energy is already so cheap that it competes with oil, diesel and liquefied natural gas in much of Asia without subsidies."

The Prime Minister and the Chancellor will be reading this over their cornflakes! The Telegraph is the newspaper of the choice of the British Conservative Party.

They will also be seeing Apple's global advertising campaign. The entire back pages of newspapers have recently featured a full colour advertisement bearing a big message: "There are some ideas we want every company to copy". The image accompanying these words is not the latest funky Apple product design. It is a simple but vast solar farm, stretching off into a sunlit hinterland. (The sunset is behind the solar panels, meaning they are facing quite the wrong way. A strange lapse of Apple's famous design rigour).

Apple are set on supplying 100% renewable power to their data centres. But why would they take out such advertisements? It crosses my mind that they must have some ulterior motive of a business kind: some venture that would require enormous amounts of solar power to be routinely deployed around the world.

Electric vehicles, I think, and then dismiss the idea immediately. Too much of a departure for a computer company, surely.

I tell the Accenture team all this and much more about life on the solar front lines, at least as I am experiencing it these days. I tell them that SolarAid's retail brand SunnyMoney has just sold its millionth solar light. I cast the achievement of the teams in Africa and London as an emblem of what can be achieved once a solar technology becomes cheaper than a fossil fuel technology. If it this relatively easy to replace a whole category of oil use, in this case in lighting, who is to say it is going to be impossible in bigger segments of oil use, once the price is right?

The Accenture team want to hear as much as I am willing to tell them about the inside story of Carbon Tracker. They are impressed with how the stranded-assets narrative is gaining traction, they tell me.

I pick my way through this, telling them everything they could pick up on the internet, little of plans in train. I am wary of going too far, given their client base.

ExxonMobil have just come out with a press release addressing the risk of fossil fuel stranded assets. Vice President of Corporate Strategic Planning, William Colton, makes the company's position very clear: "All of ExxonMobil's current hydrocarbon reserves will be needed, along with substantial future industry investments, to address global energy needs", he professes.

The company is completely ruling out the risk of stranded assets, both in its current reserves portfolio, and in any resources it may choose to convert to reserves. It is placing a bet, in other words, that growing need for its product will trump growing concern about climate change, and air quality, or the disruptive impacts of solar and storage technologies, for decades into the future. It asks its investors to be assured that there is zero risk – precisely zero risk – of any of these dangers materialising.

I know this bullish stance will not play well with investors. How strange it is, I reflect, to see the belief system of the oil heartland so starkly exposed to the light of day.

Mark Lewis, Carbon Tracker advisory board member, is now working at Kepler Cheuvreux, financial advisors to investment banks. He pulls no punches in a note to his clients. "Exxon's view of climate risk appears naively binary: the reality is much more nuanced." His own view, he explains, is that the total amount of revenue at risk in fossil fuel companies over the next two decades, if the world aligns behind a two-degree ceiling for global warming, amounts to 28 trillion dollars.

Carbon Tracker is holding its fire. We will be publishing a detailed report on the risk of stranded oil assets on May 8th. Our data will speak for us then.

Give us a preview, the Accenture team asks.

I dodge. Kashagan ought to be a pointer, I suggest.

Kashagan is a super-giant oilfield offshore Kazakhstan in the Caspian Sea. Discovered in 2000, it supposedly holds 13 billion barrels of oil. This makes it one of the largest oilfields ever discovered. By way of comparison, annual global oil production in billions of barrels is 30.

Kashagan was supposed to come on-stream in 2005, after the investment of some $10 billion by the companies involved. Almost a decade later in 2014, it is still not producing oil, and is not expected to for another two years, possibly more. It produced a little oil last October, but the oil was so acidic that it ate away the pipework, causing toxic gas to leak. The capital expenditure bill on this one oil project is running at $50 billion now – a $40 billion overspend on a time overshoot of 11 years and counting – and will clearly need to be much more if the oilfield is ever to produce oil.

If.

A senior Shell executive who trusts me has told me that within the company the oilfield is not called Kashagan but Cash-all-gone.

The situation that the companies involved face – Shell, Eni, ExxonMobil and Total – reminds me of a gambler in a casino who has lost a pile of money, but whose brain tells him to keep going so that one day he can win it all back.

That is, if his backers – his investors – bankroll him to keep going.

But suppose they tell him to give up and leave the casino before he wastes any more of their money?

I can see heads shaking around the table. I wonder if they have all heard this true story before.

Then there is the shale, I add. Here too the oil and gas industry is drilling at high cost. At least it is producing a product. But not one it can currently sell, in most cases, at a price anywhere near the cost of production. And this is at a high oil price: in excess of $100.

It isn't as though it is only mavericks like me asking questions, I point out. Bloomberg is all over this story. "Is the U.S. shale boom going bust?" a headline asked recently. "A host of geological and economic realities increasingly suggest that the party might not last as long as most Americans think." The headline in another investigation focused on the main concern: "Shale drillers feast on junk debt to stay on treadmill." One company, Rice Energy, earns only $1 of income from gas sales for every $4 it spends. It is still able to raise hundreds of millions of dollars in debt.

"People lose their discipline," a fund manager explained to Bloomberg. "They stop doing the math. They stop doing the accounting. They're just dreaming the dream, and that's what's happening with the shale boom."

And many Americans, including in high places, think that the boom will not only last a very long time, but produce so much oil that imports will become a thing of the past.

The discussion with Accenture's finest goes on for hours, as the fine wine flows. Many of the brilliant young minds there may not be advising renewables companies, but they clearly wish they were.

I have one more glass of wine than I normally would. I am quietly celebrating. Caroline Lucas has been cleared of her anti-fracking protest charges. A judge has decided that by sitting down in an unused road outside the gates of a shut down drilling site, registering a peaceful protest, she was acting within the law.

City of London, 8th May 2014

The launch of Carbon Tracker's oil report. Anthony Hobley's old employer has made available its lecture theatre, and it is packed with fund managers, analysts, and financial journalists.

Christiana Figueres has come to support us again by opening the event. The executive secretary of the climate talks was in London yesterday to give a talk on climate change in St Paul's Cathedral. The latest constituency she is targeting in her effort to build support for a meaningful Paris treaty is the churches.

Carbon Tracker has has set the tone of public discourse with its unburnable carbon analysis, she says. Please listen very hard to what they have to say today.

Today the Carbon Tracker team member taking centre stage is Mark Fulton. For months now the former head of research at Citi and James Leaton have been locked in work with their small teams on an analysis of an industry database for all oil projects around the world. We have had to pay a small fortune for access to it. Philanthropic foundations would normally run a mile at the prospect of their money being used in this way. Not on this occasion. Compiled by a Norwegian data company, Rystad, the database is used by the International Energy Agency, among others. This will be the first time it has been looked at through the type of lens that we bring to bear.

The database spans all oil reserves, and resources targeted in exploration programmes. (Reserves are oil deposits that have been mapped out, are technically recoverable, and hold the potential to be produced economically. Resources are oil deposits yet to be mapped out and be tested for economic viability). The Carbon Tracker study will look at all this oil in terms of current understanding of the cost at which it can be produced and the price it would have to be sold at to make an acceptable profit. It will slice up the data project by project, so that individual oil projects like Kashagan can be compared with others around the world. It will slice province by province, so that Arctic oil can be compared with deep-water oil, for example. It will slice oil species by oil species, so that tar sands can be compared with shale oil, for example. And of course, it will slice company by company. It will be a quite unique resource for the people in this room, and many others like them across the global capital markets. Our hope is that they will start checking much more than they have hitherto whether or not the oil and gas industry is in danger of wasting investors' money. We hope that their search for capital discipline will lead to cancellations of higher-cost projects and the beginnings of an alignment between the capital markets and the objectives of the international climate negotiations.

There are both economic and carbon reasons for this approach. High cost projects also tend to be high carbon projects. Tar sands projects, in particular, require a higher amount of energy (and therefore emissions), to boil oil from tar than is needed, say, to tap crude oil from shallow reservoirs in the Middle East. Oilfields far offshore need more steel and concrete than those onshore. And so on.

Mark Fulton is based in Sydney. He appears on a large video screen. He looks sober, speaks with total confidence and authority to the camera, using only the language of the markets. I know from months of interaction with him that he has both a ferocious intellect and a drive that holds no prisoners. It is easy to imagine him heading up the research effort of an entire global investment bank. He has a key accomplice of equal throw-weight. Paul Spedding, the former head of oil and gas at HSBC, has joined the project in the editorial phase. Wall Street seeking redemption they may be – as Mark Fulton regularly puts it – but as he lays out his case, once again I can't escape a slightly surreal feeling at having fallen into confederacy with such people.

Mark describes the new concept he and James have developed: the carbon supply cost curve. This takes all the oil projects in the world and plots them against market price on the vertical axis and cumulative gigatonnes of carbon on the horizontal axis. This result is a rising curve, kicking off with the least expensive oil at around $25 per barrel market price, to the most expensive oil, at a mind-boggling market price approaching $220 per barrel.

If we take a two degrees climate budget, Mark explains, you can only afford to burn 360 gigatonnes of carbon dioxide equivalent. That amounts to just over half the world's oil in current reserves plus resources being targeted by current oil-industry capex programmes. That budget will be blown at current burn rates of extraction by 2050. A 360 gigatonnes budget embraces all the oil from $25 per barrel market price up to around $75. Beyond that, if the two degrees budget applies, the more expensive oil can't be burned.

But would it make sense to burn it anyway, Mark Fulton asks, given how expensive much of it is against the current oil price? And what if the oil price falls, as we expect it might? Currently it is above $100 per barrel, but if supply begins to outstrip demand, it would fall, and it is easy to imagine demand scenarios not too far in the future where that happens. Suppose for example the Chinese economy slows down? The oil on the higher side of the carbon cost curve is at risk both from climate policymaking and simple economics. This is true without even considering climate change.

I scan the room trying to gauge the reactions. There is a rapt attention, a stillness, beyond the scurrying pens on pages.

Mark passes over to James Leaton, who will look at the implications for companies.

Note how the curve steepens above around $95 market price, James begins. His manner is the opposite of Mark Fulton's: quiet, thoughtful, scholarly. You could never imagine him haranguing junior analysts.

We conservatively estimate that above $95 per barrel fully $1.1 trillion of capex is at risk of being wasted over the next decade, he says. Which companies and projects sit in this zone? He talks about specialist companies in the tar sands, and the companies involved in the Kashagan project as obvious examples. There are many more. The reports we are publishing today dive into a lot of company detail.

Anthony Hobley shepherds the event into a phase where panels of experts discuss the findings. Martijn Rats, Head of European Oil & Gas at Morgan Stanley, agrees that the oil majors are coming under massive pressure on capital expenditure. That's why spending is slowing. The cost curve is actually worse than Carbon Tracker depicts, he says, because currently international oil companies need around $100 per barrel to break even. (That is to say, recover their costs before any profit. An acceptable market price for a breakeven price would be around $115 per barrel). This is because most oil companies also have substantial gas reserves, which on a barrel of oil-equivalent basis they can only realise a maximum of $35 a barrel at present, and sometimes quite a lot less. So oil needs to cross-subsidise gas.

Great, I think, better for us to be "wrong" on the downside. Morgan Stanley are saying we are being too conservative.

Martijn Rats turns to James and makes a comment that captures the whole Carbon Tracker experience to date.

I guess you have a tailwind, he says.

A question for him from the audience. So will Morgan Stanley be incorporating Carbon Tracker's analysis in its advice?

I am here in learning mode, frankly, Martijn Rats replies. What we don't say so far is that this or that project should be cancelled.

But you will, I think. You will have to.

Stockholm, 13th May 2014

The Swedish Energy Agency is launching its annual report with a day-long conference in a downtown hotel in the Swedish capital. Jeremy Oppenheim of McKinsey and I have been invited to give keynote addresses kicking off the day's discussions. We meet for breakfast to catch up on each others' lives.

McKinsey is one of the big consultancies who compete with Accenture. Jeremy heads its energy team. This year, he is on secondment to a vital project, a global commission of world leaders, chaired by the former Mexican President Felipe Calderón, who are compiling what aims to be the definitive work on the economics of climate change. The Global Commission has a brief to update the famous Stern Review of the economics of climate change, the 2006 report that is credited by many with making people realise for the first time that climate change is leading the world towards economic ruin by dint of progressively more destructive climate destabilisation, and that it makes more economic sense to abate the threat by cutting of greenhouse-gas emissions than to endure it unabated.

Just yesterday we have had graphic evidence of how ruinous the impacts of global warming can be. Two teams of scientists have concluded that the melting of the vast West Antarctic ice sheet appears now to be unstoppable: that it will collapse piece by piece until the entire sheet has slid off the continent into the ocean. This would cause global sea-level to rise ten feet over the next few centuries. The generations to come will need an awful lot of zeroes when they count the cost of that.

None of the other observational climate-science news is good, either. 2013 was the warmest year ever. Nine of the ten warmest have all been since 2000. So far in 2014 extreme weather events have continued in many places, right around the world. There have been extremes of heat in Australia, South America and Africa. Eastern North America has seen extremes of cold. The British winter has been the wettest in 250 years. The World Meteorological

Organisation needs no persuasion that these extremes, collectively, are the result of climate change.

Jeremy and I shake our heads over all this, and move quickly to what we and our allies are trying to do about it, in our different spheres of influence. Jeremy is leading the team of analysts working with the Global Commission on what will be called the New Climate Economy Report. He runs me through the conclusions that are likely to emerge when the report is released at the UN Secretary General's climate conference in New York in September. The key conclusion is that cutting carbon emissions sufficiently to keep global warming below two degrees would actually improve economic performance, not undermine it, as the incumbents so often allege. Jeremy has high hopes that the elite makeup of the Commission will combine with the up-to-date technical detail in the report to send a message out to the world, loud and clear, that tackling climate change means a much healthier global economy, and national economies.

He relays the same message to the several hundred people gathered by the Swedish Energy Agency. I deliver my latest stump speech. Then we stay on for the rest of the day to listen to the discussion.

The panel after we finish features the Swedish national chief executives of three big European utilities, E.ON, RWE and Vattenfall. They are interviewed by a Swedish television personality, sitting on high stools as though at a coffee bar. You have just heard that your business models are dead in the water and that you are heading for bankruptcy, she says, referring to my speech. What do you have to say about that?

They take turns to give versions of the same answer. The analysis is essentially correct, they agree. Some utilities are indeed going to go bankrupt. But my company is going to react to these changed circumstances. We will survive and find a way to prosper.

In the coffee break, I express my amazement to Jeremy Oppenheim.

Yes, he agrees, it is astonishing how fast things are unfolding. But can these companies change? Are they really able to do that? The cultural barriers that they face are so huge.

Chapter 9

Doom, boom, or bust

London, 12th June 2014

A conference organised by the Financial Times and the International Finance Corporation in a Mayfair hotel. A panel with Shell, GE, and Macquarie Bank. The Shell man, Wim Thomas, is chief energy advisor. The GE man, Ganesh Bell, is Chief Digital Officer. The Macquarie man, Simon Wilde, is a Senior Managing Director. The dire but typical gender balance is partially repaired by chair Pilita Clark, the FT's environment correspondent and leader of the paper's energy team. We have an interesting title to guide our discussion: "Energy: doom, boom, or bust."

Wim Thomas had his marching orders set on May 16th, in the form of a letter sent to all shareholders. "To whom it may concern," it reads, "we are writing this letter in response to enquiries from shareholders regarding the 'carbon bubble' or 'stranded assets' issue." It goes on to argue that there is nothing for an investor to worry about in these concepts. "The world will continue to need oil and gas for many decades to come, supporting both demand, and oil & gas prices. As such, we do not believe that any of our proven reserves will become 'stranded.'"

The whole concept is "fundamentally flawed", Shell concludes. And worse, "there is a danger that some interest groups use it to trivialize the important societal issue of rising levels of carbon dioxide in the atmosphere."

Carbon Tracker is not named, but it is clear who they must be targeting with this comment.

I have been accused of many things by the incumbency over the years. Trivialising global warming is a first.

The notion of stranded assets is flawed because of "a failure to acknowledge the significant projected growth in energy demand, the role of carbon capture and storage (CCS), natural gas, bioenergy and energy efficiency measures.

Energy demand growth, in our view, will lead to fossil fuels continuing to play a major role in the energy system – accounting for 40-60% of energy supply in 2050 and beyond, for example. The huge investment required to provide energy is expected to require high energy prices, and not the drastic price drop envisaged for hydrocarbons in the carbon bubble concept."

They leave a major hostage to fortune by admitting they need high energy prices to justify investment in their vision. It would be interesting to watch Shell react if oil prices drop.

There are other gaping holes in their argument. Carbon Tracker does not think that fossil fuel assets can only be stranded by a low oil price, as would be obvious to anyone who read our reports. High prices can strand assets just as well as low prices, in a world of such fast-falling alternative-energy prices. So too can carbon emission and air pollution legislation, which will be necessary to avoid catastrophic rises in global temperatures.

As for relying on climate negotiators to achieve nothing of note for decades, while fossil fuels barrel on providing the majority of the world's energy, how can Shell dismiss the risk of a different outcome? What a huge assumption to make that the destabilisation of the climate system, so evident for those with the eyes to see it today, will not at some point trigger a survival reflex in society. What an insult to the hundreds of climate negotiators from the dozens of countries who are acting in good faith at the climate negotiations, trying to find ways to agree and enforce emissions targets.

Shell's response to our oil report is very similar to ExxonMobil's in its contempt for the climate negotiations and their prospects.

Today we see unsurprising evidence that oil companies have been colluding. The industry association IPIECA, speaking for many companies, publishes a "fact sheet" on its website echoing the main points made by ExxonMobil and Shell in their responses to Carbon Tracker's work.

The members of the "doom, boom, or bust" panel take turns to set out our stalls. Wim Thomas has an interesting spin on the Shell party line. Solar will indeed become the world's dominant energy source, he says. By 2070. By then, we will still need a lot of carbon capture and storage to handle all the fossil fuels we will be using. There will be lots of gas consumption still.

In the discussion, steered by Pilita Clark quickly to the heart of the issue, I find myself in a situation I have never been in before in these four-person panels. I am part of a majority side, contesting a minority of one. Normally it is me who is the minority of one.

GE's chief digital officer, Ganesh Bell, is a creature of Silicon Valley through and through. He talks enthusiastically about the "internet of things", meaning the notion – increasingly aired in visions of tomorrow – of objects of all kinds being embedded with electronics, software, sensors and connectivity to make a network that allows exchanges of data between manufacturers, operators and/or other connected devices. The network of "things" becomes all the more valuable and serviceable because of its interconnectivity.

Wait until all this hits decentralised energy fully, Ganesh Bell says. Stand back and watch how fast the disruption of everything traditional will be: in energy, data management, economics and much else.

He doesn't spell it out, but it is clear that he doesn't see Shell factoring much of this kind of thinking into their version of the future.

Macquarie Bank's man, Simon Wilde, is little less enthusiastic about his vision for tomorrow. Where better to look for growth opportunities to invest, he enthuses. Renewables are generally way ahead of fossil fuels on his list, especially when you throw in all the smart energy opportunities. He only sees that accelerating.

Wim Thomas does his best, but as he speaks, sitting next to me, I slowly develop the impression that the Shell man is not comfortable in his own skin today. He is relaying a party line that he does not have total confidence in.

I offer my thoughts about how neuroscience plays against rational patterns of behaviour in big corporations.

Thomas is not entirely defensive, and even concedes some truth in the analysis, based on his own experience. But we will still need to depend on fossil fuels for a very long time, he adds.

And the risk of stranded assets?

There is a "slight exaggeration" in that narrative, he suggests.

So I am not trivialising anything, I am tempted to ask him. Your employer's letter to shareholders didn't seem to see anything "slight" in Carbon Tracker's exaggeration.

It occurs to me that I have been here before. From 1990 until 1997, throughout the first seven years of international climate negotiations, BP had stood full square alongside Exxon Mobil, Shell, and the other oil giants in denying that climate change is anything to worry about. The world needed "sound science" before rushing to judgement, they said, as though the hundreds of scientists in the Intergovernmental Panel on Climate Change, and their exhaustive processes of peer review, were incapable of providing it. Then, in May 1997, everything changed. The BP CEO of the day, John Browne, gave a mea culpa speech at Stanford University. It was pointless for oil companies to argue that climate change wasn't a grave problem, he said. The evidence was becoming too clear. BP would henceforth be joining the growing number of companies wanting to do something meaningful about emissions.

BP's U-turn galvanised the climate talks for the rest of that year, just at the right time ahead of the Kyoto climate summit in December, where the world's first emissions limitations were finally agreed. I describe this passage of history in my book published in 1999, *The Carbon War.*

As BP's internal discussions raged between advocates of reform and entrenchment in the run up to Browne's game-changing speech, I noticed that the BP executives I debated with in public, and conferred with in private, looking increasingly uncomfortable with the party line. They knew that, behind the company's closed doors, the blocking narrative was eroding. They gave this away in all sorts of ways.

Now, as then, evidence is becoming increasingly clear all around those who seek to entrench the incumbency in its old ways. Perhaps most notably, as they seek to justify their own performance and downplay the opposition's, the oil and gas industry's tendency for hype is being exposed on an increasingly regular basis. The most recent example from the American shale is particularly shocking. The Monterey Shale in California was promoted in a 2011 report by a supposedly independent consultancy firm working for the US government's Energy Information Administration as a huge oil play, with more than 13 billion barrels of oil recoverable. The release of this conclusion made news headlines all over the world. Since the estimate meant fully two thirds of all American shale oil reserves were now in the Monterey Shale, it was the core of the "Saudi America" mantras pushed by the industry and its cheerleaders at the time.

David Hughes, former Canadian Geological Survey veteran, fellow of the Post Carbon Institute and member of the Transatlantic Energy Security Dialogue, began painstaking work to check the estimates the consultancy had made in the Monterey Shale. He came up with a very much smaller figure for recoverable oil. On May 21st, the Energy Information Administration was forced to agree with him. They let it be known that the reserves estimate for recoverable Monterey oil needed to be revised down by a somewhat inconvenient 96%.

Let me repeat that, for fear of typos. Ninety-six percent.

This announcement, needless to say, did not make headlines around the world the way that the original report had.

On top of such revelations, and the billions Shell has written off in shale, and likely soon to write off in Kashagan, in the Arctic, and in tar sands projects, there is a growing popular rebellion to set resources off limits that the industry will need in order to deliver the majority of energy "many decades to come". The effects of this rebellion will soon become evident even to people like Wim Thomas. For example, on the same day that the EIA revised its estimate of the Monterey Shale down by 96%, Santa Cruz County in California banned fracking completely.

Some diehard incumbency defenders might take comfort from how strongly the industry's supporters in politics are fighting for them, of course. Nowhere more so than the UK. The same day of the shale-oil shock in California, the Telegraph ran a front-page headline announcing that "Ministers signal start of great oil rush" in Britain. This polar opposite to what international business editor Ambrose Evans-Pritchard is writing in the business pages of the Conservative paper these days was based on a report by the British Geological Survey, soon to be published, announcing billions of barrels of recoverable oil under the Weald of Kent, Sussex and Surrey. To access this oil, the government would apparently be allowing the industry to drill under peoples' homes without their permission, and would increase average compensation to communities undergoing fracking to £800,000.

Notwithstanding this effective bribe, residents reacted angrily, immediately. It struck me as incredible that the government thought it would be able to force this ghastly brew of undemocratic policies past their own voter base in the Conservative heartland.

On May 23rd, the actual British Geological Survey report became available, so that the government's description of it could be verified. Verification proved to be needed. There is in fact no recoverable shale gas under the Weald, the Survey concludes, and very little recoverable oil, in all probability.

Reading this, I had to take my dog for a walk and have a very deep think. There is, it seems to me, much more than failure of rational thought at work in all this drama. There is a whiff of delusional madness in the air.

At the FT debate – "doom, boom, or bust" – I wonder if I am debating with someone from Shell who might be harbouring the first faint germs of similar thoughts.

London, 2nd July 2014

The frackers look forward to public money backing them in bribes paid by the government for every British community they choose to frack in, paid from the taxes of those communities and all the others in the UK. What of the solar industry, meanwhile?

A public seminar of the All-Party Renewable and Sustainable Energy Group in the Houses of Parliament: a Conservative MP in the Chair of a panel, a Liberal-Democrat Peer to the right of him, a Labour MP to the left, and three others among them along the speakers bench – me, a government official, and Juliet Davenport, CEO of the renewable-electricity supplier Good Energy. Through the windows, strong sun on the River Thames. In the room, dark wood benches arranged like the pews in a church, with enormous paintings of robed dignitaries frowning down on them. The room would have looked much the same hundreds of years ago, but for the microphones dangling from the high ceiling, and the garb of the attendees along the bench and pews.

Juliet speaks first. Her small utility, trying to compete with the Big Six utilities using renewables, sources electricity for its customers from 800 small generators, a quarter of them solar. Solar is a fantastic technology, she says. It provides home security. It is not intermittent, as so many people falsely claim, it is variable: a big and vital distinction, because companies like Good Energy can balance it with wind. Solar is doing quite well in the UK as things stand, with more than 2 gigawatts installed, where there was next to nothing just a few years ago. So why is it that the UK government is launching its third review of the solar market in four years? Why is it that yet again the future of the industry is under threat?

I speak next, and answer the question she has teed up. Because we are doing so well, I say, despite all the obstacles. Because we pose such an existential threat to the business models of utilities today and oil and gas companies tomorrow.

Next comes the government official, the head of renewables investment at the Department of Energy and Climate Change, Hannah Brown.

We fully recognise the potential of solar, she says. We expect to see 10-12 gigawatts of it in the UK by 2020, but the key element for its deployment has to be on buildings, not on the ground. And solar must be affordable for consumers. We have a constrained budget.

As is so often the case with the thoughts of officialdom, I reflect, every statement and assertion is questionable. 10-12 gigawatts is effectively their cap on solar deployment. Why have a cap on what will be the cheapest energy technology by 2020, on current trends, and one that will need no subsidies – unlike fracked gas, nuclear and the other incumbency options? Why suddenly stop large ground deployments en masse? Why not have deployments on both buildings and the ground, on appropriate sites? As for affordability, how rich that is, when we look at the vast subsidies for the oil-and-gas and nuclear industries.

The next speaker is the Labour MP Alan Whitehead, who is a scholarly expert on renewables policymaking. The mechanism for constraining the government's budget, the so-called Levy Control Framework, is not fit for purpose when it comes to renewables in general and solar in particular, he says. The problem is that solar has become too successful for its own good. The funding for renewables is capped, yet the funding for non-renewables is not capped. Gas is favoured everywhere, by default, in the government's actions. What is more, their stated intention of a seamless transition from solar on the ground to solar on roofs is not guaranteed to work. The great majority of the constrained budget – constrained on a false premise that "costly" renewables would be a subsidy-taker rather than a money-saver for years to come – has already been allocated by the government to big offshore wind and biomass. There isn't enough left to keep growing the industry until its plunging costs mean life independent of all subsidies around 2020.

The Liberal-Democrat Peer and the Conservative Chair have their say. They don't sound too thrilled with the direction the government is taking either.

Then to questions from the audience. A degree of anguish fills the room as they unfold. Fear that the government is going to maim its domestic solar industry, whether via incompetence or design, or a little of both, can be heard between the lines of almost every polite statement. Distrust is thick in the air. The government official's voice becomes quieter as she trots out her formulaic, lawyerly, responses.

She is young. This job cannot be easy for her. She doesn't seem like a junior student of Machievelli to me. But then how can you know?

What about climate change, someone finally asks. Shouldn't we be worrying about the feedbacks and accelerating solar for all we are worth?

We should worry about the feedbacks and leaks both, I say. The American gas industry has been trying to keep civil society from measuring their leakage rates. But the few university teams that are brave enough to have a go are producing increasingly scary results. So much so that I now expect the final conclusion to be that unconventional gas production and use is actually worse than using coal.

A journalist from the Wall Street Journal seems impatient with all the British reserve on offer, and the morass of technical detail on subsidies that masks what is clearly the most fundamental of disputes.

What is it, exactly, that you are worried about, she asks me.

A repeat of 2011, I tell her. The government launched an ambush on the solar industry back then: a savage subsidy cut with only 6 weeks notice, aiming to decimate us. The gas industry was behind it. I know this because I was told so by civil servants and parliamentarians up to ex-ministerial level. My company took the government to the High Court claiming they had acted illegally. The High Court agreed with us. So did the Court of Appeal. So did the Supreme Court.

But the government still succeeded in setting the solar industry back a few years, wiping out thousands of jobs.

This feels the same. It feels like Groundhog Day.

London, 7th July 2014

It has been a long time since I have been invited inside the Prime Minister's home. The last time, in the early days of his government, I and seventeen other supposedly successful entrepreneurs were invited to have breakfast with him and George Osborne. That bizarre occasion I describe in *The Energy of Nations*. Since then I have blotted my copybook by being critical of government policy on many occasions. Sir Jeremy Leggett I shall not now be.

Today I visit Number Ten, but only the front door, to hand in a letter signed by 150 agitated renewable-energy companies and supportive institutions. As their representative, my duty is to have my photo taken, along with two representatives of the Solar Trade Association, handing the letter to the policeman on the door.

Checking through security at the entrance to Downing Street, I reflect on the differences between the leadership in Cameron's government and Obama's. The US President is still pushing his legacy. A month ago the Environmental Protection Agency unveiled historic rules reducing coal pollution by 30%. Al Gore was fulsome in his praise. The new rules were "the most important step taken to combat the climate crisis in our country's history", he said.

The Republican backlash was immediate and vituperative, especially from the coal states like Kentucky. Obama took personal charge of the pushback. "Science is science," he said in an interview with the New York Times. "And there is no doubt that if we burned all the fossil fuel that's in the ground right now that the planet's going to get too hot and the consequences could be dire."

This was the first time an American President had referred to unburnable carbon and – implicitly – the risk of stranded assets.

The Carbon Tracker team was thrilled. Speculation began immediately about whether his team had been reading our work, and briefing him accordingly. There has been so much visibility for both divestment and engagement recently, so much of it mentioning our conclusions, that it would be difficult for American officialdom to miss it. The most potentially damaging development for the fossil fuel industries has been the decision by the British Medical Association to take its pension fund out of fossil fuels. In this action, the doctors are placing coal, oil and gas in the same product-brand category as tobacco.

I wonder where David Cameron's head is on all this now. He defends the shale vision so vehemently, and yet his record on climate change is far from hopeless in other respects. He presides over a government that is completely split on the issue within its own ranks. George Osborne and the Treasury bat almost wholly for the oil and gas industry, it seems, but over at the Department of Energy and Climate Change, ministers seem to be trying hard to move the climate agenda forward. Secretary of State Ed Davey, a Liberal Democrat, is a strong advocate of wind energy. Climate Change Minister Greg Barker, a Conservative, is a strong advocate of solar energy. Notwithstanding Solarcentury's spell in court opposite him, I have few illusions about the constraints he operates under. I need little persuasion that he would do more if he had the political room to manoeuvre. Cameron could give him that, but mostly doesn't.

Almost all the Conservative ministers, save Greg Barker, have backed away from the green industrial revolution they advocated when they were in opposition. It seems such a lost opportunity, to me: a lost cause, even, when one thinks of the potential prize for the British economy. The IEA calculated last month that the world will need to spend $40 trillion on providing its energy supply by 2035. That is $2 trillion a year, up from $1.6 trillion last year. How much of this is likely to go to renewables? The REN21 renewable-energy organisation recently published the results for new power generation added in 2013. The renewables share jumped to more than half for the first time, 56%, up 8.3% on 2012. More than 22% of global power production is now renewable. In Germany it is running at more than 30%. Bloomberg has estimated that by 2030, two thirds of all energy investment will be going to renewables. The opportunity for the British economy would be huge, if the leadership grabbed its chance to accelerate the clear ultimate winner.

But instead, they elect to flog two dying horses, shale and nuclear, with an enthusiasm verging on blindness. And that is why I find myself knocking on the door of Number Ten Downing Street, with a press photographer hovering.

Signatories of the letter I clutch in my hand include Ikea, Triodos, Good Energy, Ecotricity, Kyocera and Interface. It is a document that a Prime Minister ought to be taking seriously, granting the signatories representatives just a little of his time. But no.

My comment in the press release that has gone out to the UK media attempt to explain why. "Despite all of the incredible achievements of the UK solar industry since 2010, it's still very clear that the Whitehall mindset has yet to catch up," I say. "It's time that the government woke up to the fact that, with stable support, jobs rich UK solar will be cheaper than onshore wind during the next Parliament, opening up immense opportunities for UK plc and driving down the costs of delivering the 2020 renewable energy target in the process. Far from slamming the brakes on large-scale solar, the Prime Minister should be hailing it as one of Britain's renewable energy success stories and getting behind it. Instead he prefers to push fracking, even in National Parks."

Online, 10th July 2014

Some mornings these days, I open my laptop, check the energy news and have to pinch myself to make sure that I am not still dreaming, so fast are signs of system change unfolding. Today the feeling is particularly acute. A momentous article has appeared in the Telegraph. It argues that fossil fuels are the new subprime threat to the global economy, doomed to mass stranding underground, with many billions of dollars of investments heading for write-offs.

Meanwhile, it asserts, a solar revolution is rushing up on society under the radar, the standard bearer of a transformative green industrial revolution that can no longer be avoided.

Ambrose Evans-Pritchard, international business editor of the Conservative Party's newspaper of choice is at it again.

Some wag, unknown to me, has written on Twitter: "What world are we living in when the Telegraph starts sounding like Jeremy Leggett?"

Since 2008, $5.4 trillion has been invested globally in oil, gas and coal markets, Evans-Pritchard's writes. "Little has come of it." Production from conventional oilfields has peaked. The marginal cost of new production, including from shale, is very high. Many an oil company is spending more on capex than it is generating in cash flow. And investors in this potential bubble are throwing good money after bad. "They are likely to be left holding a clutch of worthless projects as renewable technology sweeps in below [the] radar, and the Washington-Beijing axis embraces a greener agenda".

The article does not tell us what investment would be needed in solar and other clean energy to have a good chance of keeping global warming within 2°C. The answer to that question puts the vast sums heading for fossil fuel wastage in stark perspective. It is, according to US investment think-tank Ceres, in a study led by Carbon Tracker advisor Mark Fulton, $1 trillion a year, up from about $250 billion today. Ceres calls this target "the clean trillion".

What is the current $250 billion achieving? Evans-Pritchard: "...staggering gains in solar power – and soon battery storage as well – [threaten] to undercut the oil industry with lightning speed". Once the crossover point of solar costs and fossil fuel costs is reached in multiple markets, the switch to solar and storage "must surely turn into a stampede. My guess is that the world energy landscape will already look radically different in the early 2020s".

I wish this article had been written two days ago. I could have added it to the letter that I delivered to Number Ten, to increase the chances that the Prime Minister would see the message writ large in the business pages of his party's favourite newspaper.

Chapter 10

Music to my ears

London, 15th September 2014

At least three of the major oil and gas companies have full scale strategic reviews underway at the moment. Transparency disclosure statement: I am involved in the periphery of two of them. I am invited to come in and offer my advice, for what it is worth, as to what I think is going on in their markets and what they should do. Today is one such day. I sit in a room with state of the art audiovisual equipment talking to senior oil and gas executives in other cities.

Are you surprised?

I have to admit that I feel somewhat schizophrenic myself. I get these invitations to talk to them, I suppose, because they know of my past in and around the oil and gas industry, and of my status as a nominally successful entrepreneur in clean energy, which many people are telling them they should be taking more seriously these days. They know, either by past experience or by reputation, that I won't lecture them: that I will use language and a tone of voice aimed at not annoying them. I do, after all, want them to listen and consider, not close down and elect to hate me.

Of course, that's not guaranteed to stop them hating me anyway.

I have to fight to keep a sense of Schadenfreude about their multiple problems out of my voice. And I know the smallest amount of research by them will reveal what private thoughts I probably harbour.

Today is a good example. I have orchestrated a full page advertisement in the International New York Times. It is an appeal to the world's foundations and philanthropists by 160 of the world's environmental prize winners, from 44 countries. Deliberately using fake legal treaty language to evoke the Paris climate agreement, the chances of which it aims to boost, the ad pulls few punches.

"Aghast that the Earth is heading for 4 to 6 degrees Celsius of global warming, given current policies on the burning of coal, oil and gas.

Terrified that we will lose our ability to feed ourselves, run out of potable water, increase the scope for war, and cause the very fabric of civilization to crash as a consequence of the climate change that global overheating will bring about.

Devastated that our governments have not succeeded yet in slowing, much less stopping, the flow of greenhouse gases into our thin atmosphere, in the full knowledge of these risks, despite a quarter century of trying.

Aware that the December 2015 Paris Climate Summit may be the last chance to agree a treaty capable of saving civilization.

Believing that the world's philanthropic foundations, given the scale of their endowments, hold the power to trigger a survival reflex in society, so greatly helping those negotiating the climate treaty.

Recognizing that all the good works of philanthropy, in all their varied forms, will be devalued or even destroyed in a world en route to six degrees of global warming or more, and that endowments that could have saved the day will end up effectively as stranded assets.

We, 160 winners of the world's environmental prizes, call on foundations and philanthropists everywhere to deploy their endowments immediately in the effort to save civilization."

The ad was facilitated by the European Environment Foundation, of which I am a trustee. They and the environmental laureates who signed it didn't water it down, as I expected. They beefed it up.

If a critical mass of the world's largest foundations were to act on this appeal, many billions would be pulled out of fossil fuel investments, and investment in clean-energy companies would soar, as would grant-giving to non-government organisations pushing clean energy and the multiple other ways of cutting greenhouse-gas emissions. The current problems of the oil and gas industry would be compounded. This is the point of the exercise.

I do not mention the advertisement in my hour with the oil and gas executives on the screen.

The IEA has recently reported that renewable energy grew at its strongest ever pace last year. It now produces 22% of world electricity. But one needs to look beyond this, I suggest, to get a sense of the disruptiveness of the technologies that are invading fossil fuel markets. There are days of late when the solar revolution seems to be an inevitability, to people like me, and a not too distant one. If I had to pick one example from recent weeks to make my case, it would be a report by UBS on solar and storage. A big team of analysts from the world's largest private bank predicts that by 2020 it will be possible to have a solar roof, an electric vehicle and a domestic battery bank, powering everything you need in a home, with mouth-watering economics. That energy-trio purchase will be able to pay for itself within six to eight years, while giving a 7% pre-tax annual return on investment. Such household economics, UBS concludes, will change the face of the energy industry. In just a few short years, people will be queuing up at IKEA and every other retailer sensible enough to be offering this kind of energy service. People will be realising in ever growing numbers that their homes and cars are their very own power plants, wiping out fuel bills – including petrol – whilst giving pension pots a much- needed boost.

UBS are not alone. Other investment banks' analysts are saying similar things. In August, Citigroup professed that it is inevitable that solar will outperform fossil fuels in the long term, simply on pure economics, never mind any climate-change policy drivers. Ambrose Evans-Pritchard's latest Telegraph story, on August 20th bore the headline: "Oil industry on borrowed time as switch to solar and gas accelerates."

This is the bit where my Schadenfreude has to be disguised. I also have to hide, frankly, my sense of disbelief at the evidence of my own eyes.

"The props beneath the global oil industry are slowly decaying", Evans-Pritchard writes. "The big traded energy companies resemble the telecom giants of the late 1990s, heavily leveraged to a business model already threatened by fast-moving technology."

What to do about all this, I ask the oilmen and women? Think of a three-step logic chain, I suggest. First, the disruption is underway at a speed that takes almost everyone by surprise, if they are not following play. 35 years ago people like me hand-typed our theses, now we hold most of the then computing power of NASA in the palm of our hand. What will energy look like 35 years on, in the age of the internet of things, and after the massive failures of policy by the energy incumbency and its institutional supporters, with the imminent crises they entail in energy markets? The incumbency is making decisions to raise hundreds of billions assuming something close to the status quo for up to 50 years, in some cases. Ridiculous, when you think about it.

Second, many governments want zero net carbon in energy by 2050 – essentially, a total phase out of fossil fuels – as a target at the climate talks, so I am told. Others look persuadable. So why not adopt that as the end point, given logic step 1, on the principle that strategy begins with an end in mind?

Third, then you plan in 7 five-year spans, the first of which has a name like "Begin".

Working the rest out would be a lot of fun. And with a 35-year timeframe, what is there to be scared of really? We are talking about less than three percentage points of change a year.

I try to underplay the problems they are experiencing that should be driving them in this direction without being exhorted to change course by heretics like me. There are so many to choose from.

The FT has majored recently on the "terrifying" skills shortages in the oil industry. The older generation of geoscientists and petroleum engineers who were hired before the mass lay-offs of the 1980s are now approaching retirement age. What will happen to average levels of expertise in the world's oilfields and refineries when that happens? You can safely bet on more project delays, like Kashagan, and more risk taking, like on the Deepwater Horizon rig drilling on the Macondo field.

Then there is the carbon bubble. The Economist is the latest recruit to the seriousness of this issue. An article on July 18th referred to it as "the elephant in the atmosphere", as far as Shell and ExxonMobil are concerned. The conservative economics weekly did not seem at all impressed with the letters sent by the companies to their investors. "Are investors happy that – assuming the firms are right – their shareholdings will make oodles of money whatever happens to the climate? Of course not. Carbon Tracker has written to Shell taking issue with practically all its arguments."

The final thought in the article provided another of those pinch-yourself moments for me, when I read it. "If Carbon Tracker is right, then they (investors) will dump oil shares – which is what should happen if the firms are making a huge gamble that will misfire."

Carbon Tracker upped the pressure with its next move. We published a set of fact sheets on capex showing that the oil majors are gambling $91 billion on just twenty high-risk projects. Sixteen of the projects involve deep water and the tar sands.

The bad news has been building up for the shale industry as well. The dire drought in California's Central Valley is heightening sensitivities to all aspects of water policy, and in July state regulators halted injection of fracking waste into shale wells, fearing contamination of drinking water reservoirs. They ordered an emergency shutdown of eleven oil and gas-waste injection sites, and a review of more than 100 others.

Their move proved prescient. In August, new research by Stanford University showed the industry had been fracking at far shallower depths than widely believed, sometimes through drinking water sources.

Water use during the fracking process itself is also proving troublesome. On September 2nd a World Resources Institute report study found more than a third of commercially viable shale gas deposits worldwide are in areas that are either arid or have water supply constraints. Yet a single shale-gas well can require up to 23 million litres of water.

In the UK, where such considerations have yet to dawn on most citizens, a precedent was set on July 22nd that must have seemed particularly ominous to aspiring shale drillers. A county council in the Weald rejected plans by Celtique Energy for drilling a single exploratory site. The West Sussex Gazette reported residents "weeping with relief" as the announcement was made.

It the industry has this much difficulty getting permission to drill single sites, how strong do they think their chances are of permission to drill the thousands they would need if there proved to be shale-oil sweet spots beneath the Weald?

Riyadh, 18th September 2014

My first trip to Saudi Arabia. I arrive at night, and am driven from the airport into a city lit up like a fireworks display.

Six million people live here. They are burning oil in power plants to provide electricity with few constraints on waste. It is akin to shovelling their national income into a furnace, at an accelerating rate every day. If their domestic oil consumption continues to grow at the rate it has in recent years, and assuming production can be maintained around today's level – which people like me doubt, given the age and tired condition of the biggest oilfields – then they will have no oil left to export sometime between 2025 and 2030.

This is possibly the scariest fact to be found in the global energy markets. No Saudi oil exports as little as ten years from now.

That would focus a few minds on the energy transition, would it not?

The Saudis know all about this problem. Senior government officials regularly talk about it. They have a plan. They are going to replace the oil burning in electric power plants with huge programmes in solar and nuclear.

The first of those solutions is why I am here: to give a keynote address to The Desert Solar conference.

The event opens the next day. It is held in windowless space as characterless as any other hotel conference hall anywhere else in the world. The room is lit to headache level by electric light bulbs throughout a blue-sky day. The air conditioning is so low that I will quickly be shivering if I take my jacket off.

Most attendees are Saudis in desert white robes and red chequered head dress. This is the Saudi solar industry in waiting, I am given to understand. Foreign and expatriate men in suits sit among them. A few women are present, hidden from view behind screens.

I am to give the closing speech. I have two days to listen and learn before I deliver that.

I have not come completely uninformed. By coincidence, I had half a morning with senior executives from one of the Kingdom's largest companies a week before coming here. The Saudi Basic Industries Corporation, Sabic, sent a senior team to Cambridge University to study and discuss the fast-changing world. I was invited to join the faculty team describing and analysing that world for them.

My talk to the Sabic team was the same one I will be giving today. It can be summarised in a single sentence. There are two emerging megatrends on the world stage that will together create a revolution in global energy markets: solar-and-storage cost down, and oil-and-gas capex cost-up. Saudi Arabia needs to be in the front row of that.

I do not even mention climate change.

The oil capex cost-up point applies even in the Kingdom. In late August, Saudi Aramco CEO Khalid Al-Falih warned that Saudi Arabia would need to invest $400 billion over the next decade just to keep to keep oil production capacity steady and double gas production. Oil production in the old days, he made clear, didn't cost anything like this in constant dollars.

In Cambridge, I found the Sabic team receptive and quick to engage in debate. Sabic's business is the manufacture of chemicals, fertilizers, plastics and metals. Their executives were clearly worried about the implications for their supply chain of all the perturbations in the oil world. But most of the executives that day were expatriates, not Saudi nationals. In Riyadh, it is the other way round.

I have 45 minutes for the delivery of my talk. That is plenty of time in which to hang yourself, obviously. I deliver the messages politely, using a full-colour PowerPoint, and hope for the best.

The first Saudi to his feet in the Q&A is a senior gentleman with a statesmanlike air.

What you have said is music to my ears, he begins.

He explains why. He needs no persuasion of the vital strategic importance to the Kingdom of being a player in the global solar revolution. There are all sorts of reasons why this makes sense, well beyond the wasteful burning of oil in electric power plants. Among them is the possibility of low oil prices in the future, and what that will do the Kingdom's national budget.

Interestingly, the Brent oil price began falling in June. On September 6th, it fell below $100 dollars for the first time in a long time. It has stayed there for nearly two weeks now.

Chapter 11

We cannot pretend we don't hear them

New York, 21st–26th September 2014

A muggy Sunday in Manhattan. I walk along 59th Street and turn right up Central Park West as marchers gather. At the front of the column, where 59th hits Columbus Circle, are impacted communities. Pacific islanders mingle with the displaced homeless closer to home: headdresses mixed in with middle-class Americans holding banners such as: "Our homes was [sic] flooded in both Irene and Sandy".

Next come doctors: white coats the breadth of the street. Stable climate for stable health, the banners say, in different forms of words. I notice how amateurish the banners look: of necessity – the New York Police Department allows nothing sturdy to be carried.

"Another Republican who doesn't deny climate change."

Then come the children, by the family, and even class, it looks like.

"We raced to the moon, lets race to clean energy."

Then youth groups. Lots of T-shirts. The march logo is a Statue of Liberty brandishing a wind turbine.

"Earthlings unite; the Earth needs you now."

A host of bicyclists.

Students of Yale.

"Destruction of the Earth is a bad business model."

Then the inevitable dinosaur. It can't hold a candle to Greenpeace's Tyrannosaurus at the Kyoto Climate Summit: that was made of rusty oil barrels with company logos, looking like something out of a science fiction movie.

"Its not warming. It's dying."

On 77th Street I see a carbon bubble. Twenty feet across, it bounces slowly on the heads of the crowd. Wall Street meets Main Street, I think.

I join what is being called the solutions segment: people from clean-energy companies, and investors in them. Here speeches are being delivered from a PA system run by the Sierra Club. My friend Danny Kennedy, founder of Californian solar company Sungevity, is giving one as I arrive. In 2001, new electricity was 81% dirty and 19% clean, he says. In 2013 it was 58% clean and 42% dirty. The trend is clear. As somebody said on Twitter, The Mend is Nigh.

"100% renewables" signs everywhere. This the work of a single campaigner from the World Future Council, Anna Leidreiter. I am wearing a "100% Renewables" tee-shirt that she gave me in a bar last night.

I look across at the American Museum of Natural History. The current exhibition is Flight in the Age of the Dinosaurs. How rather hopefully appropriate.

Chanting now: What do we want? Clean energy. When do want it? Now!

In Melbourne earlier, the crowd chanted not "now", but "ten years ago".

Silence descends. This is the minute for the victims of climate change around the world to date.

A forest of arms held aloft as a wave of noise spreads up the column. The front of the march is moving. But it will be 15 minutes before we can, people say. And the march extends all the way to 86th Street.

I see my favourite banner so far. "I can't believe we still have to protest this crap."

I peel off the march when we get to 59th street. Back in my hotel on Times Square, as I change into my suit for the opening of the Clinton Global Initiative, I can still hear drums and cheering echoing through the canyons of Manhattan.

Bloomberg's headquarters on Lexington Avenue next morning. A Manhattan tower housing the global base of a business communications empire created by one man with an idea. Fortunately, Michael Bloomberg is a Republican who cares about climate change – indeed, is working hard to fix it. Accordingly, in the conference suite, a who's who of global finance has assembled to hear Carbon Tracker's analysts launch the second in our three-report series on the state of global fossil fuel industries: on coal. They are in town because the UN Secretary-General's summit on climate change is about to start. And Wall Street is only a few blocks away.

They are abuzz with today's news. Members of the Rockefeller family have announced that their $860 million foundation, the Rockefeller Brothers Fund, is divesting from all fossil fuels. The fund had already pulled out of coal and tar sands, but is now including oil and gas. They are doing it for both moral and economic reasons, Steven Rockefeller, a son of Nelson A. Rockefeller and a trustee of the fund, has told the New York Times. He says that he foresees financial problems ahead for companies that have stockpiled more reserves than they can burn without contributing significantly to climate damage.

This coup for the divestment movement brings the tally of institutions divesting around the world to 180, and the total divested to $50 billion. This is not a sum to shift dollar dials in capital markets investing more than $4 trillion in fossil fuels equities. But it is of enormous political significance: it tarnishes fossil fuel companies in the way the anti-Apartheid's divestment campaign tarnished companies propping up South Africa's Apartheid regime ahead of its collapse.

The Rockefeller Brothers' Fund's president, Stephen Heintz, makes an interesting comment to the FT about John D. Rockefeller's likely response to all this, were he alive today. The oilman who founded Standard Oil in 1870 would approve and would be "leading the charge" into renewable energy, he suggests. "He was an innovative, forward-looking businessman. He would recognise that clean energy technology is the business of the future."

I elect to believe that. Today, I have faith.

I watch the Carbon Tracker team at work from the back row, unable to keep a smile off my face. Not everyone is as forward looking as the modern Rockefellers. A few days ago the chief executive of the Minerals Council of Australia, speaking for the embattled Australian coal industry, accused them of being "activists. Not genuine analysts".

Mark Fulton, who leads off on the analysis for us today, is privately outraged by this. He opens with a quick reference to the grotesque smear by his countryman, turns it into a joke, and then sets out to show why the coal boss is wrong.

He, James Leaton, and their teams have been taking the same approach to coal they did for oil: constructing a carbon supply cost curve from a standard industry database. This time they have bought the data from Wood MacKenzie. The results show clearly that high cost coal producers are gambling on survival simply in the hope that prices will somehow recover. They are not likely to, our team concludes. Tim Buckley, a draftee to Mark Fulton's team for our coal project – a former Citibank colleague of his – has led the work on demand, with a sophisticated model he has developed. It shows that historical demand assumptions in the markets are unravelling. The costs of renewable technologies continue to drop at a pace faster than most experts have predicted. In parallel, governments are introducing a snowballing brew of air quality and carbon emissions measures, further destabilising coal demand. In China, peak demand for thermal coal – the type burned in power plants – could come as soon as this year, based on our reading of the data. If that happens, it will shoot holes in the seaborne coal trade. In the industrialised countries, demand is already falling.

In this demand climate, throwing additional capital expenditure at high cost production is risky to the point of being a high-probability waste of money. This is especially true for new coal mines, which typically require expensive new rail infrastructure and port facilities to get coal to market.

Over the last three years, the Bloomberg Global Coal Equity Index has lost half of its value while broad market indices are up over 30%. In the pure coal sector there is only one trend – downward; coal prices are down, returns are down, share prices are down. For around half of potential 2014 thermal coal export production, current prices fail to cover their costs.

Coal could well be a sinking ship, Carbon Tracker has concluded: an industry in structural decline.

The opening ceremony of the climate summit. Ban Ki-moon, UN Secretary General, welcomes the heads of state. He looks very different in a suit than the T-shirt he was photographed in at the demo. We are not here to talk, he says, we are here to make history.

I have finally found a way to be in two places at once. If I stay in my hotel bedroom I can watch proceedings of both the climate summit and the Clinton Global Initiative live on my computer screen.

I pretend for a living, Leonardo di Caprio tells the political leaders, sporting a Fidel Castro beard. But you do not. The people made their voices heard on Sunday and the momentum will not stop. But now it is your turn. You can make history, or you will be vilified.

After him, a mother reads a poem written for her six-month-old daughter. Kathy Jetnil-Kijiner is from the Marshall Islands. We deserve to do more than just survive, she says, we deserve to thrive.

Her poem triggers a standing ovation.

I give up following the #Climate2014 Twitter hashtag. The tweets are appearing so fast on my screen that I can't read them. This is the first time in six years of tweeting that I have experienced such a thing.

The scientists back up the poets, sadly. In the days approaching the summit many have spoken out. Their message is loud and clear. World leaders must commit themselves to holding current rises in global temperatures to 2°C. Beyond that lies mayhem.

World Meteorological Organisation secretary-general Michel Jarraud is among those leaving no room for equivocation. "Our weather is becoming more extreme due to human activities such as the burning of fossil fuels," he says. "We must reverse this trend by cutting emissions of carbon dioxide and other greenhouse gases across the board. We are running out of time."

Citizens are returning in numbers to public expressions of dismay. Yesterday's demonstration in New York was but one of 2,646 events in 156 countries. More than half a million people took to the streets in cities including New Delhi, Melbourne, Johannesburg, Rio de Janeiro, London, Paris, Brussels and Berlin. Climate change is back on the front pages of newspapers the world over.

Some of the scenes remind me of protests I saw during the Vietnam War. On Wall Street yesterday, a hundred activists were arrested. Twitter is awash with iconic images. One shows a protestor dressed as a polar bear, being taken away by policemen with hands cuffed behind his back.

Another – well, what can I say? I am often tempted to write "life, stranger than art", on the pages of this chronicle. I can only get away with that cliché maybe once. Let it be here. The NYPD has found a way to deflate the carbon bubble. I mean the one twenty feet across, carried by protestors above their heads as they poured into Wall Street. The police popped it on one of the horns of the famous statue of a bull.

In the UN, the country commitment statements begin. Four minutes apiece with a big red digital clock ticking down time on the podium.

Jose Manuel Barroso, President of the European Commission, opens. The EU is leading, he reminds everyone. Our target is 40% emissions by 2030 en route to 80% to 95% by 2050. In effect we are decarbonising the European economy. He adds the renewables target, managing to keep a straight face: 27% of the energy mix by 2030.

It is merely business as usual. British foot-dragging, and oil industry lobbying for shale, had much to do with that.

The President of the Republic of Kiribati, Anote Tong, speaks with a hoarse voice. I've been shouting so long about climate change, he croaks, that I can barely speak now.

A wry smile from Ban Ki-moon on the top table.

The President of the Marshall Islands, Christopher Loeak, elaborates. We have nowhere to retreat to. For a quarter of a century, we and the other small island states have been the moral face of climate change. You can ignore us no longer.

And so it goes on. As I sit in my hotel room, it strikes me how much the world's leaders talk of moral imperative, and how little they enthuse about the megatrends that are strengthening the case of action.

David Cameron. Will he really be able to bring himself to enthuse about shale at an occasion like this?

He can. But his speech is a score draw between shale and solar.

Finally, Barack Obama, fresh from a speech about another bombing campaign in the Middle East. One issue will define the contours of this century more than any other, he says. That's climate change. The danger is not distant, but clear and present.

He talks of the impacts America is experiencing: floods, droughts, heatwaves.

Our citizens keep marching. We cannot pretend we don't hear them. We are the last generation that can do anything about this. In America, we are acting.

He sets out the list. Included in it: we harness ten times more electricity from the sun than we did when I came into office.

Just a few minutes ago I met with Chinese vice premier Zhang Gaoli and reiterated my belief that we have a special responsibility to lead. That's what big nations have to do.

Applause, the first I have heard since the opening plenary.

But let me be honest, none of this is without controversy. In each of our countries there are interests that will be resistant to action. But we have to lead, that is what the United Nations and this General Assembly is about.

At the very end of the afternoon, the UN gives Carbon Tracker a few minutes at the microphone. Anthony Hobley delivers the speech.

Many high cost projects are also high carbon and represent financial risk now, he says. Even at a low to zero carbon price they demand a high oil price simply not to lose money. This has put over a trillion dollar at risk: the dirty trillion. In a world addressing climate change, the financial risks rise dramatically. Yet plans are there to spend over 20 trillion dollars developing these high cost projects as well as searching for more unburnable carbon to replace these reserves.

I wonder what the heads of state are making of this approach to climate. It is so different from most of what they have been hearing today.

I wonder what the energy incumbency is making of the platform Carbon Tracker is getting on the world stage.

Moscow, 14th October 2014

When I was last in Moscow for more than a stopover, in 1989, the Berlin Wall still had another few weeks of life left in it. The Cold War was cooling, but it was still glacial. In those days, as an advocate of arms control, I was in the business of trying to help melt the Cold War. (I set up a Quaker-funded think-tank called the Verification Technology Information Centre in 1986). Rather a different game from the business of trying to cool the Warming War.

At Sheremetyevo airport I am ushered to a chauffeur-driven car bearing the logo of the Russian government congress I have been invited to speak at, and we join the heavy traffic heading along the four-lane freeway into the city.

Not far from the airport we pass giant steel sculptures of tank traps by the side of the road: a monument marking the closest the Wehrmacht came to Moscow in November 1941. I remember this place from 1989. It had been pointed out to me with pride by the Soviet academicians who were then my hosts. Here the citizens of beleaguered Moscow fought the might of Hitler's Germany to a standstill, and pushed them into retreat. It was a turning point in World War Two.

I recall a tatty satellite town here then, more than ten miles from Moscow. Now modern blocks of flats loom, part of a conurbation extending all the way in to the Kremlin. An IKEA superstore overlooks the sculptures. Car dealerships line the freeway as we crawl past in four lanes congested with their former contents.

We eventually come to the inner city. I observe that the ground floor of many of the Soviet-era buildings have been converted into luxury-goods shops. I see well-dressed people on sidewalks.

What a contrast to 1989. I recall shabby queues outside shops with near empty shelves then. I remember driving around the suburbs with a host looking for a filling station with enough petrol on sale for him to fill up his battered old Lada. Now, fine food and consumer goods are stacked on shop shelves. BP and Shell stations vie with Russian oil brands along the boulevards. SUVs and Audis pull into them.

It is oil and gas export revenue that has built all this wealth in the quarter of a century since I was here last.

The topic of my panel at the government forum is germane: "The world after oil and gas: how to ensure tomorrow's energy."

I am fascinated that Russian officialdom has put such a topic on the agenda of an innovation conference, and I wonder what year they might think "after" will start in, in this context.

After a night in a luxury hotel, I am delivered to one of those convention centres reminiscent of a conglomerate of the world's largest aircraft hangers. Every city in the world fancying itself as a leader in trade, innovation and the rest of the buzzwords of business has at least one of these.

Inside, thousands of people, the vast majority of them Russian, cruise the corridors and meeting halls. In many cases they literally cruise: the complex is so vast that electric buggies and scooters ferry delegates around. The forum spans all aspects of innovation, across all sectors of business.

Before my session starts, I have time to check out other panels. I make my way to two halls where energy is the topic of the hour. One session is entitled "Innovative approaches to R&D in geological exploration and production". In an adjacent hall, "Green technologies" is the somewhat different subject.

I opt for a taste of my past, not my present.

A Lukoil executive kicks off the exploration-and-production panel by thanking the Russian government for reducing taxes. Producing oil is a very expensive business, he says, and could no longer be contemplated if taxes were too high. An executive from IT giant SAP observes that the old oil fields are becoming exhausted, and smart use of data – which her company excels at, of course – will be essential if more oil is to be squeezed out. A McKinsey consultant backs her up, criticising the oil industry for being too risk-averse. They will need to embrace technology more, he says, if they are to keep producing enough oil and gas. A Gazprom executive volunteers that unconventional gas from shale is arriving on tap just in time, because the old conventional gas fields in Russia are depleting fast.

A director in the Ministry of Energy is given the last word on this panel. Fifty percent of the national budget comes from oil and gas profits, he says, please remember that. This is a key factor for the future of this country. You have said we need a roadmap for oil and gas production in the future. It is true. We need *several* roadmaps.

I listen to it all, through rain drumming on the iron roof of the convention centre and constant echoing noise from outside the hall, including public information announcements. I do not hear, on this panel, a Russian industry brimful of confidence about business as usual. They may be confident about their ability to innovate their way out of trouble, perhaps to the point of complacency. In this there are clear parallels with the west. But listening between-the-lines, I hear a message that the Russian Federation seems to be in trouble on conventional oil and gas supplies.

Questions from the audience. Someone asks a pointed one. Will it be western or Russian technologies that will be used for all this futuristic data-dependent propping up of tomorrow's Russian production?

The questioner does not mention the sanctions that are underway at present, instigated by the United States and other western governments outraged by President Putin's apparent expansionist intentions in Ukraine. Nobody has, for the full hour and half.

The chair throws the question to Lukoil Man. In a confident and forceful tone of voice, he dodges it.

Feeling usefully prepped by this listening experience, I head off to my own panel.

Representatives of "the energy of tomorrow", in "the world after oil and gas", include an American speaking for advanced biofuels, another American for wind power, a Briton for solar – me – and a Russian for nuclear. We are asked to give three-minute opening statements. Around a hundred people listen to us. It is 5.30 already, and the congress is thinning fast.

Advanced biofuels have arrived, the first American says. Two production plants are up and running.

Wind power will include airborne turbines tethered like kites, says the second American. We have a prototype.

I major on the twin emerging megatrends of solar-and-storage cost-down and oil-and-gas capex cost-up. The business models of the big energy companies are coming under threat on account of these, on an international basis, I suggest. Russia is no exception.

The nuclear Russian opens with an assertion that his favoured technology is clean and green, offers no thoughts as to why, and then moves straight on the offensive. It will be awful after oil and gas, he says. His tone is truculent. You need them for so many things. And renewables have so many problems. Solar panels are difficult to dispose of, for example. Nothing grows around windmills, for example.

With the opening statements complete, the chair seeks immediate audience participation, and he does it in an oblique way that is probably clever, given the peer-group constraints that swirl about him.

How many here think that renewables can power the future after oil and gas, he asks?

Hardly any hands go up.

Who has visited a nuclear power plant, he asks.

Most in the audience put their hands up.

Who has visited a solar farm?

Nobody puts their hands up.

The chair nods. We form our impressions from our own experience, he observes.

He then invites questions from the audience and they come thick and fast. They are uniformly negative.

Renewables can never power the future, says one official from a government institute. How could anyone sensibly think they could?

Look at Saudi Arabia, I say. Their government is worried that they are burning too much oil domestically. What is their answer to that? Not more oil, but – they have said – big national programmes to build nuclear and solar industries. Look at Germany. They are generating 28% of their annual national electricity demand from renewables already, today.

The questioner shakes his head vehemently.

With respect, sir, I say, straining for politeness, voice level, this is a verifiable fact, not an opinion. And the share is growing fast.

The man stands, gathers his bag with a flourish, and walks out.

The chair seeks to soothe these troubled waters. Our President Mr Putin has said that the oil and gas companies would do well to diversify, he comments. And on the first day of the forum in this very building Prime Minister Medvedev said in his speech that oil is too valuable to burn.

What about exports, asks a man from an institute I didn't catch. Russia can't replace its oil and gas export revenues with renewable electricity and fuels surely?

Oh well, in for a penny, in for a pound, I think.

Let us be frank, in a comradely and hopefully constructive way, I say with a smile. Most of Europe is *desperate* to escape having to pay so much for Russian oil and gas exports, or even having to buy *any* Russian oil and gas.

At this point, I notice a few grins appear in the audience. In Russia, as everywhere, at least some people prefer plain speaking.

The preferred escape route for many European nations, I continue, is acceleration of energy efficiency and renewables, a movement that has a good chance of growing as more and more governments realise how much more attractive the relative economics are, and as solar and other clean-energy costs continue to come down. How much longer do you think Russia and other exporters are going to be able to rely on oil-and-gas export income?

The nuclear man next to me on the panel shifts in his seat with a barely suppressed harrumph.

We come to end of the session. The chair asks each panellist to make one final point.

I wonder what best to say. I notice a man towards the back of the audience in full military uniform. I decide to direct my speech to him, invoking the spirit of Yuri Gagarin.

Look, the Russians were the first into space. You won the space race. Now, we are all in a new kind of international race. You are a bit behind other nations in this one at the moment. But there is so much science and technology skill in this country, and so much of a spirit of innovation. It would be easy to play catch up. And it seems so very clear to me that it is in the Russian national interest to do that.

After most panels of this kind, I invariably find myself facing a sizeable circle of people with questions, or seeking name-cards to write me follow-up e-mails. This time there are only two.

Back in the hotel, with time to kill, I catch up on my chronicle of events in the carbon war. So much is going on, as ever. I wonder how much of it the Russians at the convention catch.

The USA will soon become the world's largest producer of liquid petroleum. The IEA reports that US production of oil and related liquids such as ethane and propane was neck-and-neck with Saudi Arabia in August at about 11.5 million barrels a day (mbd). US crude production is approaching 9 million barrels a day: still lower than either Saudi Arabia's at about 9.7 million barrels a day or Russia's at 10.1, but having soared from 5 mbd in 2008 as fracking has boomed. US imports are expected to provide just 21 per cent of US liquid fuel consumption next year, down from 60 per cent in 2005. It is very easy to see why this narrative is so beguiling in America.

But can it continue like this in the face of the dire economics? Both Bloomberg and the FT continue to question whether drilling can continue while it simply builds up debt. Investors are losing confidence, they report. And the casualty list continues to build. Sumitomo of Japan is the latest big name to pull out of American shale, after a disastrous two-year foray in search of oil. Its write-down of $1.6 billion has wiped out almost all its earnings this year.

The industry's systemic hype barrels on. A Bloomberg investigation has found that 66 out of 73 US shale drillers report higher reserves to the public than they do to the regulator, the Securities and Exchange Commission.

My collaborator Mark Lewis is among the analysts predicting pain ahead. His latest report for Kepler Cheuvreux, with the graphic title "Toil for oil means danger for majors", concludes that oil prices will have to rise again, after the current spell of falling prices works its way through to less drilling. When they do, there will be little comfort for survivors. High capex requirements plus renewables cost reductions will mean that oil can be unattractive whether the oil price is high or low.

Others are more apocalyptic. On specialist websites the analysis persists that shale drilling is a giant Ponzi scheme, akin to the sub-prime housing boom.

Is this really the model that Russia wants to import to replace its ageing conventional oil and gas fields? I can't imagine it would be, if its advocates were following play in America closely.

In the UK, David Cameron continues as a shale advocate with blinkers fully on. His government has rejected 40,000 objections from voters saying they have concerns about the potential for fracking below their homes. UK fracking companies can go ahead, the government says, and in their forthcoming Infrastructure Bill will say that they the oil and gas companies can use any substance they like.

If this is the undemocratic lengths that a democratic government is prepared to go to, I think, how bad will it be in Russia if full-scale shale operations like those in Texas and North Dakota begin?

And then there is the Arctic. We have plenty of oil and gas there, Russian advocates of the status quo would say. Rosneft and ExxonMobil have just announced a discovery of 750 million barrels in their $700 million joint well in the Kara Sea, have they not? Igor Sechin, Rosneft's chief executive says that the company could invest $400bn in the Arctic in the next 15 years. He expects to open a new oil province there, "with reserves comparable to the developed reserves of Saudi Arabia," or so he told Der Spiegel.

Well yes, but there are two problems. The first is geopolitics. Exxon is being forced to wind down its collaboration with the Russians as western sanctions bite over Ukraine. That means technical problems for Rosneft. There will also be financial problems, if western capital continues to be unavailable from western banks. Sechin has already had to go to Putin, cap in hand, to plead for $42 billion in the short term as western sanctions have bitten his company.

More importantly, in the long term, there is the economics. The cost of developing and producing Arctic oil is too high. The oil produced, by the time it is produced – well over a decade from now – will not be able to compete with alternatives in energy markets. Any reserves found will be stranded then.

That assumes the drillers do what they do when they say they can do it by, at the capex costs they currently report. But nobody has ever drilled a large oilfield far offshore in the Arctic. Who is to say that the experience won't be akin to Kashagan? A $40 billion over-run on capex, and an eleven year over-run on schedule, and counting.

Greenpeace's campaigners – long since released from their Arctic jail cells, thank God – may have failed to stop Russia's Arctic drilling with their campaigning. But economic and operational factors will very likely be doing their job for them.

Chapter 12

Seven percent by 2035

Freiburg, 21st October 2014

A visit to the biggest renewables research institution in Europe, the Fraunhofer Institute for Solar Energy Systems in Freiburg, a beautiful German city awash with solar installations. A trip designed to top up my batteries both in knowledge and morale. 1,300 people work here, on a mix of federal and industry funding. There is nothing remotely like it in the UK.

Eicke Weber, friend and director, drives me in his zero emissions hydrogen car. It takes three minutes to fill up and can drive 400 km on a tank. He is a fan of batteries as well, but he loves the idea of hydrogen. The gas can be produced by solar in summer, he says, by electrolysis, then stored in salt caverns. Anyone who says storage is a problem for renewables is just flat wrong.

I tour the labs. I am shown a solar concentrator cell, using lenses to focus the sun's rays, with a world record of 46% efficiency. I am walked through pilot thin-film manufacturing lines – joint ventures with big-name solar companies. I hear about the latest spin-off start-up companies the lab has underway, and in planning. This is the technical front line in intelligent system control for decentralized energy systems: not just solar, but combined heat and power plants, and others forms of micropower.

I meet researchers from all over the world, including the States, France, and the UK.

In Eicke's office, we talk solar politics. This is a man I am instantly on the same page as, every time we meet.

We compare notes on the incumbency pushback in the UK and Germany. I am surprised by how downbeat he is about the status in Berlin of the Energiewende. It seems the massive demonstration I went on, and consistent popularity of renewables in opinion polls, do not impact the political world as much as big energy lobbying does.

Not even the Greens speak up for the programme these days, Eicke complains. The problem lies in media coverage. The big energy companies put the press through a kind of schooling. They are bought. The energy giants use their advertising budgets to ensure key papers and magazines play soft with them: they tell the media that if they report too favourably on the Energiewende, they will take their advertising elsewhere.

What about German politicians, I ask. Are people on the Green side, or in the SDP, not emerging to replace Hermann Scheer? Do you not have progressive conservatives of the kind we have in the UK?

Eicke laughs. SDP politicians often end up working in the energy industry, he replies. This is something nobody talks about here in Germany.

London, 24th & 25th October 2014

A two-metre wire fence surrounds the grass square opposite the Houses of Parliament. It is lined by a mix of police and private security guards wearing black caps, facing outwards. They stand side by side, looking bored, reminding me of the kind of paramilitary one sees on the streets of Latin dictatorships. They far outnumber the protestors they are barring from the grass. These, the Occupy Democracy group, sit and stand on a narrow strip of grass beside the main road. They have a bedraggled look. Their banners, facing the heavy traffic on the road, are makeshift.

The banners speak of emerging threats to democracy. I didn't vote for privatisation of the NHS, says one. I didn't vote for fracking says another. Hoot if you love democracy, says a third.

There is a lot of hooting from passing cars.

The tacit support is unsurprising. The excesses of the banking industry go on and on, one scandal rolling out after another, with a wholly inadequate regulatory pushback. Recently it has emerged that six major banks have been caught rigging the currency markets. They face a fine of more than a billion pounds for that, but as ever no jail time for executives. We have also learned that since the financial crisis the banking industry has paid more in remuneration of staff, especially of course senior staff, than it has in dividends to shareholders. Before the crisis, the reverse applied. It is a wonder to me that there are no investors at this demonstration, alongside the students.

I am with the mother of a protestor, an old friend. We are delivering food to the night shift. She is not sure if we are breaking the law. So many things have been banned, under the weird law rushed out by the coalition government to ban protest on this space opposite parliament. Police confiscated tents that protesting students wanted to erect on the square itself for the week-long period of their protest. They arrested a member of the London Assembly, Jenny Jones, for simply turning up to monitor police behaviour. Green MP Caroline Lucas was threatened with arrest if she passed food to a protestor who had climbed the statue of Winston Churchill. Students deprived of their tents, including my friend's daughter Daisy, sat on a tarpaulin instead. They were arrested, carted off in vans to Brixton Jail, and incarcerated for ten hours.

How can a modern democracy suppress protest by its disenfranchised youth in this way? What kind of future does that point to?

Daisy has a double first from Cambridge. Her friends are also Cambridge graduates.

The police aren't going to know what hit them, says my friend, when they confront this coven of ferocious intelligence in court.

My friend was no student protestor herself, in her youth. But her daughter's treatment is plainly radicalising her. She is particularly outraged at how the mainstream media is ignoring the demonstration.

We sneak the food in, and I am introduced to Daisy and half a dozen young protestors. They tell me they will be sleeping rough here tonight, lying on the grass. The police have told them that if they use so much as a pizza box to rest their heads on, they will be arrested.

I notice the students are all wearing strips of blue tarpaulin, pinned to their lapels.

The next day the square opposite the seat of democracy sits empty under beautiful autumn light. I am scheduled to speak to the protestors about solutions: the solar revolution and community power, people-power capital, all that. I am going to find it difficult. The police have confiscated the PA system.

If I were more pragmatic – wiser, you might say – or had an eye on the trappings of the Establishment, such as the New Year's honours list, I probably wouldn't do this kind of speaking engagement. But I have far more empathy for these young people than I do for the system I am a creature of.

I look around the hundred or so gathered close around me to listen, hearing the din of the traffic, and the regular hooting. Expect a sore throat tonight, I think.

The talk is being live streamed on the web, as they all are. Be frank, but be careful, I tell myself.

Before I talk about solutions, I begin, I want to say something that I think is very important. Modern capitalism is broken. I say this as a nominally successful capitalist. My experience has shown me that the system is suicidally dysfunctional. It is on course to destroy both itself, and more importantly the viable civilisation people of my generation have a moral duty to hand over to our children and grandchildren.

There are many reasons why I believe this, but let me take just two. The first is climate change. How can a system not be broken, when it leaves us on course for four maybe six degrees of global warming: a future where we lose the ability to feed ourselves and our access to drinkable water. The second is financial crisis. How can a system not be broken, when it allows bankers to take home more in bonuses after the 2008 crash than they have returned to their shareholders?

This broken form of capitalism will imperil democracy, if we let it. In its death throes, it will bring in fascism.

But it can be changed. It can be fixed.

That is why what you are doing here is so important. You may feel like a minority, at four o'clock in the morning, outnumbered by the paramilitary guarding the grass. But you have so much support out there.

They clap and cheer.

I hope what I say is true.

But the first person to come up to me is a man with cropped hair and a thuggish air. My first thought is "undercover policeman", and not too cleverly disguised at that. Media reports of late make it clear that the Metropolitan Police uses officers as snoopers extensively within protest groups. In several cases, they have gone to bizarre excess. Just yesterday a court ordered the Met to pay £425,000 damages to a woman who gave birth to a baby fathered by a man she didn't know was an undercover policeman. He was spying on her and her animal-rights-activist friends.

Miscreant bankers and energy executives do not get this kind of undercover attention from officers of the law, it seems.

But it turns out that this man is from something called the English Defence League. He speaks in a voice dripping with choked-back aggression.

What gives you the right to stand up in fron'a all these people, and talk for 'arf an hour? You just wanna sell solar panels, don'tcha?

London, 29th October 2014

Last week Occupy Democracy. This week Oil and Money. I lead a strange life, it sometimes occurs to me.

Four hundred and fifty suits and twin sets from oil and gas companies in 40 countries assemble in the Park Lane Intercontinental to hear the captains of their industry hold forth on their latest preoccupations. Today, there is a clear and present danger, it seems. The oil price has fallen to $80 a barrel. I am going to be very interested to see how they deal with this little wrinkle in their market.

But first, a tragedy. Christophe de Margerie, charismatic CEO of Total, has been killed in an air crash in Moscow. The conference organisers call for a minute's silence.

I take the time to remember my own memories of de Margerie. All I experienced of him was a panel debate at an oil industry retreat in Norway. I had the firm impression of a man who was his own master. And I would have liked to have known him better – to try and understand how his brain worked. Why? Because of all the oil bosses, he seemed closest to an honest appraisal – as I see it – of both oil-depletion threats and solar-energy opportunities. Total has talked openly of pressures on global oil supply, and have invested billions in solar, where ExxonMobil, Shell, BP and the rest talk of a "new age of oil" and badmouth solar despite all evidence and converts to the contrary. Total adverts festoon Heathrow at the moment: "Committed to Better Energy," the headlines read. The strap line, in smaller print, makes it clear what is meant by "better": Oil, natural gas, and solar energy – 100,000 women and men." It's a start.

First up is the OPEC Secretary-General, Abdal Salen el-Badri. He is very clear on what $80 oil means. Fifty percent of US shale oil production is so uneconomic it will have to be shut down.

I always learn so much at these industry conferences. I am particularly looking forward to the first discussion of Kashagan this morning. Eni's Franco Magnani provides it. We are going to have this kind of problem again & again, he says.

Not if investors spit the dummy, I think.

Bob Dudley, BP CEO, speaks just before lunch. He is the first speaker to even mention climate change.

He gives BP's view of the most likely energy mix by 2035. Oil, coal and gas will each be at 27% of primary energy by then. Renewables are 2% today, and will be 7% by 2035. This is not the most desirable outcome, he says, but the most likely.

Bob Dudley has offered a 30-minute recipe, I tweet, for torpedoing the global climate, global economy, oh and of course BP itself.

BP's retreat from renewables has been so dismal to watch. After the high hopes of the Browne years came a shameless retrenchment to carbon under Tony Hayward's leadership, notwithstanding hardening scientific evidence on climate change. Now, under Bob Dudley – who once headed BP's solar unit – BP is on a dedicated mission to emulate ExxonMobil's worst.

They manage to get their heads around this abrogation of responsibility – to shareholders, staff, customers and citizens alike – by saying that although climate change is indeed problematic, the responsibility for dealing with it lies with government. Tony Hayward was crystal clear on this when I debated with him at another FT event in September 2012, as described in *The Energy of Nations*. BP and Shell foot soldiers echo the message at events like this.

Of course, ask virtually anyone in government not looking for BP largesse in the next phase of their career, and they will tell you that BP lobbies hard to stop any government efforts to escape oil-and-gas dependency. So when Bob Dudley says 7% renewables in the global energy mix by 2035 is not the most desirable outcome, but the most likely, he is being disingenuous. BP lobbyists are doing their very best to make the undesirable outcome the most likely.

Over lunch, a protest. Two women who do not work in or around oil and gas have found a way through security. They turn on a rape alarm and chant about how the assembled oilmen and women are raping the planet. They are soon bundled away by security guards.

After lunch comes a session on oil, gas and security. There is strong interest in this.

The final session of the afternoon is on oil, gas and climate. I am on the panel. Half the audience leaves before the session begins. Climate change has nothing to do with security, I guess.

In the Q&A session I am asked how I reconcile my bullish view about clean energy with BP's view about its role.

In business schools across the world, I reply, students will soon be laughing at what Bob Dudley said here today.

Chapter 13

Winning dirty in an endless war

London, 30th October 2014

Twenty solar industry executives gather for wine and dinner in a conference room above an office on a crowded West End Street. The price of their supper is a pitch from me. The Solar Trade Association has kindly convened this early-evening gathering so that I can try to organise some solar-industry lobbying beyond the kind that they themselves do.

My idea is simple. I want ten companies to contribute £10,000 each to fund a basic effort to counter the oil and gas industry's efforts to sabotage the solar industry's interests. In particular, I want us to develop a voice in the fracking debate. Why should we not, when the wonders of shale are so often espoused by incumbency cheerleaders as a reason for putting solar on a back burner or worse?

The latest evidence of what we are up against has appeared this very day in the New York Times. Someone in a roomful of oil and gas industry executives has developed a conscience and leaked a tape of a presentation by a top incumbency lobbyist. In it, Richard Berman, a veteran fossil fuel defender, tells his audience that they should regard themselves as being in a state of "endless war" with environmentalists and others who oppose fracking and advocate clean-energy deployment. The oil and gas industry cannot "win pretty", he says. They will have to "win dirty". By this he means the use of tactics such as digging up embarrassing information about environmentalists and liberal celebrities, and exploiting emotions like fear, greed and anger, and turning them against the opponents of fossil fuels.

The executive who did the leaking told the paper that he did so because the black-arts approach Berman suggested left a bad taste in his mouth.

This is one of the reasons why I think of the battle of ideas on climate and energy as a "war". A civil war: fossilistas versus cleantechistas, incumbents versus disrupters, the dark side versus the light side. I am criticised for this metaphor by some of my friends, especially by women friends. Come on, they say. There are no bullets anywhere. You had too many toy soldiers when you were a little boy. Don't be so melodramatic.

They are correct, both about the bullets and the toy soldiers, and I do hold some reservations about the metaphor. But a civil war is how it *feels* to me. There may be no bullets, but people have started to die who wouldn't have died so early otherwise, especially in the developing world. That casualty list will mushroom unless we act. Plus the belief systems breed hate as they skirmish. I try to suppress it in myself, but I have seen it clearly in the eyes of coal advocates as they jab their fingers at me. It can be seen in many a shouting match in the US Congress. And like any civil war, you find believers in the two warring belief systems under the same roof: in the same government, political party, company, household even. And for many on the light side, it is an existential struggle. We lose the struggle, we lose civilisation. What other metaphor works better than global civil war, given these stakes?

Berman was seeking a $3 million top up of his already-considerable backing from the fossil fuel industries that day, back in June in Colorado Springs. This evening in London I am seeking £100,000 from the companies present. And there will be no question of playing "dirty". I am simply recommending that we make a concerted effort to get the right information about shale into the right hands at the most opportune times.

A little effort by a few good people, I argue, can go a long way in this regard. Take the events of just the last week. The already-bad economics of US shale drilling is beginning to look increasingly shocking as the oil price has fallen to around $80. This has wiped at least $158 billion off the market price of drilling companies, according to a Bloomberg analysis of 79 of them. Meanwhile, doubts continue to emerge about the industry's transparency based on its demonstrable tendency to hype. The most recent analysis comes from the Post Carbon Institute, written by David Hughes, the geologist who correctly warned about the vast overstatement of reserves in the Monterey Shale. His reading of company data suggests that both shale oil and shale gas production from the main US drilling regions will peak by 2020, and drop steeply thereafter. The two most productive shale oil regions, the Eagleford of Texas and the Bakken of North Dakota, will peak next year, he suggests. His story is dramatically discordant with the narrative-of-plenty pushed by the companies and the US government's Energy Information Administration. The EIA claims production of both gas and oil by 2040 will be at levels far above those Hughes sees feasible.

Environmental problems continue to build for the industry alongside the woes arising from economics and overoptimism. The latest episode to join the growing catalogue is the discovery by an environmental watchdog of benzene, a cancer-causing chemical, in the liquids drillers pump underground during fracking.

The public is picking up on the escalating drip of toxic news. A poll in New York State has shown a remarkable eight in ten voters in favour of the state's current moratorium on fracking, an opinion that spans all political divides. Shale is demonstrably beginning to feature as a negative issue for American voters.

But how many UK parliamentarians, policymakers, journalists and investors know about all this, I ask the solar executives. They are not reading Bloomberg reports every day, much less the specialist websites where most of information languishes. We need a concerted campaign to present them with a digest of such material, all reliably sourced. Perhaps then they would be less keen on default support for the alluring image of abundant cheap oil and gas right under our feet in British shale, and more enthusiastic about the alternative of a green industrial revolution.

Around the table, the disappointingly thin audience of solar-company representatives looks and sounds keen enough this evening.

Why is it that I've never seen these horrible aerial images of fracking hot spots in America before, one asks.

Why indeed.

But previously unseen horrible images or not, I wonder if their boards will be up for even the low level of funding for lobbying that I ask for. Just to make a start on the basics.

Online, 2nd November 2014

When the Intergovernmental Panel on Climate Change completed its First Assessment Report in 1990, almost a quarter of a century ago, I worked for Greenpeace International as a climate campaigner. For the previous decade I had been a university scientist, on the faculty of the Royal School of Mines at Imperial College. As an earth scientist, my research had been on the geological history of the oceans. That was what made me worried about climate change. Based on my studies of oceanic sediments, I figured I knew about the natural rhythms of the planet. I didn't like what I was seeing in the first climate models to incorporate the build-up of greenhouse gases in the atmosphere. I quit my life as academic and consultant to the oil and gas industry and became a campaigner. Or a turncoat, as most of my ex-colleagues saw it.

Then, as now with the Fifth Assessment, the overall IPCC Assessment Report was drafted based on three reports from working groups. In May 1990, the working group on climate science completed the first. "Race to save our world", the headline in one British newspaper read the next day. The headlines in other papers were all versions of that. The world's climate scientists had found dire rates of global warming and sea-level rise. Even Mrs. Thatcher, the then UK prime minister, was worried.

The second report, on impacts, and third, on solutions, came out during the summer of 1990. The impacts report sat uncomfortably beside the scientific report, as though its authors had been instructed by their governments to tone things down. The solutions report was completely discordant. Many governments could not bring themselves to say that the deep cuts in emissions advocated by the scientists could be achieved with renewable energy, energy efficiency, and the rest of the solutions toolkit.

I gathered a group of eminent scientists and policy experts together who did believe in the feasibility of deep cuts in emissions. We persuaded Oxford University Press to publish a "shadow" IPCC report: one where the projected impacts and solutions would fit the climate science. It came out in August 1990, just in time for the World Climate Summit in Geneva, the meeting that kicked off the global climate negotiations that have been running ever since.

We echoed the IPCC's climate assessment, but said it did not go far enough. The IPCC projection of global warming was a "best estimate". Policymakers needed to hear a "worst case", based on the probability, as we saw it, of positive feedbacks – natural amplifications of warming – coming into play.

We were accused of scaremongering at the time. But climate scientists today routinely discuss the feedbacks we described, and how they are very much in play.

We argued that the impacts would be worse than the IPCC argued in 1990. Again the march of events has proved us right.

Nobody likes people who say "we told you so." I have had to bite my tongue many times over the years since then as scientists and policymakers caught up with the IPCC "shadow report".

We described how the world could be run on renewables and energy efficiency. Few believed us. In 2013, more new generating capacity came on-stream around the world from renewables than from fossil fuels and nuclear combined.

That share will continue to grow. Ever more people believe that a zero-carbon energy system is feasible today. They can see the evidence of their own eyes in countries leading the way.

The Fifth Assessment has already been making headlines around the world, ahead of its actual release. As usual, leaked copies are in circulation. Climate change will inflict "severe, widespread, and irreversible impacts" on people and the natural world unless carbon emissions are cut sharply and rapidly, the report concludes. In terms of emissions budget, if the world is to stand a better than even chance of limiting warming to no more than 2°C, only about one trillion tonnes of carbon dioxide can be emitted from 2011 onwards. In terms of the investment needed in clean energy in order to achieve this, the report reinforces the IEA's analysis that an extra $5 trillion of energy investments will be needed in the 22-year period to 2035, relative to its base-case scenario.

I watch the launch press conference online to see how the UN officials representing the panel fare in putting across their stark message.

The first question is for the UN Secretary-General, Ban Ki-moon. The impacts are so severe and obvious that we wonder why are you still an optimist.

The head of the UN elects to emphasise the engagement of the business community in climate change, of late. What he saw at the September summit in New York was very encouraging, he said. As was a follow-up meeting with industry leaders yesterday.

The second question: why will you succeed where you failed in Copenhagen?

World leaders were not ready then. The times are different now. This synthesis report will make a big difference. Again the Secretary-General moves to the pivotal role for business. He speaks directly to investors. Please reduce your investments in the coal- and fossil fuel-based economy, he urges, and move to renewable energy. He is resolutely upbeat. I am confident we can reach agreement in Paris 2015, he says.

Michel Jarraud, head of the World Meteorological Organisation, chips in. Ignorance can no longer be used as excuse for inaction, he says. Decision-makers will be held accountable for decisions they make today. That can happen in 50 years from now, he says. Then corrects himself. Even 30 years.

City of London, 1st December 2014

Accountability. Yesterday was one of those days that breathed credible life into that rather loaded word, where defence of fossil fuel profligacy is concerned. Two pinch-me-to-check-I'm-not-dreaming developments hit the headlines one after the other. First, the FT reports that the Bank of England is launching an internal enquiry into whether or not fossil fuel companies pose a threat to global financial stability. The Bank failed to spot the threat that the financial sector's excesses posed to the capital markets in 2007 – or at least, failed to sound an alarm. Evidently, it doesn't want to risk repeating the act, less than a decade later, in the case of the energy sector.

Governor Mark Carney is clear on why. A key risk in need of appraisal involves stranded assets. If climate policymakers begin delivering on their promises to cut carbon emissions to sub-danger levels, most coal, oil and gas reserves will need to remain underground.

Second, Germany's largest utility E.ON announces it will be retreating from coal and gas to focus its future growth on renewables, distribution and customer services including energy efficiency. Legacy assets from the old regime – coal-, gas-, and nuclear plants, energy trading and exploration and production – they will park in a separate company.

So the first major energy company has instigated a U-turn with its business model, firmly in the direction of energy transition. And the Bank of England, by recognising that there is a case the energy incumbency threatens financial stability, has cast a long shadow over the prospects of all fossil fuel investments going forward.

Nobody will be told whose arguments carried the day in these milestone decisions. But the Carbon Tracker team can surely be allowed some quiet self-congratulation about their role in the Bank's decision. And Hermann Scheer's parenting of the German feed-in tariff for renewables can certainly assume posthumous credit for E.ON's.

What a day, then, to check out the state of the conservative brain on climate change and energy. The Spectator, a British magazine of great popularity among the political right, is running a conference on energy. Carbon Tracker, desirous of appealing to all political tribes, is co-sponsoring it. But the colleagues who dreamt up this collaboration are away on the day: an exhausted Anthony Hobley on paternity leave, Mark Campanale and James Leaton on the fundraising trail in Denmark. Mr Chairman has to hold the fort.

Our PR agency decrees that I need some coaching in "how to talk to the Right". I protest, but not too strenuously. One Spectator issue ran a cover headline exhorting readers to "Relax: Global Warming Is All A Myth". I need coaching in how not to have a heart attack, if I am to be faced with stuff like that.

Laura Sandys, a delightful Conservative who actually wants to conserve things – important things, including the planet – is assigned the task. She is a serving Member of Parliament but also sits on Carbon Tracker's advisory board. I listen, and find her perspectives fascinating.

The strategies of the Right include a concerted effort to create cultism, she tells me, with disarming frankness. They and their media supporters, of which there are distressingly many, seek to paint believers in global warming as old-fashioned flat-earthers who are anti development. They view this as their strongest message. They are very professional and co-ordinated in pushing it. They focus on heart and pocket, not head. They are not frightened of being controversial and emotive, and they use a wide range of messengers.

The worst thing, she emphasises, is that they may be winning. In 2005, 82% of the population professed concern about climate change. Now only around 60% do.

There is no point engaging on climate science, or even using that many facts, Laura argues. Feelings are what work in this pseudo-debate, and what we need is a new route to arouse them.

Somewhat consistent with my belated education in the recent discoveries of neuroscientists, then. But depressing nonetheless.

We need to paint a desirable picture of a necessary transformation, Laura continues, using innovative new technologies, ones that offer greater productivity, efficiency and profitability; that are modern, progressive and secure.

Jeremy, she says, I think that the most powerful arguments are about modernity versus old fashioned business models mixed with the cost of not addressing climate change. None of these guys want to look behind the times.

That's it, I think. That is what I will major on. Telling the rightists that they risk being on "The Wrong Side Of History." The developments at the Bank of England and in E.ON's Stuttgart headquarters have surely made my job easier for me.

Andrew Neil, former editor of the Sunday Times, one-time favourite of arch-conservative media mogul Rupert Murdoch, now chairman of the Spectator, chairs the event. He introduces the panel I am on, the first of the day.

James Ball, a director of the first company to export LNG from the US, speaks first. He is a trenchant advocate of the shale narrative. He recites the litany of benefits this boom has bestowed on America. As he reaches the opposition to fracking, his tone becoming truculent. None of this backward thinking can he understand.

Or, clearly, tolerate, I observe.

There will be no fracking in the UK because of ignorant nimbyism, he asserts.

But Ball ends with what turns out to be another pinch-me moment. There is one brand new caveat that holds the potential to bring the whole shale boom in America to a halt, he says, his tone edged with incredulity. It is the phenomenon of the city of Denton. Denton was the first urban area to be fracked in Texas. It is a thoroughly Republican city from top to bottom. Yet its citizens have voted to ban fracking going forward. This bizarre turn up for the books is a wild card that could undermine the entire shale boom, he concludes.

Trying to conceal my shock at this seemingly massive new setback for the incumbency, I duly offer my perspectives on being on the wrong side of history. As Laura Sandys advised, I major on the upside potential for prosperity of a green industrial revolution, where I can speak with a degree of authority from the front lines. British Conservatives were once rather good, collectively, at articulating this vision, I say. Before they were in government. Companies have started to adopt it. Take E.ON's recent decision.

Andrew Neil responds as soon as I finish. E.ON is a basket case, he says, not an example to be held up of things to come. It is the subsidies for renewables that have killed their business model.

His argument, it seems, is that the big utility has been sunk because foolish politicians have eschewed the pure free markets that rightists hold so dear by giving subsidies to expensive and generally useless technologies like windmills.

And conventional energy needs no subsidies? I respond – the obvious retort. Far more subsidies go to the incumbency than to renewables. Ask the IEA.

I watch him articulating his counter punch. He doesn't look at me. Like Ball before him, he radiates an aura of intolerant contempt when debating with me.

Chapter 14

From carbon capitalism to climate capitalism

Lima, 3rd–14th December 2014

This is the first annual climate summit of the UN Framework Convention on Climate Change that I have been to since the bitter disappointment of Copenhagen in 2009. Senior diplomats have been telling the press that this session of the climate talks, and the follow-up through to the Paris climate summit a year hence, is the best chance in a generation to tackle climate change. One of the main reasons for their optimism is the announcement of a bilateral deal between the US and China, the two main emitters. On November 11th, after months of secret talks, Washington and Beijing unveiled new emissions targets of their own, and committed to lead the global effort to combat climate change together. Presidents Obama and Xi Jinping announced the initiative in Beijing. The US will emit 26% to 28% less carbon in 2025 than it did in 2005: a target double the speed of the previous reduction commitment. China, starting way behind the US in terms of emissions commitments per capita of course, will finally come to the emissions-reductions game with a commitment to stop its emissions growing by 2030. By that time, President Xi says, 20% of China's energy would come from solar, wind and other renewables. Energy, not electricity, note. That includes transport. The Chinese must have EVs in mind.

In late October, EU leaders agreed to cut greenhouse-gas emissions union-wide by 40%, by 2030. In Berlin on the eve of the Lima summit, Angela Merkel's cabinet agreed extra German cuts from the power sector to meet a 40% target by 2020.

The EU commitment, and Germany's renewed push for leadership within the union, is nothing if not a big reminder to E.ON that it has made the right call with its U-turn towards clean energy.

While I am squirting carbon into the air from an Airbus, slumped over my computer, I exchange e-mails with my Carbon Tracker colleagues. Issue of the day is that the expert group of the Norwegian Government Pension Fund has delivered its recommendations on how to treat fossil fuels in the fund. We have to respond with a press statement.

The news is mixed. On the one hand, advisors to the world's biggest sovereign wealth fund advise engagement with investee companies on climate, and divestment from the "worst" offenders, in terms of carbon pollution. In this they advocate the approach favoured by Carbon Tracker. On the other hand, they seem to envisage substantial amounts of fossil fuels being burned for decades to come, an assumption which allows them to assume limited scope for stranding of assets.

Meanwhile, in the early action in Lima, the Alliance of Small Island States leads the first group of governments at the climate talks to directly support a complete phase out of carbon pollution by 2050, or zero net carbon is it is also called. The Alliance of Independent Latin American and Caribbean Countries supports them. So does Norway.

The environmental organisations are quick to remind negotiators that zero net carbon has to mean 100% renewable energy by 2050. Carbon capture-and-storage and nuclear lead the non-renewable technologies that put themselves forward as contributors to zero net carbon. They cannot be realistically considered because they are both proving themselves to be costly illusions: the very antithesis of renewables.

It surprises me that the Norwegian government supports this zero net carbon target. It will be interesting to see how they deal with the report from their Pension Fund's advisors then.

Many Alliance of Small Island States delegates are tabling their arguments at the climate negotiations now for the sake of the planet, not their homelands. In the Republic of Kiribati, far away in the South Pacific, the president, Anote Tong, gives a heart-rending interview in which he describes how the rising sea floods his Pacific atoll nation increasingly frequently, shrinking drinking water supplies. Even strong action isn't going to save his people, he says. They will have to migrate at some point.

4th December: A ride through Lima's suburbs to a city of temporary tent-like buildings in the grounds of the Peruvian Army's headquarters. I walk through the throngs. More than 12,000 people from nearly 200 countries are here in this small city of giant tents. I used to know so many people at these negotiations. Now, it is an hour before I see anyone I recognise.

A Korean student is the first person to speak to me. She and a cluster of her peers pass pamphlets to delegates advocating the need for universal love to underpin the global climate deal that must be done in Paris. They have made green paper wallets to give out to as a token of this love. With great inappropriateness, I feel like hugging her.

Three women in spectacular saris wander past, deep in conversation. Are they representatives of a citizen group, or Indian government negotiators? There is no way of telling.

A tatty life-size polar bear lies dead on the ground nearby. A notice pinned to his chest announces "Arctic Methane Crisis".

My first meeting is with a UN official: a session to plot how the carbon bubble argument can best be injected, in and around the negotiations, to boost the prospects of success in Paris. He describes the shape of the deal that the UN is angling for.

Countries intending to be parties to the eventual Paris treaty must submit their commitments for emissions reductions in 2015, ideally by the end of the first quarter, but if not then, in plenty of time for the UN to add up where the sum of commitments leads in terms of the amount of global warming it commits the world to. The Secretariat considers it highly unlikely that the sum total of Paris commitments will put the world on course for two degrees in one leap, given the tortuous history of the negotiations. But that does not mean the community of nations will have failed in the quest to contain the problem, so long as several other vital things are captured in the Paris treaty. Critically, governments must commit honestly to review their collective progress regularly, every five years ideally, with the avowed intention of ramping up emissions commitments until the world is on course for staying below two degrees. The UN Intergovernmental Panel on Climate Change's work informs the negotiators that they have a fair chance of keeping the world below two degrees provided net greenhouse-gas emissions are phased out altogether by the second half of the century. That goal will become ever more credible, so the UN believes, simply with the passage of time, as the feasibility of a zero-carbon future becomes ever clearer to governments and populations from the evidence of their own eyes. What this means is that governments must view their Paris commitments – Intended Nationally Determined Commitments, to use the UN jargon (let me simply call them national emissions commitments in the rest of the book) – as minimum first steps, not the final word.

As for the legal framework of the treaty, negotiators are these days much influenced by the intractable problem of climate denial in the US Congress, especially on the Republican benches. Emissions-reductions commitments binding in international law are very unlikely to be ratified by any conceivable US Senate, because of this denial. Hence, there is now a movement at the climate talks to wrap the commitments as a less formal layer round a more formal core: to make the rules of the treaty – the review period and process – binding in international law, and the emissions commitments subject to national law. That way, the reasoning goes, the treaty has a better chance of being ratified by the US and enough other crucial parties to bring it into force.

The bottom line in all this is clear to me. The UN is shooting for a signal to be sent from Paris that is strong enough, despite the inevitability of imperfections, to create a positive direction of travel, one that can keep alive the possibility of staying below two degrees of global warming.

I can buy into this line of reasoning without too much difficulty, believing what I do about the potential for disruptive system change inherent in the two emerging megatrends of fossil fuel cost-up and clean-energy cost-down. The right direction of travel in climate policy would make it more likely that clean-energy markets can accelerate to the extent needed for eventual victory in the carbon war, and renaissance in the world.

So I tell myself.

But as the UN official and I discuss the prospect of success amid pragmatism and realpolitik, elsewhere on the Lima summit site old problems of multilateral diplomacy play out as they have so often over the quarter of a century the climate talks have been in progress. China tells the summit that climate-finance pledges from the rich nations to help the poor nations adapt to climate change, and avail themselves of low-carbon technologies, are so far inadequate. Head of delegation Su Wei's intervention has the feel of warning shot – one that has been fired many times before by China, but also India and Brazil.

In Copenhagen, the rich nations agreed that the appropriate sum should be $100 billion a year by 2020. They have pledged only $10 billion so far.

This $100 billion a year sum will inevitably be the cause of much arguing over draft treaty text at this summit, whatever the big game plan for Paris. Rich nations, mired in domestic austerity as many are, will be reluctant to sign up in concrete terms. They are much more focussed on the need for emissions-cuts commitments from all nations. Yet, it strikes me, this sum, shared between many nations, is precisely twice the capital cost of a single oil field, in the case of Kashagan. The contrast is one of those snapshots of how crazily skewed global priorities are. $50 billion to access new oil to cook the planet with: few questions asked. $100 billion a year to help the poor nations adapt to climate change and use new technologies that won't cook the planet: endless all night bickering at the United Nations.

The United Nations Environment Programme releases a report updating its estimates of how much adapting to a warmer climate could cost: up to $300 billion a year by 2050, even if global warming can be kept to a two degrees ceiling, the agency says. This is a sum almost three times as much as previously thought. The conclusion deepens the problem of the rich nations' adaptation funding for poor nations. But matters could be worse yet. I am always suspicious of global-warming damage or adaptation estimates. How can they cost in the potential for amplifying feedbacks in the climate system, and the synergistic effect of multiple climate impacts doing their worst all at once? An abiding concern, harboured ever since the two-degrees target for maximum global warming was first mooted at the climate talks, is that it won't prove to be a strong enough to hold at bay the natural amplifications of global warming that might be triggered along the way. Many governments share that view. At the 2010 climate summit at Cancun, Mexico, one of the pledges in the closing statement was for a review by 2015 on whether the objective of two degrees needs to be strengthened in future to a 1.5°C goal, on the basis of the best scientific knowledge available.

These days, almost all the talk is of two degrees. But the possibility that we will need a 1.5°C target is still there. That was what the poor dead polar bear was worried about. And methane release from the Arctic is just one of the potential amplifying feedbacks that we know about.

I tell myself this. Just focus on playing your bit part in looking for ways to help a clear direction of travel to emerge: an accelerating transition. Once the clean-energy revolution is really rolling, there is every chance that it will amaze people at its speed. Then many options open up. Including, just maybe, keeping global warming below 1.5°C.

6th December: I suppose we should be flattered. The oil industry has organised a special side meeting at the summit to rubbish Carbon Tracker's arguments. It is truly remarkable that in their one set-piece press conference, at a global climate summit, they avoid every other topic germane to oil and gas – not least the still-falling price of oil and the huge issues it raises – simply to focus on trying to undermine the arguments of a small team of analysts based in London.

Their event is in a room deep in one of the larger tents, a structure that is creating its very own greenhouse effect. Fans blow warm air noisily, creating a virtual sauna, making it difficult both to hear and to concentrate.

The companies represented on the platform today are Chevron and Eni. Robert Siveter of IPIECA, the global oil and gas industry association for environmental and social issues, opens the event. IPIECA represents half the world's oil and gas production, he says. We contest the concepts of stranded assets and a carbon bubble. We do so for five reasons.

First, there is no evidence of a carbon bubble. More than 80% of company valuation is by proved reserves. Resources may not be used for 15-20 years, and barely impact the valuation of companies. How can there be a bubble if there is no threat to valuation?

I am sitting among journalists. Calm is good. But how on earth do they think it is relevant to mix up the bubble of potentially unburnable carbon that Carbon Tracker talks about with the way oil companies are valued? Even if we accept the premise that valuation of oil companies is all about current reserves, there must at least be *some* risk that some of those reserves will be stranded by policy. And anyway, if oil companies go on forever recycling vast amounts of profit into capex investments to turn resources into ever more reserves, as their business model requires them to do by default, in what way is this not inflating a pool of assets at risk of being stranded at some later point? And a growing pool of overvalued assets is a bubble, is it not?

Second, all energy forms will be needed going forward, Siveter says. So the International Energy Agency tells us.

Yes, I think, if you selectively choose from their policy scenarios. But the IEA has a long history of promoting fossil fuels. It was set up by governments to do so. Why not at least consider what the Fraunhofer Institute, or the Rocky Mountain Institute, to name just two centres of excellence, tell us about the future energy mix? They tell a very different story.

Third, not all fossil fuels are the same, says Siveter. Gas burning emits around 50% less carbon dioxide than coal burning for power generation.

Oh for heavens sake. Only if you ignore all the fugitive emissions everywhere along the gas infrastructure chain from the drilled wells to the gas hob in the home. And even if it were true, a two-degrees global-warming target requires 80% of fossil fuel reserves to be left underground, including a lot of gas. Even the IEA will tell you that.

Fourth, Siveter says, the industry manages many risks. We use a shadow price for carbon, for example.

Yes, you do use a shadow price for carbon (an assumption in economic calculations that the carbon price will end up higher than its current market value). For your operations. Not for the use of your product. As for the other risks you manage, let us hope the industry will manage climate risk better than BP managed blowout risk in the Macondo well, or than Shell managed weather risk when it towed its drill rig through stormy seas to the Arctic and crashed it on the coast of Alaska.

Fifth, innovation and technology, Siveter says. Energy-efficiency and carbon capture-and-storage are vital – and we do invest in renewable energy.

After many years of carbon capture-and-storage advocacy, you and the other energy companies have innovated your way to just 13 operational plants in the world. When there should thousands, at industrial scale, if it were really to keep meaningful amounts of carbon out of the atmosphere.

As for the oil industry's investments in renewables, please, give me a break.

After they finish, I ask the first question. I elect to avoid the catalogue of disagreements I could plunge into, and try and fire one big torpedo.

We disagree on stranded assets and the carbon bubble, obviously, I say. Please see our website for details of our arguments. My question is on the Bank of England announcement that they are holding an enquiry into whether or not fossil fuel companies pose a threat to the stability of the global financial system, citing the possibility of stranded assets. On a scale of 0-10, how confident are you that the arguments we have heard this morning will persuade the Bank that they have zero need to worry?

Chevron's man, Arthur Lee, responds first. I haven't heard of that statement by the Bank of England.

I hope the journalists clock that one. A week has passed since the announcement. Could it be that the oil industry, or Chevron at least, is that badly informed? Or perhaps that it doesn't even take the Bank of England seriously?

But in any event, Lee continues, we have the fact sheets, we have what the IEA tells us. Our case is strong.

So that, I say, would be 100% confident then?

I'm not answering that question, Lee responds.

Eni's man, Renato De Filippo, is equally sanguine. I hadn't heard about the Bank of England's enquiry either, he says. But we are sure and calm about our case.

7th December: I marvel at the complacency of the big oil companies. “Our case is strong. We are sure and calm about it.” Really? And your investors? And the journalists who cover your arguments, other than the paid-up re-hashers of your press releases?

The headline in the Financial Times this morning suggests another view of Big Oil’s case. “UN climate talks call future of energy majors into question”, it reads. “ExxonMobil and Shell would cease to exist in their current form in 35 years under measures UN negotiators are considering for a legally binding global climate pact to be sealed in Paris next year.”

The substance of this article is about the target of zero net carbon emissions by 2050, as espoused by a growing number of governments.

IPIECA did not address this possibility at all in their press conference yesterday. I suppose it was because they didn’t think it would be politically wise to tell nearly 200 governments that the oil majors think they have precisely zero chance of agreeing a treaty with any teeth, much less a tooth as sharp as zero net carbon just 35 years from now.

One component of the complacency inherent in this view is the failure to consider that at some point governments might tip en masse into a mood for strong action simply because of the course of events with global warming. Today a million Filipinos have had to flee their homes as a ferocious typhoon, Hagupit, devastates the nation. This is the third climate summit running where a “megastorm” has coincided with negotiations.

There will be more such extremes, of storm, flood, drought and fire. An elevated global thermostat drives the surfeit of extreme weather events around the world, the World Meteorological Organisation tells us. And the WMO has come to Lima armed with the latest data for global temperatures. 2014 is set to be the hottest year ever. Fourteen of the hottest fifteen have fallen in the 21st century. Global average temperatures are 0.57°C above the average of 14°C for the 1961-1990 reference period.

10th December: “Something huge is happening this week”. So reads an e-mail from the online activist group Avaaz to its 40 million members. In Lima, governments are about to set a goal to cut carbon pollution completely. But it is at risk. The e-mail appeals to Avaaz members to get busy signing one of their multi-million-signature international e-petitions.

Young digital campaigners are not alone in exhorting action. A group of Catholic bishops, spanning every continent, publishes a demand for an end to use of fossil fuels. Their statement calls for 100% renewable energy, and a strong focus on finance for adaptation.

Today the Lima summit will witness an all-too rare fighting alliance among the renewables industries, in support of the 100% target. The Global Wind Energy Council gives a joint press conference with the European Photovoltaic Industry Association. Steve Sawyer and I, representing the two trade bodies, have the mission of sending a signal to negotiators that they can be confident of a target of total decarbonisation in the energy sector by 2050. It is the second week of the talks. Ministers have arrived at the summit. Now is the time to state our case loud and clear.

In the UN media centre, Steve Sawyer opens with the thought that negotiators are generally 10 to 15 years out of date in their thinking on renewables. I profess that a 100% renewable-energy future is feasible far sooner than most negotiators at these climate talks appreciate, and certainly well before 2050.

Sawyer attributes much of the technical ability to reach 100% renewables to advances in the solar industry. Five years ago, he notes, I would, and did say, the majority of renewables would come from hydro and wind power. But with the dramatic up-take in the solar industry in the last five years, and the cost reductions in that period of time, solar is already making a substantial contribution and rising faster than anyone thought possible.

I spell out the key facts: that solar prices have fallen by two-thirds in six years, and according to the International Energy Agency, prices will come down much more.

Elsewhere in the world, our enemies counter attack. Wind and solar are "just not ready for prime time", Exxon's chief strategist Bill Colton tells journalists. The two technologies will only be providing 4% of the world's energy by 2040. Fossil fuels will still be providing the great majority.

Harlan Watson, former lead US climate negotiator in the Bush years, and now a fossil fuel lobbyist, echoes this here in Lima. "I don't think you could get from here to there in development terms just with renewables. They're still very expensive and intermittent," he tells the press.

In Germany, at around noon on the day of our press conference, 29 gigawatts of wind power and 5 gigawatts of solar provide 46% of the nation's electricity.

In the afternoon, a call from the special advisor to the UK Secretary of State for Energy and Climate Change, Ed Davey. Davey, head of the British government delegation in Lima, is looking for a story to hit the UK press back home, where he has so far found it difficult to attract attention. He is thinking of doing a speech on the carbon bubble that evening, at an event at the British Ambassador's residence. The Spad, as special advisors are known, wants my opinion. He is very specific. What can possibly go wrong if the Secretary of State majors on this?

That is code for will Carbon Tracker support him.

I can't think of anything, I respond. Rather, he would surely look statesmanlike on a cutting-edge issue.

Thank you, says the Spad. Your invitation to the event will be in your e-mail in-box in just a minute.

I take a taxi to the residence, in the hills above the city. It is a modern mini-palace. I drink champagne and chat with luminaries of the Peruvian business community. Ed Davey sits in a corner of the luxuriant garden, surrounded by the entire British media corps – more than a dozen of them. They are all scribbling in notebooks.

Later he gives his speech. The theme is "from carbon capitalism to climate capitalism". It is one of those titles you wish you had thought of yourself. The content is all about the carbon bubble. Carbon Tracker is the only organisation he namechecks other than the Bank of England.

The Secretary of State tells his audience that he has written to the Governor of the Bank of England asking him two things, essentially. First, what will the Bank be doing, during the course of its investigation of the threat that the fossil fuel industries might pose to the global financial system, to safeguard British pensioners' investments? Second, what will the Bank be doing to engage other regulators around the world in its enquiry?

The Secretary of State does not leave in the balance the question of whether or not risk exists, the way the Bank carefully does when talking about the enquiry. He clearly assumes that it does.

I am granted fifteen minutes talking with him solo afterwards, uninterrupted in the dark behind a tree. I have a few thoughts about the role of shale, I tell him, in the context of all the excellent points he offered in his admirable speech.

11th December: US Secretary of State John Kerry is in Lima. The plenary hall is full beyond capacity for his speech. Delegates unable to get in crowd around TV screens all around the site.

The US will take a lead, Kerry says, but every nation must act. We simple don't have time to sit around going back and forth. We must take giant, clear, measurable steps forward. That means concrete actions.

But with 24 hours left of official time in Lima, the shocking news from the negotiators is that they have agreed only one paragraph.

We are going backwards, civil society groups lament. The slow progress is costing lives, Oxfam's boss tells journalists. Winnie Byanyima accuses negotiators of having detached their human feelings from the negotiations.

Every day, first thing, the non-government organisations meet to co-ordinate, in two separate meetings: one for civil society groups, one for business groups. In the first decade of the climate negotiations the business meetings used to be a forum for co-ordinating attempts to sabotage the talks. No more. Too many business organisations want positive action. These days, the business meetings are simply a forum for the neutral exchange of information and insights. The fossilistas keep their sabotage ammunition dry for their own, very private, co-ordination meetings.

At the business event this morning, I sit in the back row, smiling at my memories as the bland proceedings unfold. At the Kyoto climate summit in 1997, the first I attended as a business delegate, the business caucus was chaired by the representative from Texaco, the American oil giant later merged with Chevron. He and the representatives of ExxonMobil and the US coal industry dominated proceedings, acting as though they spoke for all companies, in all industries, in all countries. When first I walked in and took a seat, he tried to have me thrown out. UN officials told him he was not within his rights: I was a solar industry representative.

Today I see two of the carbon club's hitmen from those dark days in the room, still in the game. They are Brian Flannery of ExxonMobil, and Harlan Watson, mentioned above. My book *The Carbon War* chronicles their antics in the run up to the Kyoto climate summit.

I stick my arm up at the appropriate point and say that I would like to add a point of information to the discussion. I just want to check that fellow business delegates have checked the press coverage from the UK Secretary of State's speech on carbon asset-stranding risk last night, I say. I offer a sense of the messaging in all the British papers.

After the event, I elect not to speak to my old enemies. I prefer the idea of lunch. I stick my head into a working group en route. Hundreds of diplomats and officials sit along tables under whirring fans. Big screens along the walls and at the front of the hall are full of red text.

12th December: The condensed draft text tabled first thing this morning risks seriously inflaming poor countries, and encouraging any other country secretly hoping the talks will fail. The original text has ballooned to 60 pages. It is dangerously weak, civil society groups say. A coalition including Venezuela, Iran, Saudi Arabia and India is now professing that no deal would be better than a bad deal. Talks could resume in early 2015, they say.

This is not good. This is derailment.

Pope Francis sends a message to Lima urging delegates to agree a strong deal. The time to find global solutions is running out, says the head of the Catholic Church.

The Lima summit is extended by a day as the bickering reaches boiling point.

13th December: As negotiations drag on, workers start dismantling the tents on the summit site. With frustration turning to fury in some quarters, many negotiators seem incapable of escaping old agendas. The nations negotiating have been divided, for a quarter of a century now, into "developed" countries and "developing" countries. This distinction has dominated proceedings, even as it has become increasingly out of date with the growth of economies such as China's, India's and Brazil's. The bloc of 133 countries classified as developing is insisting that those classified as developed in 1992, when the Framework Convention on Climate Change was signed, should pledge emissions-reductions commitments before the 133 do. The US, EU and the others classified as developed way back in 1992 are unwilling to continue with the distinction.

Last minute attempts to inject new text are taking the talks backwards. With the clock now running over by 25 hours, the US lead negotiator, Todd Stern, warns that the Lima climate talks face "major breakdown".

14th December: I go to sleep deeply depressed, thinking that – despite all the opportunities open to them – the governments of the world have let their citizens down badly. I wake to find that overnight they have pulled a form of success from the jaws of failure.

Let me start with the good news. The Lima Call For Climate Action commits all nations, for the very first time, to emissions cuts. Going forward there will be no binary classification of parties to the treaty into developed or developing nations. Parties will simply be a community of nations – 196 of them, nearly all the nations on the planet – who say they will come to Paris with national commitments of different kinds to cut greenhouse gas emissions.

The US and China lived up to their commitment to lead. A compromise was imperative. Todd Stern met with his Chinese counterpart Xie Zhenhua to seek it. The solution they came up with might seem like pedantry to anyone unfamiliar with the history of the climate negotiations. A line from the recent bilateral US – China climate agreement would work, Stern and Xie felt. The UN document should refer, like the US – China one, to the principle of "common but differentiated responsibilities and respective capabilities" of nations. Instead of listing all the rich countries based on a classification dating from 1992, contributions would be "in light of different national circumstances."

Six simple words. But they made the difference. When America and China speak together, others do well to listen.

Now the bad news. The outline text of the treaty has pushed many of the difficult questions for resolution into 2015. It extends to 35 pages, and includes as "options" much language that key nations would not sign up to, as things stand. On this basis, the prospect of a clear collective target of zero net carbon by 2050, supported by over 100 nations according to one report, remains present, but far from certain of a place in the final Paris treaty.

Among the key absentees in the text is a pledge for more funding from the rich nations to help poorer nations adapt to climate change, beyond the paltry sum committed to date. Similarly, poorer nations wanted adaptation commitments included in the national commitments along with mitigation commitments. That has been left optional.

Other vital rules of engagement are unclear. The draft, as it stands, would not guarantee regular reviews of the collective adequacy of commitments to stay below the two degrees Celsius global warming ceiling. That is a crucial factor for ultimate success.

The negotiating race will clearly be intense during 2015. But it is still on the track. The planes out of Lima today will be full of massively sleep deprived but grimly satisfied negotiators.

I take an American Airlines flight to Miami. Todd Stern is in the seat behind me. He looks like a ghost. I manage to resist the temptation to ask him even a single question.

Beside a log fire, 31st December 2014

As the logs spit and hiss their way towards 2015, I review the order of battle in the carbon war over what remains of a bottle of wine.

If the US and China can perform on climate in 2015 the way they did in 2014 then hope of a good outcome in Paris must be very much alive. They are doing passably well without any help from "events". But where events are concerned, the goalposts are hardly likely to remain static. Consider key developments since I returned from Lima.

Religion continues to emerge as a potentially key theme. The Pope announced immediately after the Lima summit finished that he will be convening a summit of religious leaders in 2015 in support of the Paris climate treaty. He says he will also send a lengthy message, a so-called encyclical, to all 1.2 billion Catholics. He leaves people in little doubt what flavour it will have. Catholic climate deniers everywhere will be running for cover, it seems.

The Financial Times headline about the fossil fuel companies facing crisis because of the carbon bubble has caused predictably severe ructions in oil and gas land. Shell CEO Ben van Beurden wrote a letter to the FT on 15th December. It managed to convey both conciliation and defiance. "We recognise that an energy transition is under way", van Beurden professes. But the answer still has to revolve primarily around more gas in the energy mix, and carbon capture-and-storage.

Never mind climate as a reason for energy transition, what about pure economics? The oil and gas industry's problems continue to deepen as the oil price slides. The price of Brent crude fell below $60 for the first time in more than 5 years in mid December. A day later, Goldman Sachs released the results of an appraisal of 400 oil and gas fields, essentially repeating the Carbon Tracker analysis of May. We had argued that over a trillion dollars of capex would be at risk above a breakeven oil price of $80. Goldman Sachs concludes that the picture is worse than we painted it: nearly a trillion dollars of what they call "zombie" investments above an oil price of $70.

And I repeat, the oil price is at $60 and falling.

The BBC reports that in the North Sea the oil industry is "close to collapse". The Chairman of the independent explorers association has told them that almost no new projects are profitable. A third of quoted UK oil and gas companies face bankruptcy, Company Watch estimates: 70% of them are unprofitable, racking up losses.

Chevron has dropped its Arctic drilling plans indefinitely. Economics plus regulatory requirements have put them off a target of drilling a first well as far away as 2025.

The debt-laden oil industry now poses a new threat to banks, the FT reports. As much as half of outstanding financing may be stuck on banks' balance sheets. Bond investors are becoming skittish over the plight of big energy companies in emerging markets. Once popular bonds of Petrobras, Pemex, and Gazprom are looking shaky. Meanwhile, credit derivatives are becoming a headache. The bill here could exceed $100 billion if the worst drillers default, the FT reports. The Bank of England has warned that lenders to US shale drillers may not be able to repay their loans. This could have a knock-on effect in the wider junk bond market, Governor Mark Carney worries.

The International Monetary Fund also warns that there is risk of contagion. This is how crashes occur, as we saw in 2008: fast-rising panic about losses actual and potential, and infectious loss of confidence in markets, leading to sudden collapse of value.

In Russia, the government is having to prop up both ailing energy companies and banks. Rosneft's precarious finances have become part of a blame-game raging in Moscow. The first Russian bank has been bailed out, suggesting echoes of Northern Rock in the days leading into the 2008 global financial crisis. Russia's Central bank has handed Trust Bank $530 million. The ruble has fallen to dramatic new lows as a Kremlin-imposed interest-rate rise fails to halt sliding stock values.

In Brazil, Petrobras sits on $170bn of debt, by some estimates the most indebted corporation in the world. It professes to need $221 billion in capex for drilling in the sub-salt over the next five years. (I marvel at the 1 in the 221. Such precision in financial forecasting.) But it risks pariah status among the investors it needs to attract. With billions creamed off the company's contracts in blatant fraud by executives, PWC has so far refused to sign off accounts.

The US Securities and Exchange Commission is investigating. So are Brazil's own authorities.

In Saudi Arabia, the question is how long can the Kingdom keep trying to show that only the fittest survive in oil markets by drilling at record rates despite low oil prices. The Saudi game is now clear: they are trying to protect market share by driving high-cost American shale drillers out of business. But in the face of the low oil prices they have created their national income has dropped to such an extent that their deficit has reached $39 billion, 5% of GDP. They are having to cut the wages of government employees and are dipping into their reserves of more than $700 billion in foreign exchange in order to keep their national budget afloat.

In the US shale, drillers have idled the most rigs in two years. Beyond their problems with debt, there are the environmental issues. The reasons for the fracking ban in the Texan city of Denton are spilling out in the media. Too many of the 277 wells were drilled near children, in homes and schools. Drillers displayed callous disregard to parental and other public concerns and have paid the price at the ballot box.

New York state is not making the same mistake. Health officials are speaking out, saying potential risks from fracking are too great to tolerate. A University of Missouri team has found that fracking chemicals could pose risks to reproductive health. They recommend that people living near frack sites should be monitored. In the face of all this, Governor Andrew Cuomo has made the decision to ban fracking state wide. There is no prospect of the 6-year moratorium ending now.

This is a huge blow for the oil and gas industry. Much of the Marcellus Shale, so productive in Pennsylvania, lies under New York State.

The Governor's ratings have gone up as a result of the ban. One in three voters think better of him since. Health groups are confident the New York fracking ban will now spread to other states.

Yet in the UK David Cameron has waded into the green groups for what he calls their "religiosity" for being anti-fracking. He says that all subsidies for onshore wind will end, and claims that tax breaks for fracking are not subsidies. To what extent, I wonder, will his blindness to the direction of events in America expose him in the British General Election, to be held in May, should energy and shale feature as an issue?

Reviewing 2014, as the logs hiss on New Years Eve, it seems clear to me that "events" will have a power all of their own in 2015. As the multiple crises in incumbency energy markets unfold, any seriousness of intent of politicians and policymakers at the climate negotiations could well be amplified by the emerging expediency of a shift away from fossil fuels for reasons other than climate change. On the other hand, should energy events conspire to trigger another global financial crash – as seems entirely possible – the sudden plunge into worldwide economic crisis could easily divert attention from climate change to much more immediate concerns.

Whatever happens, great drama can be assured in the year ahead.

Staring into the fire, I have an idea. It is to write a live history of the drama, as I see and experience it, as events unfold. I will edit my diary into a hopefully readable true story, and publish it as a free-download monthly serial.

I will call it "The Winning of The Carbon War".

Because I have come to the view that we might just be starting to win it.

Chapter 15

In search of the oil-and-gas U-turn

London, 8th–12th January 2015

Less than a year to go to Paris, and time to clear the decks to make more space for climate campaigning. If not now, as Christiana Figueres often says of the climate-change end-goal in her speeches, then when?

Among the changes, I have elected to step down from chairing the Solarcentury board. I will stay on as a director, but that role requires less time.

This change needs to be explained carefully. "The market", as business people like to say, needs to know, and trust, that this is my idea, and not a coup by venture capitalists out to unseat me, dismantle the culture at the company, withdraw funding from SolarAid, and whatever else enthusiasm-heavy and detail-lite journalists might dream up.

I elect to offer Terry Macalister, Industry Editor at the Guardian, an exclusive briefing on what is afoot. He is a veteran pillar of sensible reporting, though the nuclear industry may harbour a different view.

In the Guardian's headquarters at Kings Cross, Terry and I sit and talk for an hour. It is the morning after the Charlie Hebdo terrorist murders in Paris. A single unarmed guard stands at the door of the UK's flagship liberal newspaper. Through glass floor-to-ceiling windows I see a packed morning editorial meeting. The editor and his sector chiefs will have much to discuss in our unsafe world. They too have irreverent cartoonists.

Terry turns out not to be that interested in the latest news at Solarcentury. The oil price has fallen under $50. There is a rout underway in the global oil market. He wants to know what I make of it.

It probably means that an oil and gas major will copy E.ON and turn its back on fossil fuels in the near future, I venture. By this I mean it would become the first in its sector to do a U-turn with its business model, and focus all growth on renewables and other clean energy, operating legacy fossil fuel assets only for cash until they run down.

I base my case on the speed at which two megatrends are squeezing the industry, I explain. Capex has been soaring since 2000, as the industry drills ever more expensive prospects. The pips were beginning to squeak at $100 oil, but since the oil-price drop began in June, capex has become a clear and present danger. Wherever you look in the world, billions of dollars worth of projects are being cancelled: Shell in Qatar, Chevron in the Arctic, Premier off the Falklands, Statoil off Greenland, the list goes on. Norway is about to hold crisis talks on its oil industry. In the American shale, rigs are being being withdrawn from operation at a rate never seen since records started. The first oil company has gone bankrupt, unable to raise more cash. It is a small Texan driller, but analysts warn there will certainly be others. In the US shale, a whole segment of an industry seems to be in the process of going bust. The shale boom looks like it may be set to collapse like a house of cards.

But then the price will just go up again, Terry says, won't it?

Almost certainly it will, I reply – unless there is some apocalypse involving demand – on the principle that if you don't explore, drill, and produce, you must at some point constrain supply to the extent that it struggles to meet demand. But it is no use to argue that the oil industry's fortunes will turn around when it does. Analysts who know the industry well make the point that once you cut expenditure and staff, switching back on is far from easy. Moreover, if the price goes too high, the cost-down in clean energy becomes ever more alluring. The oil and gas companies lose at high prices as well as low prices. This is especially true when the disaster of oil-capex-up plus oil-price-down spills over into the financial sector, as it is in the process of doing. The financial pages have been full of danger to banks of loans to the oil-and-gas industry turning toxic.

I watch the shorthand hieroglyphics speed across Terry's pad with some wonder.

The impacts of the low oil price are spilling over into stock markets, polluting the value of quoted companies, I add. The S&P 500 benchmark is down at a time when consumers should be enjoying a bonus from lower gasoline prices. Too many companies have made investments on the premise that oil will always be expensive.

Next I give him my views on the second arm of the vice: solar and battery cost down, and why this increasingly means trouble for the twentieth century oil-and-gas business model in the twenty-first century.

The story about my prediction appears on the 11th.

My inbox the next day is a thing to behold. Most e-mails are in the vein "I hope you are right".

Some disagree with me, including my friend Jonathon Porritt. His long experience of trying to argue for an oil-and-gas business-model U-turn, via his think tank Forum for Future, suggests to him that the task is futile. No major oil-and-gas company can commit to renewables in the near future, he writes, because their internal culture is too robustly vested: they are forced to adopt "intricate patterns of denial and self-deception".

I agree with him about the culture. It is dysfunctionally defensive at best, deliberately malign at worst. But what about the market dynamics? Is the double whammy of capex-cost-up solar-cost-down really escapable now?

If Terry had asked me to define what I meant by "near future" for the business-model U-turn, I would have said three years.

One e-mail, from a former senior BP executive, surprises me. "That's some bet", it reads, "but probably good within, say, 5 years."

Copenhagen, 12th–14th January 2015

Denmark is one of the nations supporting a zero net carbon target for 2050 at the climate negotiations. It is not too difficult to see why. Denmark generated 39% of annual electricity from wind in 2014: the fourth consecutive annual record. Copenhagen has reduced greenhouse gas emissions 40% since 1990 with smart heating grids and other techniques. Chinese policymakers flock to study the Danish capital's low-emissions route to heating cities these days.

A Carbon Tracker team is in town today to try and persuade Nordic pension funds to squeeze the fossil fuel companies harder than they are. Mark Campanale, James Leaton and I will have board-level access over the two days.

In the first meeting, Mark focuses on oil and gas. The top 20 sell-side analysts will tell you that all the oil and gas can be burned, he says. Ask them why they don't see the carbon risk, and they say its because their clients haven't asked about it. They have to be forced to change their assumptions and tell clients what fair value looks like. We recommend that you tell them that the profitability of their legacy assets will give cash for 15 years, and if they cancel risky capex they can pay you dividends. If they do that, they are contracting in size, but their share price will go up. We are talking about transition and contraction versus business as usual. What you don't want is these oil and gas giants wasting your money in marginal projects.

I watch the fund directors mulling it over. It sounds so blindingly obvious to me. Why are they even remotely hesitant about acting? The only answer I can think of is the neuroscientists' point about the innate humane tendency to resist change – *any* kind of change.

The next day, Nordea organises a forum. They are the largest Nordic asset manager, with well over $200 billion under management. Many of their peers attend. Former Danish Environment Minister Connie Hedgaard gives a keynote speech. She has just stepped down as European Climate Commissioner. She tells the assembled fund managers, with visible relish, that her freedom from the constraints of diplomacy will allow her to speak with a candour she has not been used to today.

She is full of excitement about the progress being made by the divestment movement, she says. She explains why, listing some of the cities, universities and other institutions who have divested. There are not enough pension funds on the list, she observes.

She has a question on climate for the hundred-plus investment professionals in the room. You're not sitting on the fence, waiting for the outcome, are you?

I am sitting at the back of the room, watching the body language in the audience. Too many heads are down in smart phones and iPads.

I have noticed, Connie continues, that some of the fiercest advocates of free markets are the first to sit back and wait for important issues to be resolved by the politicians. So please, act yourselves. Change your investment portfolios. You won't lose money: this investment model doesn't underperform compared to other investments. What are you waiting for?

Even more heads down.

I catch a very senior fund manager from the City of London over coffee afterwards. He had better remain nameless.

He tells me he enjoyed Connie's speech: thought she was bang on target.

We must become more professional, he tells me. Fund managers tend to think of themselves as above the sector. They aren't.

London, 15th January 2014

The Secretary of State at the Department of Energy and Climate Change, Ed Davey, is delayed in Parliament. But he more than makes up for it by giving me a whole hour in his office. He has a meeting fixed with the Bank of England, he tells me, to follow up on his letter about the Bank's internal enquiry into the impact of the fossil fuel companies on global financial stability. I relay the latest news from Carbon Tracker. It feels strangely like a meeting with a collaborator, but for the four largely silent officials listening to everything we say. It is difficult to imagine that Solarcentury is involved in a court case with the government over their latest solar subsidy-cut decision.

Just keep off that subject, I remind myself. Grown ups know how to operate glass walls and not take things personally. Oh, and avoid the government's decision to go ahead with the Hinkley Point nuclear reactor. What you want to discuss, apart from the carbon bubble, is the evidence of the fast-unfolding renewables revolution and its implications, and the emerging shale fiasco and how it adds to the imperatives for renewables.

I have been talking to sources inside E.ON, I tell Ed Davey. The situation is fascinating. The Board's decision to effect a U-turn has been very popular with staff. A re-energised workforce is full of executives keen to work in the clean-energy growth company. But the strategy is still very much work in progress. The walls at the utility's German headquarters must be one huge collage of flip-charts and post-it notes. What is increasingly clear, though, is how right they were to make the decision to change course. Deutsche Bank, for example, has just come out with a report saying that solar power will be at grid parity in 80% of the world by 2017, assuming power prices keep inflating at the rate they have of late.

Other utilities fear what is they are increasingly calling a "death spiral" unless they make a similar change of course. In America, an Edison Electric Institute paper warns of a situation analogous to mobile telephony, and what that particular set of disruptive innovations did to the business models of the wire telephone companies.

How will this play in the UK? E.ON of course is one of the Big Six utilities that Ed Davey has to deal with on an almost daily basis. He is no friend of theirs, and has on at least one occasion called for them to be broken up unless they are more public-spirited towards their customers. I suggest to him that there are now a lot of easy wins that he can leverage from the utilities, given E.ON's change of course.

On to shale. I run through the news from the US, checking to see how the Secretary of State is being briefed by his officials. Debt that can never expect to be serviced much less repaid is piling up now that the oil price has fallen below $50. This is a house of cards very likely to fall down, I tell him, even conceivably before the British General Election in May.

The environmental news from American shale regions makes matters worse. Advocacy groups are suing the US Environmental Protection Agency for public access to the toxic chemicals used in fracking, for example. Goodness knows what unpleasantness will hit the press if that eventuates.

I pitch to him that there should be a Plan B, wherein if the house of shale does start collapsing in the US, the coalition has somewhere else to go by way of retreat from the Cameron-Osborne illusion of shale nirvana – a clear vision of an accelerated green industrial revolution based on renewables, smart demand control, all the rest – so minimising the damage of the inevitable U-turn. A Plan B of the kind that E.ON is actively considering as we speak, in other words.

I wonder if you could pitch that to George Osborne, I conclude: get him to agree at least a hedge bet, and work with him on it.

I make my case without much hope in my heart. The Secretary of State for Energy and Climate Change and the Chancellor are famously unimpressed with each other. They mostly seem to communicate through the national newspapers these days.

There will be a lot of communicating to be done on shale in 2015, if I and people like me are right.

Paris, 16th January 2015

The heart of the Parisian business quarter in La Défense. Corporate towers soar above the plaza. Electronic billboards announce "Je Suis Charlie." The EDF and Areva buildings are among the most impressive. I wonder who will occupy these towers in ten years time, given how badly the nuclear-power saga has been playing out for those companies in recent years.

Total's 36 floor edifice is the most impressive of the skyscrapers. A screen in the bustling foyer reads "Committed to better energy", in English, Arabic and Chinese. Earnest young people in hard hats – a league of nations both masculine and feminine – appear fleetingly among imagery making clear that "better energy" does not entail oil necessarily, at least in the collective mind of Total's marketing department. Solar features strongly.

I am here today to talk to the Sustainable Development & Environment team, both about the big picture of the clean energy family replacing fossil fuels worldwide, and the much smaller but emblematic picture of solar lighting replacing oil-burning lamps across Africa.

High up the tower, with a distracting view of Paris splashed across the windows, five executives, all French, invite me to state my big-picture case. I suggest that oil-and-gas capex-up plus clean-energy cost-down, plus-or-minus climate policy, means progressive undermining of the utility business model today and the oil and gas business model in the not too distant future. I don't think today's clean energy companies can execute the energy transition on their own, over the 35 years that a hundred-plus governments want it done in: that it will take one or more of the utilities and one or more of the oil and gas majors to execute U-turns with their business models. I profess that I want to collaborate with those who do, or are most likely to, literally as part of my life mission. That's why I am in Paris with them today.

The team leader, Jerome Schmitt, tells me he appreciates the sentiment and can easily understand why I hold it. We certainly are on the same direction of travel. But Total doesn't agree with my premise of 100% decarbonisation by 2050. It will take longer than that.

I say that we don't need to argue the case, though it might be fun, because of the direction of travel, and because crucially – if my analysis and that of the many people like me is correct – the goalposts will move fast in the face of the capex pressure building and the disruptiveness of solar and storage technologies.

I leave a pause for him to disagree with that, but he doesn't.

I suggest we move to solar lighting and Africa. The background is this. SolarAid's retail brand SunnyMoney is the lead retailer of solar lights in Africa, with more than 1.5 million sold to date, most in the last two years, most in just two countries. Total is the number two retailer, with over a million sold to date, from petrol forecourts all over the continent.

The Total goal is to reach 50 million people with solar lighting by 2020s, a member of Jerome's team tells me. Currently, they reckon their million lights have reached c 5 million people. The company takes this very seriously, and there is a lot of pride in Total at what they are achieving.

Their target is laudable, I say, but SolarAid's view is that a much tougher target could be entertained. We want to play a lead role in replacing all kerosene lanterns, everywhere on the continent, by 2020. That would mean around 200 million solar lights sold – by SunnyMoney, Total and all the others that would be in the market five years from now if we were to succeed with the mission.

Could there be a way of working together on this audacious goal, I ask them?

Chapter 16

We want to play our part

Abu Dhabi, 18th & 19th January 2015

I fly overnight from London on the biggest aircraft in the world to the sixth biggest oil-producing nation, where somewhat counter-intuitively some of the most impressive efforts anywhere in the world are being made to accelerate the renewable-energy revolution.

My first stop is the annual assembly of governments party to the International Renewable Energy Agency, the UN agency for the promotion of renewables, which is headquartered here. I arrive just in time to hear Adnan Amin, Secretary General of IRENA, give a speech to a plenary of ministers and officials from 150 nations, and representatives of a hundred institutions. The 2014 figures for investment in renewables are just out. Adnan shows the chart proudly. Investment is up 16% on 2013 to $310bn, 29% of it in China. More than a hundred gigawatts of solar and wind were added in 2014, around half of each, up from 74 GW in 2013. New renewable generation has exceeded all other forms of electricity generation combined for the second year running. All this comes just in time for the climate mission in the run up to Paris, he says. Let us hear about that now from Christiana Figueres.

On the huge screens in the hall, Christiana betrays none of the exhaustion I imagine she must fight every working day. A murderous travel schedule, the constant meeting and greeting, an e-mail inbox I can barely conceive of. And all the time, hanging over her, the knowledge of the sacred mission she must deliver on, if she can, in Paris.

$90 trillion will be invested in infrastructure in the next 15 years, Christiana says. Fifteen, not fifty. Most of that will be in the developing world. That infrastructure *must* involve the energy technologies of this century, not the last. An effective Paris agreement is vital to that. Please, she urges the ministers, make the targets you table in the run up to the summit as renewables-rich as you can. There can be no excuses now. We dare not fail.

Ségolène Royal, French Environment Minister, speaks next. I echo all that, she says. In France, as host of the summit, we have a particular obligation to increase the share of renewables in the energy mix, and are responding to that responsibility, decreasing the nuclear share and substituting with renewables.

She speaks with passion, using no notes, like Christiana. It strikes me that two powerful women will have key roles in the endgame at the Paris climate summit. So much better a prospect than so many of the grey aging men I could easily imagine might have been appointed in these pivotal roles.

National interventions follow: Germany, South Africa, USA – all positive. But you would hope they would be, by definition, at the gathering of an international renewables agency.

Then comes an intervention from the business community: Steve Howard, Chief Sustainability Officer at IKEA. Our company's target is 100% renewable energy right across our global operations by 2020, he says. We are one of the biggest retailers in the world, and our target is just five years from now. As for the low oil price, we are not fazed by it. We have seen this movie before. The oil price goes down. It also goes up. And amid all this volatility, bad as it is for business, what role do our renewables play? We know the cost of the energy from all our wind turbines and solar panels will be zero tomorrow, the next day, and in 20 years from now. This is good for business. We have started to sell solar panels in our stores. This is good for consumers. A solar roof system is the best investment a householder can make.

He works up an impressive rhetorical flow as he describes IKEA's vision of a renewable-powered future, and the role of the business world in delivering it.

Now is our moment, he concludes. We must seize the day.

The applause is enthusiastic.

After this session, I have a catch-up meeting with Christiana. We sit in the marbled lobby of a five-star hotel as famous professional golfers wander by wearing their caps and suntans. The Abu Dhabi HSBC Masters is on its final day.

Steve was great, Christiana agrees. But why do we have a silent majority in business? We need many more to speak out like him in the year ahead. We all need to work harder to make that happen. What are you intending to do about it Jeremy?

Evening comes, but there is no rest for the participants. They are bussed to the airport, where a hanger has been decked out for a Question Time-style debate sponsored by the Financial Times, and a reception fuelled only by fruit juice. The hanger contains the Solar Impulse aircraft – the 72-meter wingspan solar-powered marvel that will attempt to fly round-the-world on a series of legs between May and July. The designers and pioneer aviators, Bertrand Piccard and André Borschberg, intend their world-first to demonstrate the wonders of solar cells, modern batteries, and state-of-the-art design with low-weight high-strength materials. They are, they say, an example of what can be done on a global scale in the energy transition.

Pilita Clark of the Financial Times supervises the panel, on which I sit. All five participants share the view of the Solar Impulse pioneers: a clean-energy revolution is on the way. It is more of a mutual agreement ceremony than a debate.

The next day, the World Future Energy Summit kicks off with the awards ceremony for the Zayed Future Energy Prize. SolarAid is on a shortlist of five, in the category for non-profit organisations, distilled down by a committee of global luminaries from over a thousand applicants. A million and a half dollars goes to the winner, and SolarAid's wholly-owned retail arm SunnyMoney has just passed the milestone of 1.5 million solar lights sold. Believers that fortune mirrors coincidence are having a hard time managing their expectations.

The video played to the thousand-plus attendees at the opening ceremony surprises me. The origins of all the wealth in this nation of glittering desert towers is passed over almost apologetically, to be followed by an impressive catalogue of renewable-energy projects: bankrolled and executed, budgeted and planned, both domestic and international. Welcome to the future, is the message, a renewable-powered future which the oil-rich nation of Abu Dhabi is going to play a lead role in shaping.

Dr Sultan Ahmed Al Jaber, founder of Abu Dhabi's low-carbon city Masdar, is the architect of much of this focus on energy transition in the Emirates. He now gives a speech. The low oil price won't stop renewables because they have come of age, he says.

A Lifetime Achievement Award. Al Gore wins it. Climate change is real, says the young version of Al in a video clip. President Obama joins him with a soundbite from a recent speech. We don't have time for deniers, he says. The flat-earth society, he calls them.

I am by this point amazed at what I am seeing in Abu Dhabi's videography, and choreography. I look around again at the ranks of so many hundreds. Maybe half of them are wearing Arab clothing. If I am surprised at this projection of mix of imagery of a future replete with renewable energy, and language about deniers and flat-earthers. What must many of them be thinking? I know the Saudi Arabian climate envoy is in the room, for example. I am meeting him later.

Today's Al now takes the stage: the man who has become a global legend in his post-Vice-President years. A member of the Abu Dhabi royal family presents him with his award.

The world is going to say yes to the renewable future, he says in his acceptance speech. The only thing that is missing is political will. And that too is a renewable resource.

In Abu Dhabi today, I find it easy to believe him, and to believe in our hosts.

I sit nervously near the front of the vast audience. A million-and-a-half dollar prize would save SolarAid and SunnyMoney a lot of heartache this year. The truth is that our breakneck sales growth in 2013 has evened off in 2014. We budgeted for growth, we bought solar lights in volumes expecting growth. Too many of them sit now in a Dar es Salaam warehouse. As a consequence, our cash is dangerously low.

The situation is worse than this. We have not had a CEO in the field for SunnyMoney since May last year, when start-up specialist Steve Andrews returned to the UK after a four year tour of duty. This is probably my fault. I had thought it would be easy to replace him with a world-class growth-phase retail specialist, simply using my network. That has proved wrong, indeed – I now think – complacent. The first tranche of candidates did not work. I delayed too long using head-hunters. A team of three has been holding the fort leading SunnyMoney in Nairobi, led by John Keane, solar-lighting and development specialist, a veteran of SolarAid from the outset. They have done their level best, but with proper solar lighting markets now in place in Kenya and Tanzania, facing strong competition from well-resourced new entrants following where SolarAid and SunnyMoney led, their task has been difficult in the extreme. We all know we need a retail expert with strong African experience.

I could do with a major lucky break to cover my shortcomings today.

The suspense is an interesting experience. The organisers have kept tight confidentiality. None of the five short-listed finalists for the non-profit organisation category of the Zayed Prize know who has won. We sit together and watch short films about each others' achievements on the huge screen. The massive audience claps at the end of each. The SolarAid film features an interview with a member of the Kenyan SunnyMoney marketing team, Olivia Otieno. She radiates a controlled passion for the mission, with perfect fluency. How I would love SolarAid to win this prize, for people like her.

I sit and reflect how lucky I have been in the past, with Solarcentury, when times of existential pressure arose from mistakes I made. The answer is very lucky indeed. But can that luck hold?

It doesn't. In the seat in front of me a young social entrepreneur from the Philippines throws his arms in the air and marches beaming to the stage.

I tap out an e-mail to the SolarAid and SunnyMoney teams immediately. All the five short-listed projects are brilliant. So are many of the thousand-plus others considered, no doubt. We can be happy that so many people are working so hard for the future-energy transition, and proud that we came so close.

I have no time to contemplate the near miss, and the difficulties SolarAid and SunnyMoney could have avoided in the year ahead. I return to my hotel, another marbled five-star palace, for the next meeting on my agenda. I sit alone for an hour with Khalid Abuleif, a senior official at the Saudi Ministry of Oil who is the Kingdom's envoy at the climate negotiations, and his lieutenant.

They have agreed to swap perspectives on the carbon bubble and the solar revolution. I tell my story of relative advantage for the Kingdom, as we see it. Saudi oil is low on the carbon cost-curve, and will be among the oil of choice in a world transitioning to sub-two degrees. The Kingdom's scope to become a hub in the global solar revolution is substantial, and the need pressing, given domestic consumption of oil. There is already Saudi leadership in the region. A Saudi company, ACWA Power, has just installed the cheapest solar energy in the world in Dubai. The 200 megawatt plant will produce electricity at 5.85 cents per kWh, a world record for low cost solar, cheaper even than a gas fired power plant.

Khalid is a quietly spoken man with degrees in environmental engineering. He knows all that I tell him, of course, and much more besides.

I wonder about the role of batteries in your narrative, he says. Is the scope to disrupt oil's transportation market as potent as you say? Not based on what I have seen in China.

I am not a technologist, I reply. I have to rely on those around me who are. I may be subconsciously cherry picking my sources, I admit. But let's see. So much of this will play out fast in the short-term, as with the solar cost-down. Apropos which, are you able to tell me if your ministry has modelled the transition to a sub-two-degrees policy regime, and how the Kingdom would fare in such a world?

It has, he says. We don't think it can happen as fast as you say.

I don't ask him to elaborate. I know from press reports that he thinks a hundred-percent-renewables by 2050 target is impossible.

Can we introduce our experts to each other to compare what numbers we can, I ask? Maybe Carbon Tracker analysts could visit Riyadh? Or Saudi officials meet them in London?

By all means, Khalid Abuleif replies. And please, don't think we are being obstructive on climate change. We know the global community has a problem that needs to be dealt with, and we want to play our part.

London, 11th February 2015

Another panel in the Question Time format, this one at the Royal Society of Arts. Like the one in Abu Dhabi, none of the panellists are going to be doing much disagreeing with each other. I sit on the stage and look out across the audience. There won't be much disagreement here either, I suspect.

I wonder about the usefulness of this exercise. We need to be thinking of ways of carrying the climate message to the unconvinced, and awakening as many as we can of those who would prefer to ignore or even deny the problem rather than deal with it.

But there is the problem. Ignoring or denying makes it easy to avoid giving up an evening to come and listen to something like this.

No sooner have I had this thought than the chair gives me cause to reconsider. He asks for a show of hands before the debate begins, on two questions.

How many here fear climate change, he asks.

The vast majority of hands go up. As I thought then.

How many think we will defeat it?

Maybe thirty percent.

So there is this evening's challenge, I tell myself. See if you can help instil some hope and belief in the 70%.

Positive momentum continues to build in both the political and business worlds. President Obama has extended his climate campaign to India, where plans for coal burning are just as worrying as China's, maybe more so given the recent Chinese efforts on air quality. At the end of the visit he and President Modi promise to back an ambitious climate treaty in Paris. Back home, the American President is strong on climate in his State of Union address. He mocks the recent Republican use of a line "I am not a scientist". The obstructionists would do well to listen to US government employees who are, he says.

The US Senate refuses to accept man's role in climate change. This, to a European, is extraordinary. Twenty seven years after NASA first attributed global warming to human emissions, with an ocean of further positive proof in the interim, senators defeat two measures that try to make the same attribution in Congress. Yet an American poll shows that half of all Republicans support climate action. Two-thirds of Americans say they are more likely to vote for political candidates who campaign on fighting climate change.

There could be no better demonstration of the power of political funding from fossil fuel companies than this willingness of Republican Senators to ignore majority opinion among the electorate.

Meanwhile, in China, in the face of government air-quality regulation, coal production falls for the first time this century. This is an entirely unexpected positive development. Many people have grown accustomed to projections of growing Chinese coal use far into the future. The Beijing government also puts a moratorium on new coal mines in the eastern region of the country. They are attacking both supply and demand.

Business leaders continue to contribute to the momentum. Members of the "B team", convened by Richard Branson to act on climate change, call for a zero net carbon target by 2050, ahead of a session of the UN climate talks currently under way in Geneva. Unilever boss Paul Polman and Ratan Tata of Indian conglomerate the Tata Group are among those joining Richard in the appeal. "The net-zero emissions by 2050 target is not only desirable but necessary," the Unilever CEO tells the Financial Times. "This is not going to be easy, but the earlier we act, the greater the economic opportunities will be."

As for the French hosts of the Paris summit, just ten months from now, they are doing nothing to take pressure off themselves. French Foreign Minister Laurent Fabius tells the negotiators in Geneva that "the survival of the planet itself is at stake".

Online, 13th February 2015

Global Divestment day. 450 events in support of pulling capital out of fossil fuels, in 60 countries, spanning six continents. In the UK, rallies and protests include London, Bristol, Oxford, Nottingham, Swansea and Edinburgh. Around the world, 181 cities, universities, pension funds and other institutions are now committed to divest.

Nordea, the largest Nordic asset manager, has announced it will pull up to 40 coal-mining companies from its $228bn of assets. It is not clear whether the Carbon Tracker representations in Copenhagen had anything to do with this. In Norway, the oil fund divested from 40 coal mining companies in 2014, it turns out. Its first ever report on responsible investing, published without fanfare, lists 140 companies dropped on environmental grounds, including five operating in the tar sands.

The health sector joins in. A coalition of medical organisations says its sector should divest from fossil fuels on ethical grounds, repeating the stand it took on tobacco. The Wellcome Trust disagrees, saying it prefers to remain invested and engage with fossil fuel companies. But then it cannot give an enquiring journalist a single example of how it engages.

I suspect this would prove to be a common failing among big institutional investors advocating engagement over divestment, should journalists do a little follow-up questioning.

One of the sessions of climate talks preparing for Paris has finished today in Geneva. Almost 200 countries have agreed a draft text for the Paris treaty negotiations. It runs to 86 pages. A zero net carbon commitment by 2050 is still in the text.

Achieving near-zero emissions in the second half of the century will be central to the Paris deal, Christiana Figueres says at her press conference closing the Geneva session.

Singapore, 28th February 2015

Five major companies want to stake out a new frontier for corporate action on climate change. Some have made very good starts on their own. But now senior executives have gathered in Singapore for two days to see if they can find a way to do even more, by working together.

I am invited, to give the first presentation, on the first morning. They want me to be what they call the "firestarter". I have fifteen minutes.

I know what some readers will be thinking at this point. Has this guy ever sat down and asked himself how much greenhouse gas his flying puts into the atmosphere? Does he really think his words and actions, in all the places he goes to, can somehow offset all those tonnes of carbon?

Well, the answers are yes I have, and no not yet. The jury, I very much hope, is still out on the second question, however. This trip is a case in point. I can see a huge potential unrealised prize for the climate – indeed potentially a series of them – that would go a long way to making the frequent-flier carbon-offset equation balance for me.

For the first time in all my travel, jet lag does not allow me a wink of sleep the night before the first morning of an event. I have been sleeping quite well of late, and decided to leave the chemicals at home. Big mistake. I face the 15 minutes I have been allotted feeling vaguely ill with fatigue.

As ever, there is so much material to sift, assemble into a narrative arc, and convey. A crystal-clear head would be good. Committed as these executives are, I am confident they will know little of what I can pick up and stitch together, in my 3 to 4 hours a day following events as they unfold. How can they, with full-time jobs, and so much in the vital action in the carbon war absent from front pages – indeed often absent from mainstream media altogether?

Potentially the biggest news of all, in the last month, failed to make the mainstream, that I saw. Apple wants to start mass producing electric vehicles, so a trade journal reports, and as soon as 2020. It has been working in secret on EV design, with a team of 200 people, and busily poaching Tesla- and battery-company staff. For its own part, Tesla plans to unveil battery storage for the residential market soon. CEO Elon Musk says details will be available within a month or two.

Storage sits today roughly where solar sat six or seven years ago in terms of cost-down, analysts are saying. German company Belectric, trialling grid-balancing storage for big solar installations in Germany and the UK, talks of storage bringing a "new era for renewables".

This latest news in the highly-disruptive partnership of solar and storage is a good place to start my 15 minutes. There is so much more, just from the last month, to pick from and weave in. Apple and Google are clocking up renewable generation fast. Both have invested over a billion dollars. Google has made its biggest solar bet yet, investing $300 million in a $700 million fund for SolarCity roofs. Their total investment in renewables is now $1.8 billion. In the biggest-ever solar power-purchase agreement, Apple is buying nearly $1 billion of power from a First Solar project. In a move that is nothing if not timely, regulators are allowing Californian utility PG&E to provide their customers with 100% renewable electricity supply, and as soon as the end of 2015. This is a utility with customers in Hollywood.

Meanwhile, more grounds for cautious optimism on global carbon emissions. They flattened in 2014, preliminary data from the International Energy Agency suggest. This encouraging new development mostly reflects the fall in Chinese coal-burning, it would seem. Hopefully this can continue, given the right policies and pressure from governments and progressive corporations like those I am with in Singapore. Since 2010, two coal plants have been cancelled globally for every one built. Beijing is to shut all of its four major coal plants, the last in 2016. They will be replaced by gas plants. More than 2,000 smaller Chinese coal mines will be closed by end 2015.

In the US, the coal sector is clearly now in structural decline. The latest report from my prolific Carbon Tracker colleagues charts at least 264 mines closed in the period 2011-2013, a period in which 76% of value has been lost in coal companies. This trend, Carbon Tracker concludes, foreshadows similar pain for the oil and gas industry in the years to come.

At the climate talks, Switzerland becomes the first country to submit a Paris commitment on carbon emissions: a 50% cut by 2030, on 1990 levels, 30% at home, 20% abroad or in carbon markets. The EU then pledges 40% by 2030, on 1990 levels. In the UK, David Cameron, Nick Clegg and Ed Miliband all sign a strong climate action pledge. They call for an unabated coal phase-out worldwide and 2°C binding treaty in Paris.

It is good to see politicians set aside their tribalism, on such a vital issue. Cameron, in particular, must have exercised a degree of courage in signing, giving the extent of climate denial or obstructionism in the ranks of his own Conservative party.

No such luck in the USA. The Republicans are trying to rush a bill through Congress aiming to give a green light to the Keystone XL pipeline bringing tar-sands oil south from Canada. President Obama vetoes it: only the third veto of his Presidency, and the first of the second term. It will not make much difference to American oil imports, he says, and it will make the climate-change problem worse at a time the world is trying to cut emissions. The Republicans vow to fight on.

Oslo becomes the first capital to join the global divestment movement. No more coal, the city decides, joining 40 other major cities. The UN itself backs the divestment campaign, saying it sends a signal favouring a good outcome in Paris.

The Guardian becomes the first major newspaper in the world to launch a divestment campaign. "The argument for divesting from fossil fuels is becoming overwhelming", says editor Alan Rusbridger.

Meanwhile, divestment or engagement by investors notwithstanding, the shale narrative-of-plenty in the US continues to unravel. With the American oil price having fallen as low as $45 – a drop in excess of 50% since last June – drillers are retreating from North Dakota oil shale at their fastest rate yet. The rig count has now fallen by 30% since October. Cash-starved oil producers are trading their pipelines for money. "At some point they all get desperate enough," one fund manager tells Bloomberg.

Global layoffs in the oil and gas industry have passed 100,000. The US shale regions are hit the worst, but the pain is global. Oil boom towns are beginning to turn into ghost towns.

"Oil collapse could trigger billions in bank losses," the Telegraph worries. The surge in junk loans to the industry in recent years is coming home to roost, it seems.

Where will the oil price go next? This has now become a very critical question. Lower still, say some analysts. Right back up again, say others. BP boss Bob Dudley goes to Davos and tells the world to expect low oil prices for up to 3 years. The IEA forecasts $60 oil for another two years. "The price correction will cause the North American supply 'party' to mark a pause," the agency says. "It will not bring it to an end."

Bullish as ever. Let us see.

OPEC Secretary-General Abdallah Salem el-Badri has a very different view to those of BP and the IEA. He suggests the oil price will soar to $200, without giving a timeframe, as investment shrivels. My collaborator Mark Lewis argues in the Financial Times that when the rig count translates into oil output reduction – after a lag because of a backlog of completed wells yet to be produced – the oil price will rise in the second half of 2015. That sounds like the best guess to me.

I have fun in Singapore asking the progressive-company executives to imagine they are running oil and gas companies. Discoveries of new oil and gas reserves dropped to a 20-year low in 2014, I say, the fourth consecutive year of falling discoveries, with no discoveries of giant fields. Knowing this, and all the other setbacks accruing for the oil and industry, it is tempting to sense a whiff of incipient panic in decisions they are taking now. Exxon says US shale will drive its growth in the future: it will double production in the next three years, creating cash for overseas projects even at $55 a barrel, because of the gains it has made in the efficiency of fracking operations.

Really? Warren Buffett wouldn't seem to agree. The legendary investor sold all his ExxonMobil shares in the last quarter of 2014. He also dropped ConocoPhillips.

Shell, having written off billions of dollars in failed shale investments, opts for focus on the Arctic. Despite $15 billion of capex cuts in the next 3 years over the rest of its operations, CEO Ben van Beurden announces plans to resume drilling in the icy north. It is important "not to overreact," he tells the press. I wonder what Total will think of that, having announced that Arctic drilling is too dangerous to contemplate as an option.

As for Petrobras, the most indebted oil company in the world, intent on the oil frontier below the thick salt deposits far offshore on its continental margin, the Financial Times writes that it is "bringing chaos" to Brazil. An editorial suggests that the scandal-ridden company is "too big to fail and too corrupt to carry on as it is."

In the UK, the shale gas revolution that the Prime Minister and Chancellor hope for is falling flat. A Guardian analysis shows only 11 new wells, and 9 scheduled for fracking: 8 new wells and one existing well. And the industry seems to be facing setbacks with every move it makes. Lancashire County Council planning officers have recommended rejection of Cuadrilla's latest application for permission to drill for traffic and noise reasons. MPs have called for a UK-wide fracking moratorium. The all-party Environmental Audit Committee warns the government that it is pushing through "undemocratic" laws and will undermine climate goals.

David Cameron rejects this call. The US shale revolution can be repeated in the UK, he insists once again. George Osborne asks the Cabinet to fast track measures to free up fracking. A leaked letter to ministers requests they all make this a "personal priority".

To no avail. Labour forces a U-turn. The fracking free-for-all measures are embedded in the government's Infrastructure Bill, currently in Parliament. They have to accept restrictions to avoid a defeat at the hands of democracy.

Meanwhile, the Scottish government announces a moratorium on fracking. It wants a full public consultation and commissions an enquiry into health impacts.

I weave as compelling a kick-off presentation as I can for the progressive executives in Singapore, with my extracts of all this, and the dramas that have gone before in the period covered in this book. I then stay on for the two days of their retreat, essentially as a member of faculty. I have plenty of opportunities for further interventions. Somehow, I go 48 hours without sleep at night, jet lagged to my core, without falling asleep by day. The atmosphere is too exciting for that. And I grow more encouraged by the evidence I see of corporate commitment to change with every hour I am here.

Chapter 17

Give us moving pictures

London, 1st March 2015

A day about communication in the battle of ideas on climate energy. I am launching my serialised book at a public lecture in the London School of Economics. It will be videoed, as so many presentations are these days, and posted for endless posterity on both the university's website and my own. The audience is mostly students, and I will have a responsibility to offer them both hope and as much entertainment as I can.

I elect to use a PowerPoint but make it image-heavy and chart-light. And I do something I have never done in a talk before: I include a short video clip.

The point I want to emphasise, at the end of the talk, is that young people today have an advantage, when it comes to communicating in the battle of ideas, over the aging men who for the most part run the PR for the oil and gas industry. The youngsters understand the zeitgeist far better.

The stakes are high here. The world's biggest PR agency has just parted ways with the American Petroleum Institute, in the face of withering criticism on social media. After more than $300m in billings from Big Oil, Edelman ends a relationship that is now giving it a PR problem, especially with the young thinkers that it sees as vital to its future.

Fossil fuel companies have been hitting back at the divestment movement with crude videos and websites bearing names like "Big Green Radicals", pushing an extreme defence of the status quo rooted in simple messages and smears. One attack website, funded by the oil-billionaire Koch brothers, is focused solely on solar. Its rationale: "Shining a light on the dark side of solar power. Billions blown on solar offer little bang for the bucks."

Another, run by the lobbyist exposed by the New York Times for his exhortation of oil-and-gas executives to wage "endless war" on advocates of transition, Richard Berman, offers a particularly inane video defence of fossil fuels. It depicts an oil barrel as a young woman, complete with pink hair ribbon, lipstick and red high-heeled red shoes, who has a boyfriend who has made the mistake of dumping her. She finds a splendid new boyfriend. The ill-advised one freezes in the dark, absent the oil barrel in his life.

Is this seriously the best they can do?

A group of young women in the UK Youth Climate Coalition decide to put together an immediate response. In its devastatingly understated humour it immediately strikes me as a perfect illustration of how the communications battle is likely to play out as the carbon war enters its endgame.

A series of young women speak direct to camera.

"My ex was a fossil fuel", admits the first. She and others then explain why they have dumped their fossil fuels.

"One day he was up, the next down. So volatile."

"He didn't exactly scream long term to me."

"Before I knew it, there would be some new disaster that he would never take any responsibility for."

"He said he was going to clean up his act, but it was a total lie."

"Whilst it might make life quite convenient" (this one is washing dishes as she speaks reflectively to camera), "and he might go on about how I could never live without him, actually" – she pauses for effect – "there are plenty of alternatives."

And that is how I close my LSE talk, on that killer sentence, from a splendid young amateur actress, campaigning for climate action.

There are plenty of alternatives.

London, 5th March 2015

In the headquarters of the BBC, I sit talking with veteran environment correspondent, Roger Harrabin. I am accompanied by my Solarcentury colleagues Frans van den Heuvel and Sarah Allison. We want to explore with Roger whether there are ways that solar energy can be better covered on television.

Hundreds of journalists and producers sit at long tables in an open-plan hall, staring at screens and tapping at keyboards. There are other floors just like this, visible through the glass that the whole building seems to be made of. All the BBC's many outlets are based here: television and radio, national and international.

We are crammed in a small soundproof bubble where meetings can presumably be held without disturbing the long rows of people. I wonder how much shouting BBC journalists and producers do at each other, under the stress of their deadlines and storylines. I imagine it is not inconsiderable. Hence the bubble, maybe.

Roger opens our meeting. I am very aware that there is an immense economic upheaval underway in society, he says, a complete energy transition, and that we are not covering it at all well. I read a lot of things about how clean energy is exploding, and I get it about the crossover into storage and transport. I accept that there is a major running story around the carbon bubble too, as yet largely untold on television. But to tell these stories on the news, I need moving pictures. It can't be solar farms or solar lanterns. It can't be rooms full of investment bankers.

You seem to be saying that we are at a newsworthiness disadvantage, I say, because we can't do stuff like exploding oil rigs, burning oil trains, and oil-caked pelicans.

I can sense Sarah wincing beside me. She is Solarcentury's head of press. But I have known Roger for a long time. We have a frank relationship.

And so we brainstorm, searching as hard as we can for things that will look interesting as they move.

I was in this building yesterday, on a different floor, talking about the solar revolution on radio, for a business programme on the BBC World Service. They wanted to explore my idea that a solar revolution is inevitable. The interviewer was sceptical to the point of hostility, which I always welcome: antipathy, fake or otherwise, helps me get my points out.

With the recording complete, she showed me a different face. I hope you're right, she said.

So do I. But whatever, when it comes to the news, it looks worryingly as though solar, with its lack of moving parts, will be confined to the radio.

London, 9th March 2015

I forgot the Solar Impulse, sitting in its hanger in Abu Dhabi. Today it begins its epic flight around the world. The event has captured the imagination of the mainstream moving-picture media, on an international basis. It is easy to see why, looking at the elegant plane taking off en route to Oman on the first leg of its circumnavigation.

The phone rings early, and I am invited into the BBC again. This time it is for World Service television. Pictures of the amazing plane, with a wingspan greater than that of a jumbo jet, but lighter than a car, fill the screen. As it flies, I talk about solar energy and its cost down, how batteries of the kind packing the plane are also plunging in cost, and the implications of all this.

I try to give a sense of how fast things are moving. Just in the last few days we have seen the National Bank of Abu Dhabi produce a report saying that oil can't compete with solar any longer, and that the great majority of global power investment will be in renewables in the years ahead. Another report, by Deutsche Bank, says solar is closing in on coal-fired power in price terms. Prices could fall a further 40% by 2020, going on to create $4 trillion of value in the next 20 years.

As for storage, Tesla's charge continues. "We're going to spend staggering amounts of money," Elon Musk tells analysts on the EV company's latest earnings call: $5 billion for its battery factory alone. They are not alone. SunEdison has acquired a storage company. The ambitious American solar giant aims to match SolarGrid Storage's technology with its own capabilities in solar and wind.

City of London, 10th March 2015

A conference centre full of money people. Solar money people. The industry is attracting a significant subset of conventional finance now. The room before me looks like any other finance conference I have ever attended.

But my job, another opening presentation, is not to talk about solar at all. I have been asked to talk about the overall energy-market context of the emerging solar revolution. I am finding increasingly that investors in solar want to hear not just about solar cost-down, but incumbency cost inflation and other troubles of the fossil fuel industries. Both themes encourage the belief that money invested in solar now is money well invested, and the synergy between the two amounts to more than the sum of the two.

I run through the latest economic bad news for fossil fuels. In the US, the oil and gas rig count is now down more than 50 per cent. It has fallen off a cliff in 2015. "One day I expect the punchbowl will be taken away," an investor in US shale tells the Financial Times. He explains that he wants companies who can produce their product for $1 in and sell it for $2, not the other way round.

I invite the investors in this room to imagine the situation if solar companies were manufacturing and installing for $2 and selling on for $1. How long could we stay in business? How long would you stay invested?

Yet somehow the rules of the junk-debt feast in the American shale allow a different model. I wouldn't call it a business model.

And it is a problem increasingly extending to oil operations outside the US shale. The oil majors are piling on record debt, on their various mostly unprofitable frontiers. They have hiked borrowing 60% in the first two months of the year.

As for the environmental bad news, new figures for spills, leaks, and ruptures across three states show that fracking operators ran up 2.5 violations a day between 2009 and 2013. These breaches were hidden from the public.

It is often said that for fossil fuels to prosper, democracy and regulatory oversight need to be put on the shelf. Nowhere is this clearer than with shale and fracking.

London, 13th March 2015

Tim Yeo, Conservative chair of the all-party House of Commons Energy and Climate Change Committee, is stepping down today. A conference on UK energy policy has been organised in his honour in Bloomberg's London HQ. Tim is a good man, a long-serving advocate of environmental action. He is someone who fully appreciates the climate problem, and who is not scared of telling his colleagues who don't that they are wrong. Maybe this is why he has been de-selected as a Conservative Member of Parliament.

But now he is telling his audience, in an opening speech, that the next government should put shale oil and gas at the centre of its energy policy. Opponents of fracking are misguided, he asserts.

I sit in the audience silently sighing. I honestly wonder if he could hold that view were he to read the reports that I read every day. He clearly can't be seeing them. He would surely be wary of looking foolish if he were. "Misguided" is a big word to use about people who hold a different view to yourself. Then there is the risk of flogging a dead horse. This very day, aspiring UK shale driller Celtique Energy has dropped its fracking plans in West Sussex. Restrictions in the Infrastructure Bill are the reason for backing away from its proposed sites at Fernhurst and Wisborough Green sites, it says. The threadbare list of potential fracking sites compiled by the Guardian looks even thinner.

Nick Butler, former BP executive and now energy columnist at the Financial Times, chairs a panel after Tim's speech. Most sensible people think fracking is safe, Nick says in his introductory remarks.

Sensible. Another big word.

I have to wait my turn to reveal myself as misguided and not sensible, on a later panel. Nick Butler chairs it. David Hone, a Shell executive, is on it.

Shell has been most upset by Ed Davey's support for the inconvenient and flawed notion of a carbon bubble. They and other companies have protested the Secretary of State's talk of "risky" assets. Oil & Gas UK has written a letter of protest to the Secretary of State saying the industry is "deeply unsettled" about risk to investor sentiment. Shell boss Ben van Beurden has given a speech to an industry gathering on the subject. "For a sustainable energy future," he says in it, "we need a more balanced debate. Fossil fuels out, renewables in – too often, that's what it boils down to. Yet in my view, that's simply naïve."

Naïve. So many big, derogatory, words.

Then this: "Yes, climate change is real. And yes, renewables are an indispensable part of the future energy mix. But no, provoking a sudden death of fossil fuels isn't a plausible plan."

I am not aware of any advocates of a global energy transition who think there is any remote feasibility of a "sudden death" for fossil fuels. That's why we call it a transition, after all. Non fossil fuels still occupy a small minority of total global primary energy as things stand.

At the Bloomberg conference, I run through the solar upsides and the shale downsides in my opening statement. The government needs to get its shale gas U-turn out of the way so we can focus on the green industrial revolution, I suggest.

David Hone counters that with the standard oil-industry mantra: fossil fuels are essential. You can't solve these problems with a few solar panels on roofs, he adds.

That's not what I'm saying, I respond. I am talking about transition away from oil and gas to clean energy over a period of thirty-five years. You and your CEO should not resort to straw-man arguments pretending that those you debate with seek a "sudden death" switch from fossil fuels. I don't know of a single person across the table from you who does.

I say this with a small element of what I hope is well contained passion. These are, after all, matters of life and death. Hone sits listening, not looking at me, a small smirk on his face.

We don't want this to turn into a Chelsea match, says Nick Butler.

London, 1st April 2015

A speaking request of the kind I have never had before. At the Solar Trade Association's all-day state-of-play conference, held in PWC's offices, I am asked once again to give the opening speech. But this time, I am told, it is April 1st. Perhaps I would care to spring an April Fool's stunt on my audience?

It strikes me that I could.

I tell the audience that I am going to run through 24 amazing developments in energy over the last two years. Twenty three will be amazing-but-true. One will be amazing-and-false: the April Fool. I invite them to see if they can spot it.

I know from many conversations how little busy solar practitioners tend to pick up of the run of events, notwithstanding their roles on the front lines of the carbon war. They are simply too busy in their day jobs. And as I list the 24, I can see smiles and shaking heads in the audience. Many of the things I describe would have to sound literally incredible, if you rarely read a newspaper.

Many big energy companies are telling governments to end all subsidies for solar, of all kinds.

Really? I thought they wanted to work with us. And what about their own subsidies?

The Bank of England has launched an internal enquiry into whether or not the fossil fuel companies pose a threat to the stability of the capital markets.

How could that ever happen? I mean, much of the value of the FTSE 100 derives from fossil-fuels.

When Bloomberg first asked whether the shale boom was going bust, because the industry was feasting on junk debt, the oil price was more than $100.

Bust? And the oil price is half that now, right?

The oil industry as a whole has lost many billions of dollars in US shale to date. Shell alone has written off $2 billion and counting.

But isn't it supposed to be a boom?

A US government estimate of recoverable oil in the Monterey Shale, once considered to hold two thirds of US oil reserves, had to be revised down by 96%.

Don't be ridiculous. Who makes mistakes that huge?

The geologist who discovered this inconvenient fact predicts a production peak in the main US oil shale regions, wait for it, next year.

When? But isn't America supposed to be rivalling Saudi Arabia in oil production?

Global conventional crude oil production peaked in 2005 and the main sources growing global supply since have been US shale oil (greater share) and Canadian tar sands (lesser share).

2005!?

Apple has a secret team aiming to mass manufacture electric vehicles by 2020.

Come on, they are a computer company.

E.ON has announced a 180 degree turn in business model, becoming the first utility to break ranks and focus all growth on renewables and energy services.

Yes, this one we have heard about: true.

Shell is announcing this very day that they will be the first oil-and-gas company to break ranks and copy E.ON.

Well, perhaps then.

And so on.

At the end, I tell them the Shell item is the April fool. Not only has there been no U-turn, but rather a defence of the ruinous status quo of such truculence and disingenuity that a former senior British Foreign Office official, former climate envoy John Ashton, has recently described their approach, in an open letter to Ben van Beurden, as "narcissistic", "paranoid", and "psychopathic".

True. And he is no Chelsea supporter.

I drift around in the coffee break, trying to gain a sense of how many delegates spotted the April Fool. On a small sample, none. A selection of the other amazing facts seemed to them more likely to be the deliberate lie.

One otherwise very sensible person – one of the most knowledgeable solar experts in the UK – actually tweeted that people should expect a major announcement from Shell today.

Chapter 18

Dallas, Kent

Online, 9th & 10th 2015

The prospect of a full working day at home. I have been looking forward to it. A chunk of writing early, while fresh. The sun rising over the pond. Check the press before breakfast. Run the dog before coffee. And all the while, blackbirds in song, spring in the air, London's fumes and queues a world away.

But my ideal day is not to be. Leading the BBC TV news is the supposed discovery of 50 to 100 billion barrels of oil below the Weald. A small oil company drilling a single well near Gatwick Airport professes that up to 15% of this vast "find" may be recoverable. This 15% would be equivalent to the entirety of Brazil's reserves, including its enormous offshore sub-salt oilfields.

Their CEO is interviewed. He explains that this terrific news is based on one well, with no flow rate measurements. He is a tough looking Australian. He shows not a hint of shiftiness or shame as he trots this rubbish out to camera.

Any geologist will instantly know that this is rootless hype. You cannot possibly base such figures on a single well. You could not even base it on multiple wells without knowing the flow rates.

I wait for the counter view to come: the expert who will say that this is an exercise in hype and explain why, hopefully adding that even if if were true, the oil probably couldn't be produced at a price that would make the project economic anyway. Especially if the UK's famously strong environmental constraints on on land drilling were to be applied. Look what is happening in North Dakota, etcetera etcetera.

No such expert appears. The BBC relays the whole story as though it is reliable fact.

There is another item on the news. Air pollution over southern England is at record levels. Polluted air from Europe has combined with fumes from British cars to send particle and toxic gas counts through the roof. People are advised not to take exercise outdoors. Dogs too, I have to suppose. Especially 45 miles per hour ones like mine.

Nobody on the breathless BBC refers to the irony of the supposed oil find and its implications for future air quality, if true. And as for climate change implications?

So much for my run with the dog, to release the stress of 100 billion phantom and/or unaffordable, emissions-who-cares, barrels.

I expect the madness to be corrected by the lunchtime TV news bulletin. It isn't. This time the BBC interview an enthusiastic professor from Imperial College who essentially backs the claims. This one at least has the decency to look shifty.

The share price of the company soars during the day.

In the early evening, a call from a radio station, LBC. Will I come on air, they ask, and give my opinion on how best to spend all the Dallas-like riches that are now going to descend on the UK as a result of this amazing discovery.

I say I won't, because they have have swallowed a – how can I put this politely – beguiling myth.

Ah, says the producer. That's interesting. Why don't you come on air and say that then?

Bright and early the next morning I do.

You mean you don't want to exploit this oil?, says the presenter, radiating incredulous hostility down the phone line at me. You want ambulances to run out of petrol?

London, 14th April 2015

Dinner for fifty in a luxury location where leading lights of the Democratic Party often go when in town. The guest list has a climate and finance theme. Al Gore is going to give us a speech on that subject after the main course.

I am an early arrival. Champagne in hand, I chat with with David Blood, co-host, former CEO of Goldman Sachs, now managing director of Generation Investment Managers, the fund-management company he co-founded with Gore. Their creation is now a frontrunner in clean energy investment.

Al wanted to call their creation Blood and Gore, no doubt thinking of his enemies in the climate-denial business at the time. David was less keen.

Al also arrives early. To my astonishment, he remembers my name without prompting by David Blood. How does he do that? All the thousands of hands he has shaken over the last quarter century, mine only half a dozen times among them.

I first met Al Gore in 1991, in the Japanese Parliament, where the then Senator had organised a gathering of parliamentarians concerned about climate change. I soon found that he not only knew what a methane hydrate was, but he could talk about their potential role in amplifying global warming as though he were a climate scientist himself. (Hydrates are ice-like mixtures of water and methane, formed in sediments with high organic matter content at low temperature and / or high pressure. Increase the temperature, and the ice melts, releasing the methane. There are hundreds of gigatonnes of methane hydrates in the Arctic, where of course temperatures are rising faster than most of the rest of the planet. Hence the concern: even more so now than back in 1990.)

Gore impressed me then like no other politician had done before or since. I kept in touch with him through staffers in his office, feeding in information that I thought might be of interest, and we met periodically at other climate meetings, including the Rio Earth Summit in 1992. When he became Vice President, e-mail contact dried up, unsurprisingly. After his eight years in the job, and his narrow defeat to George Bush in the infamous "hanging-chad" Presidential election, I met him again a few times during his reincarnation as *spiritus rector* of global climate campaigning.

These days, his energy in advocacy seems no dimmer than it was a quarter of a century ago. He has founded an organisation called the Climate Reality Project that trains trainers, on a global basis, in the science and policy implications of climate change, and contests climate contrarians, exposing both the errors in their arguments and the energy-incumbency origins of their funding.

I chat with Al now for a few minutes. I tell him that in all these years, I have never felt so bullish about our chances.

I agree completely, he says.

We trade a few reasons why we should feel this way, especially knowing what we do about methane hydrates.

I mean, I say, when Apple is intending to be mass-producing solar-charged electric vehicles by 2020, what other signal do we need that disruptive system change, beyond the imagination of most people, might be just around the corner?

No comment, he says with a grin.

I had forgotten. He is on the board of Apple.

London, 16th April 2015

BP's Annual General Meeting, at the Excel Centre. An exercise, for Carbon Tracker, in helping to rack up the pressure on the company from its shareholders. A consortium of those shareholders, led by Helen Wildsmith of the CCLA, fund managers for the Church of England, has submitted a resolution which will be the focus of shareholder action today. Special Resolution 25 reads "To direct the company to provide further information on the low carbon transition." It contains a range of climate transparency goals. The hope is that in forcing the company to conduct an open analysis of the risks to its business from climate change, it will be encouraged internally by staff, and externally by investors awakening to the problem, to join E.ON and position for energy transition.

A similar resolution has been tabled at the Shell AGM. To the surprise of campaigning shareholders, both companies have agreed ahead of the AGMs to support the resolutions themselves. The general expectation is that they did not want the negative publicity that would come with rejecting a simple, reasonable, request, preferring the option of agreeing to report, then stalling and obfuscating during the process.

Outside the Excel Centre in London's Docklands, I pose for a picture with a team from Share Action, another shareholder campaign group much involved in planning today's exercise. We stand behind a banner. "Give climate change a seat at the table", it reads.

It is time to enter the building. I people-watch as I queue at security. Very many of the shareholders are over 70, I observe.

I see obvious private security types, with tell-tale earpieces, cruising the crowds in some numbers. One smartly dressed young woman is taken by the arm by an earpiece man and a woman with a clipboard, removed from a queue, and walked towards the exit. I ask who she is and what is going on. She is Sue Dhaliwal, and she protested about climate change inside an earlier AGM. She will not be allowed into this one, I am told.

A man who might be Samoan is also attracting attention from the earpieces and clipboards. I ask who he is. Derrick Evans, resident of Turkey Creek, Louisiana. He represents coastal communities affected by the Gulf of Mexico oil spill.

When I enter the main hall, BP CEO Bob Dudley is giving a speech to the hundreds assembled. Fifteen directors sit with him on the stage.

Dudley does not say anything you wouldn't expect a CEO of modern BP to say.

Questions from the floor. Shareholders queue at the microphones. Some are unquestioning supporters of the company, whose lame queries are thin excuses for compliments. Others have issues. I listen to them come and go, fended off with calm patience by the chair, Carl-Henric Svanberg, and the CEO.

How urbane BP's leaders are. How easy it would be to believe everything they say, if you were of a belief system that wanted to, or knew little or nothing of the background to the issues.

The Gulf of Mexico is recovering faster than expected, they assert. They show a short of video of an American employee assuring a camera that it is.

Derrick Evans has a different view. For starters, you leave out the tremendous health impacts both of the oil and the dispersant you used: on your clean-up workers, on fishermen, on residents of the Gulf. Metabolic disorders are widespread over five states.

Thank you Derrick. Our dispersants were approved by the US Coastguard. We will discuss this with you further.

You say your investment in Russia, a 20% share of Rosneft, is good value for shareholders, says another questioner. It isn't good value in the signal it sends to Mr Putin. The company should pull out.

International commerce is important in relations between nations otherwise in dispute.

In Columbia, people are being killed if they raise questions about oil operations you are involved in. In Azerbaijan, where you are so proud of your gas operations, the government has a horrific human rights record.

We are a company, we don't do foreign policy.

The questioners are as polite as the executives are unruffled. Only rarely is there is a sense of the underlying passions.

Native communities around tar sands operations like yours are suffering dreadfully from the pollution.

I have visited tar sands communities, says Bob Dudley. I found otherwise. I think only a minority are unhappy.

That, Sir, is very far from the truth.

We come to Resolution 25.

Helen Wildsmith and Bill McGrew, an executive at CalPERS, introduce it. It covers five areas of transparency on climate. There is to be one question on each theme. Mine is the second.

The first questioner asks the board whether BP will consider adopting hard emission targets.

We are not yet ready to set targets, Carl Svanberg says. It's not that we are unwilling. It's just that it could be counter-productive. His example of counter-productive is the additional coal burning that has been imposed on Germany by shutting nuclear.

My question. I stand at the microphone for a moment surveying the sixteen characters on the stage. Only four of them have said anything so far. The CEO and Chairman have done most of the talking. I wonder what the rest are thinking, how they are selected, how they are remunerated, the degree to which they harbour any doubts at all as BP ploughs right on throwing gasoline on the fire of global warming, despite all the transparent concern in the outside world.

The International Energy Agency has a global energy scenario for 450 parts per million of carbon dioxide in the atmosphere, I say. It is very different from the future energy scenario that you favour. It involves much more renewable energy, much less fossil fuel. It offers some chance of a ceiling to global warming below two degrees Celsius. Would you be willing to include a portfolio stress test for the 450 scenario? If not, may we know why? And what might the company's contingency plan be if forced to follow the lead of others?

Carl Svanberg now waffles a little. This is his least impressive answer so far. If I understand the gist of it, it goes like this. The IEA's outlook for demand is virtually the same as BP's. The demand scenarios for the next five decades don't change dramatically. It is hard for the company to plan for anything other than what we believe in at the time being.

The final questioner challenges BP about its lobbying. Despite all the lip service to the climate problem, and its exhortations to government to do something about it, the company remains a member of numerous European lobbying groups which take positions against climate action, such as Business Europe, CEFIC and Fuels Europe.

Svanberg says that they didn't necessarily align with the lobbying position of these groups. Membership doesn't necessarily signify agreement.

Bob Dudley makes an observation on Paris that strikes me as potentially encouraging, potentially ominous. BP got together with Shell, Total and Saudi Aramco in Davos this year, he says.

This I didn't hear about.

We decided we should form a common view for governments. We're not there yet, but will be by Paris.

New Forest, 18th April 2015

A hundred people sit in a community centre having given up a few hours of spring sunshine to talk solar energy. They and three hundred others have recently built a 2.5 megawatt solar farm in a field nearby, and financed it all themselves. I am here at their invitation to tell them how I see the wider context of their project.

This is my second AGM in two days, I tell them. Yesterday I was at BP's.

They laugh politely.

You are a microcosm of a revolution that is unfolding at a speed many people can't imagine, I begin. Including the board at BP.

But they don't look like a revolutionary vanguard. They look like a pensioners' club of some sort. Yet they raised £2.3m in equity and £241,750 in bonds to make their project happen. The gross return for members so far has been just over four percent. The prospectus they signed up to said it would be 2.4%.

I wonder what the 70 year old shareholders at BP's AGM would make of those numbers, if they knew them. BP's return on equity for the quarter to December 2014 was minus 4%.

Lausanne, 20th April 2015

Now this is going to be a challenge. The cocktail party is over. Three hundred delegates flood into the dining hall and head for their designated round tables. The hall isn't big enough for them all – an annex is full of tables too. They move to their seats afire with conversation. Much champagne has already been consumed.

This event is the Davos of the commodities industries. All the major trading houses are here: Glencore, with the vastness of its coal assets. Vitol, the secretive oil trader who must have tankers full of oil all over the world right now, sitting and waiting for the oil price to go up before cashing in. Trafigura, who in 2006 shipped toxic waste to Africa that was banned in Europe, and tried to cover up the environmental disaster it caused.

The delegates are in buoyant mood. Business is good. Oil traders are celebrating a market rout resulting from low oil prices. The more volatility the better, for them. Glencore's famous CEO Ivan Glasenberg, who must be at least a little worried about coal, talks nonetheless of a "blowout year".

Their bankers are here in numbers too. The Financial Times has reported yet more malfeasance on their part of late. Petrobras bribes have been paid through Swiss banks, a former executive has told the FT. He claims he laundered a hundred million dollars through a web of accounts, including at HSBC.

I sat with a very senior lady from HSBC at dinner last night. We avoided Petrobras, and the criminal case the bank faces in France over tax fraud.

The Financial Times tells me the commodities industry is trying to clean up its act. But that is not the reason they have invited me here. I am the after dinner speaker. They want me to offer the traders a different view of what their energy world might look like, ten years from now.

I make my way to the top table. I cannot remember the last time I was this nervous.

Tony Hayward, just in from the airport, greets me. The former BP CEO is chairman of Glencore these days. We shake hands and catch up as people do who first met more than a quarter of a century ago, and not often since. We were graduate students at the same time: we researched the same types of rock, at different universities. It is very difficult to take people seriously who you knew as a fellow long haired student and shared a drink too many with. I expect he feels the same way as I do about that.

Lionel Barber, editor of the FT, sits between us. He tells me I am no longer after-dinner speaker, but pre-dinner speaker. There are demonstrators outside, agitated about carbon and the rest of it. Security is worried that if we wait until after dinner there will be too much din for me to be heard.

But the din I fear more is inside the dining hall. I have seen dinner speakers at conferences literally drowned out by a rising tide of chatter from tables, even if they are saying things the diners might normally want to hear – which these traders and bankers will most certainly not be hearing from me. Guests at occasions like this so much prefer talking to people they are sitting with: especially if they have been both cooped up all day in sessions and have been quaffing champagne before dinner.

Sure enough, as soon as I am a few sentences into my talk a table at the back of the room starts up an audible conversation.

I press on, now expecting that contagion to spread to the nearby tables, and on around the hall. As for the tables outside, where diners can't even see me other than on a TV screen, I am amazed they are not talking already.

I recall some advice I was given as a young university lecturer, by an old professor of petroleum geology, a colleague at the Royal School of Mines. If faced by a class of students up to collective subversive mischief like tapping their twinkle toes in unison under the desks, drunk on their own devilment, you keep going, you maintain confidence at all costs, as though completely unaware. Life is, after all, theatre. So you act. You put on your best performance for the one person in that room who might be wanting to listen to what you have to say, or at least be open to it.

Full marks to the Financial Times for thinking that the commodities industry might need to be shaken up a bit on energy. I will do my level best to execute their brief.

I tell the whole story as a series of short stories, concentrating ferociously. From the carbon bubble being popped on the horns of the bull on Wall Street, to a President telling world leaders "we cannot pretend we do not hear" the protestors on his streets. From the doubt about stranded assets at the Bank of England in 2013 to a British Secretary of State talking fossil fuel crisis to an entire national press corps on an embassy lawn in 2014. From Shell admitting that they know there is an energy transition underway, to an Apple strategy that suggests "way sooner than you think, man". From the world's largest private bank predicting solar-and-battery prosperity for all-comers by 2020 to the world's most respected furniture retailer announcing it is going 100% renewable itself by then.

I take them back to Fifth Avenue at the dawn of the 20th century, and how it took just 13 years to go from 100% horse-drawn carriages to 100% horseless carriages. Will it be much different this time? I invite them to Poznan in December 2008 and the palpable fear of IEA officials about global crude production collapsing from 72 million barrels a day in 2005 to, ahem, 28 in 2035, and on to the Saudi-America nirvana-rhetoric whipped up by the shale drillers just a few years later. I try to worry them with the tale of the Monterey Shale hype in 2013 and the – er, really? – 96% downgrade in May 2014. I take them to Riyadh and the Saudis' fears about their own soaring domestic consumption. I tour the frontiers of the late hydrocarbon century: the trillion dollars of zombie investments Goldman Sachs sees at an oil price $20 higher than today's; Kazakhstan and the Kashagan Cash-All-Gone loss-fest; Brazil and the sub-salt requirement for – cue hysterical laughter – $221 billion dollars by a company the FT calls "too big to fail, too corrupt to continue"; the Arctic, and Total's insistence that the danger of drilling is unmanageable versus Shell's blind desire to rush through drilling seemingly oblivious to everything, even economics. I remind them that the first utility has jumped ship: a U-turn away from fossil fuels, in search of profitability. How long, I ask, before the first oil and gas company joins them?

As I rattle out the story at machine-gun pace, I clock that the talk-among-ourselves syndrome is not spreading from the one table at the back of the hall. People are listening. Bemused looks and smirks aplenty, but they are listening.

I conclude. Profound change I can guarantee you, these next ten years, in energy markets. The specifics are difficult to predict, given the multiple variables, and anyone who pretends otherwise is delusional. But the generalities are clear. For the fossil fuel incumbency, problems will grow, opportunities will decline. For the clean-tech insurgency, problems will decline, opportunities will soar. I invite you, therefore, to join the revolution.

I return to the top table. Tony Hayward shakes my hand with a smile, though he must have hated almost every word I uttered.

Chapter 19

Must do, can do, will do

London, 29th April 2015

Another soiree, more champagne, this time served in the glassy exhibition centre owned by Siemens in London's Docklands. I talk to two executives of the company, who tell me proudly about their sustainable cities work. Their exhibition centre is itself an exhibition: its carbon dioxide emissions per square metre per year are more than 65% lower than comparable office buildings.

The Guardian Sustainable Business Awards ceremony is held in a lecture theatre that can seat hundreds. Guardian Deputy Editor Jo Confino takes the stage as master of ceremonies. Last year, he tells his captive audience, you were not well behaved. This year we thought you ought to be seated.

Jo is well loved by all sectors involved in sustainability. Our confinement, and the temporary absence of alcoholic beverages, is taken in good heart.

The first award of the many Guardian Sustainable Business awards to be announced in the hour that follows is for Innovation in Communicating Sustainability. Carbon Tracker wins it. Anthony Hobley and Mark Campanale are called to the stage to shake Jo's hand.

In the last fortnight the remarkable reach of the team they lead has stretched to new horizons. Carbon Tracker has released a new report, the first in a series we label "Blueprint", as in a blueprint for companies seeking a sustainable transition from carbon-based energy to clean energy. Mark Fulton led the drafting. Based on the first Blueprint, HSBC has written a private report warning clients of the growing likelihood that fossil fuel companies may become "economically non-viable", as they put it. The report, inevitably leaked and covered in the press, recommends one of three approaches: divesting completely from fossil fuels; shedding the highest risk investments such as coal and oil; or staying invested and engaging with companies on their capital discipline.

Investors who stay in fossil fuels, HSBC observes, "may one day be seen to be late movers, on 'the wrong side of history.'"

Anthony Hobley tells Newsweek "it's incredibly important that a mainstream financial institution is effectively taking our narrative on the carbon bubble, analysing it and then producing a research report that reinforces our conclusions. I think we are at the beginning of a very important reframing of this issue, and of climate risk being understood by the mainstream financial markets."

It seems increasingly likely this is the case with every passing week now.

But we do still have a long way to go. The Asset Owners' Disclosure Project (AODP) has found that 85% of the world's largest 500 asset owners – the majority of them pension funds – are doing nothing on climate risk. If they continue to be inactive, AODP says, they will face a new danger: that of being sued. AODP itself, acting with the activist lawyers' group Client Earth, threatens to lead the process.

Meanwhile, in the fast-spreading divestment campaign, the School of Oriental and African Studies becomes the first London university to divest from fossil fuels. It will pull out of them all within 3 years. An FT survey of the rising tide shows Prince Charles and Richard Branson shunning fossil fuel investments, and the National Trust and Church of England among those reviewing their options.

On the same day that the Carbon Tracker team accepts its award, a major new development unfolds. It emerges that the G20 powers are investigating carbon-bubble risk. The G20 has asked the Financial Stability Board in Basel to convene a public-private inquiry into the fall-out of stranded-assets risk, modelled on the Bank of England's enquiry in the UK. All member countries have agreed to co-operate or carry out internal probes. France championed the idea, so the Telegraph reports. The hosts of the Paris climate summit are obviously coming to see finance as a critical lever for successful delivery of a treaty in December. Among the nations who have agreed to the review are the United States, China, India, Russia, Australia, and Saudi Arabia: major fossil fuel producers and/or consumers all.

Now this is going to be interesting.

Online, 2nd May 2015

Yesterday in Silicon Valley perhaps the most remarkable businessman of the 21st century so far launched a product he hopes will change the world. Today, a video of his 18-minute presentation is available on YouTube, and I am getting on for the millionth person to watch it.

Welcome everyone to, basically, the announcement of Tesla Energy, says Elon Musk. An audience sitting in a dark conference room claps and cheers.

This man is at the heart of the solar and storage revolution. The electric-vehicle manufacturer he founded, Tesla Motors, is galvanising the electric-vehicle market. It is worth billions today. Now he is taking the lessons his company has learned in the use of lithium-ion batteries, and applying them to solar-powered buildings.

What I am going to talk about tonight is a fundamental transformation in the way the world works. This is how it is today, Musk says.

An image of fossil fuel burning appears on the screen.

It sucks, he continues, with a stifled giggle.

He has a diffident, almost casual, manner. He is not the typical silver tongued Silicon Valley entrepreneur-visionary orator, but everyone watching knows the towering scale of his vision, and – crucially – the track record of spectacular success that he backs it up with.

The upward curve of rising atmospheric carbon dioxide concentrations appears.

The solution to this is in two parts, Musk says. Part one, the sun. We have this handy fusion reactor in the sky called the sun.

Laughter from his audience.

You don't have to do anything. It just works. It shows up every day and produces ridiculous amounts of power.

To make the point, he shows a map of the United States with the area needed to produce all US electricity as a tiny blue square, occupying a small minority of northern Texas. Mostly that will be on rooftops, Musk says. You don't even need to disturb the land.

On to the need for storage.

You will see one red pixel in the blue square, he says. This is the area of batteries you would need for the United States to have no fossil fuel generation. This is a very tiny amount.

Clapping.

Now the issue with existing batteries is that they suck.

More laughter, including from Musk.

They are really horrible. There is a missing piece and that is what we are going to show you tonight.

The Tesla Powerwall appears on the screen now, as dramatic music swirls. It is a simple wall-mounted flat box, albeit a beautifully designed one. Attendees lift smart phones to take pictures.

A normal household can mount this in their garage or on the outside wall, Musk explains. It gives you peace of mind. You don't have to be worried about power outs. If you want to, you can go completely off grid. And the cost of this is $3500.

Clapping and cheering.

You can stack up to nine of them in a home. That's 90 kilowatt hours.

And very importantly, this is going to be a great solution for remote parts of the world, he suggests. We're going to see what happened with cell phones and landlines, where cell phones leapfrogged landlines.

You can order the Powerwall right now on the Tesla website.

Whooping. I imagine people reaching in the dark for iPads to place their orders at the head of the queue.

We're going to start shipping in approximately 3-4 months. Next year we will ramp up as we transition to the Gigafactory in Nevada.

This is a reference to the enormous Tesla battery factory currently under construction.

What about something that scales much larger? For that, he continues, we have the Powerpack.

An image of this second product appears, a wardrobe-sized box.

It's designed to scale infinitely, Musk enthuses. You can go gigawatt hour or higher.

Now would be a good time to transition the power we are using in the building to batteries, he says. A dial appears, showing that the entire launch event has been powered by batteries: batteries entirely charged by solar on the roof of the building the event is being hosted in.

His audience is by now ecstatic.

What's really needed to scale the world entirely to sustainable energy? Is it insurmountable, impossible? It's not. With 160 million Powerpacks you could transition the entire US. With 900 million you could transition the world's electricity – all renewable, primarily solar. It would take two billion Powerpacks to do all the world's transport, electricity and heating. That may seem like an insane number. Its not, in fact. We have approximately two billion cars and trucks on the road. And every 20 years that gets refreshed – there's about 100 million cars and trucks made every year.

We can do this. We have done things like this before. Tesla has started, with Gigafactory One. What we are really designing in the Gigafactory is a product of Tesla, a giant machine. There will need to be many Gigafactories in the future. Tesla won't do it alone. Other companies will need to build Gigafactories, and we hope they do. The Tesla policy of open sourcing patents will continue.

This is a reference to his remarkable policy of open-book design. All the more remarkable because competitors are already forging into the battery industry. Bill Gates is into the game with an investment in liquid-metal battery company Ambri, who are currently testing their first field deployments. Bloomberg has surveyed the storage scene and reported a clutch of other technologies under development. This is no one-horse race: Musk's preferred option of lithium ion is not yet guaranteed to prevail.

Elon Musk returns to the carbon dioxide curve, showing a variant of it now where carbon dioxide concentrations flatten, all because of solar panels and batteries.

That is the future we could have, he says. That is the future we need to have. Solar and batteries are the only path I know that can do this. And I think its something that we must do, and we can do, and we will do.

Berkshire, 8th May 2015

I have a guilty secret. I am a closet golfer. These days I am very occasional, and highly erratic, but I enjoyed a misspent youth on golf courses. I could once hit the ridiculous little ball straight and the right distance most of the time.

Golf, played badly or tolerably, is a great way to forget about the rest of the world and its ways for a while. Today is a useful day for that. Yesterday was the British general election.

As I wait for my companions to arrive at The Berkshire Golf Club, I am offered a glass of champagne by a Marilyn-blonde lady of the shires, a hostess on what turns out to be a golf day of the Countryside Alliance. She and her happy companions assume nobody in a Home Counties golf club could be doing anything other then celebrating a majority Conservative government elected by little more than a third of the voters on two thirds of the eligible turnout.

In fact, I am apprehensive in the extreme. The broken British electoral system has replaced a government propped up by 50 Liberal Democrats broadly supportive of renewable energy and action on climate change with a government propped up by 50 to 60 right wing Conservatives tending to frothing climate-change denial and visceral hatred of new fangled ways of generating energy.

My companions arrive and we head off. I have much to mull as we wend our way through the pines and heather.

The city of Gothenburg announces their prize for sustainable development this week. Previous winners include Al Gore, Kofi Annan, Paul Polman, and Gro Harlem Brundtland. Yet this year I am to be named as one of three recipients. The judges say they have given me the award for my work with Solarcentury and SolarAid.

I find myself more uncomfortable than flattered. There are three reasons why. First, Solarcentury is no superstar of the solar age like SolarCity, the company Elon Musk chairs, or Tesla, the company that he runs. We are an unfinished story: no more than a healthy middleweight survivor-so-far.

Second, although SolarAid might be describable as remarkable, both it and Solarcentury have achieved their substantive growth under teams that I did not lead. Derry Newman and now Frans van den Heuvel have led Solarcentury, other than in the chaotic early years of my own leadership. Nick Sireau led SolarAid from day one, then Steve Andrews, now Andy Webb. They and their teams did the execution. I mostly sat in my office in Kent, working on what I tend to call "wider context". It is all these people – the past and present teams – that should have won this prestigious award.

Third, and most importantly, the cash-flow crisis at SolarAid is worsening. An existential threat hangs over the organisation I founded. We are entering a make-or-break phase, and it cannot last for more than a few months. By then, unless we can find a solution, our cash position may well be untenable.

Big questions swim in my mind, day and night, about this situation. Have I been pursuing the right proposition with SolarAid as it is currently configured? Am I doing the right thing in arguing so hard for a non-traditional business approach on the solar-lighting frontier? Am I serving the climate- and development causes best, in this particular microcosm of the carbon war?

I review the basics constantly. I have long thought that making 100% global renewable energy credible would require breaking the mountain of transition down into short climbs. The first climbs should be the theoretically easiest. Replacing oil used for lighting is an obvious candidate. Kerosene costs so much in the poor nations that avoiding it by using solar lighting saves households money within weeks: a no-brainer sale, even to some of the poorest people in the world.

I figured Africa might be a good place for an adventure in social entrepreneurship to test this idea. If fossil fuel could be knocked out in one sector there, on purely economic grounds, knocking other sectors out as clean-energy costs fall would seem increasingly credible.

I founded SolarAid to chase the mission, a charity to be funded with 5% of the annual profits from Solarcentury, the much more conventional solar company I had founded in the developed world. Conventional capital like the venture capital backing Solarcentury would be unlikely to pioneer new African markets in solar lighting fast, I reasoned. My experience was that it would be too risk-averse to face African frontier conditions, and that investors would sound the retreat too quickly in the face of setbacks. I wanted to test the idea that a different financing model – a mix of philanthropic donations and low-interest debt, mostly crowd-funded – was the best way to crack open the first mass markets.

SolarAid set up a retail arm, SunnyMoney, to experiment with the new model in 2008. The idea was that all profits (once they eventually materialised) would be recycled back into the mission, and the retail operation, although functioning like a conventional business, would remain wholly owned by the charity. We began in Kenya, Tanzania, Malawi and Zambia, but our mission was to play a lead role in ridding every African country of kerosene lighting by 2020.

Progress was at first slow. Then the teams in the field found a model that worked. It involved schools. SolarAid makes available the details of how this model works on an open-source basis.

Sales rocketed in 2013 and 2014. SunnyMoney has now sold 1.7 million lights, making it the biggest retailer of solar lighting on the continent. Most sales are in Kenya and Tanzania, where proper markets are now functioning as venture capital and impact investment flow in behind SunnyMoney's start, as we always intended.

Our 1.7 million lights have improved the lives of ten million Africans so far. Over the three-year product lifetimes, we are in the process of saving $360 million, averting 890 tonnes of carbon dioxide emissions and creating two billion extra homework hours. That is just where the social benefits begin.

The question now is whether we can replicate this market kick-start model in other countries, particularly the "big three" in terms of population: Nigeria, Ethiopia and Congo.

The venture capital and other equity financing raised for companies coming in behind us in Kenya and Tanzania exceeds $150m. On one level that sounds a lot, but venture capitalists expect most investments in their portfolios to fail.

The estimated SolarAid funding requirement, if we are to lead the way on new frontiers with enough working capital, would amount to not many millions of dollars over the next five years. For the oil industry, the sum would be a rounding error in a single oilfield. For us, it is a series of "big asks" that will begin to be made of potential foundation and corporate partners over the summer. We will see then whether the SolarAid experiment can work in more than two national markets.

Heading off climate change is all about timing. It seems increasingly likely, as fossil fuel costs rise and clean- energy costs fall, that a great energy transition will sweep the world. But can it happen quickly enough to save civilisation? If we leave the clean-energy frontiers in developing countries only to traditional forms of capital, I have my doubts. This summer we will see if others agree with me.

Chapter 20

A reminder to the powerful of the Earth

Paris, 19th May 2015

Climate Week in France. Six months to go to the Paris summit: a good week to sit on a Eurostar and review the state of play in the carbon war's multiple theatres. We have just learned that atmospheric carbon dioxide concentrations topped 400 parts per million in monthly average for the first time in February. They have risen 120 ppm since pre-industrial times, half of that since the 1980s. The worrying impacts of this global-thermostat boost continue to unfold. The British Antarctic Survey reports that an ice shelf the size of Scotland is now at imminent risk of collapse into the ocean. Warming seas are causing ice loss below Larsen Ice Shelf C, they say. As for the rampant droughts underway in California, Australia and elsewhere, Swiss researchers find that three in four such heat extremes are now ascribable to human activity.

Some progress is being made on emissions. China's coal imports have fallen 42% as the government's air pollution action bites. The government is extending the ban on coal burning in cities to the suburbs.

The divestment movement gathers more steam, with the Church of England deciding to end investment in coal and tar sands. The Church's £9 billion fund has "read the last rites" to the two industries, jubilant campaigners tell the press. Oxford University will also divest from coal and tar sands, though it stopped short of excluding all fossil fuels, to the vocal disappointment of many present and past students and faculty members, myself included.

Glencore insists that there is no chance of its fossil fuel assets being stranded, including its coal. CEO Ivan Galsenberg says that what he calls energy reality means there will be no measures to stop full utilisation of his commodities.

The Bank of America is a new addition to those that disagree. It crafts a new coal policy deeming investments increasingly risky. "The dynamics around coal are shifting," one of the largest coal financiers now says.

In the face of this kind of development, the floundering coal giants come over as increasingly renegade. Peabody Coal's CEO Gregory Bryce professes that global warming is "a crisis predicted by flawed computer models", and that the real crisis is the poverty that absence of coal will inflict.

How many key influencers does he honestly expect to believe such mantras today?

Those clinging to the tar sands look no less isolated. Canada has reneged on past emissions commitments. Their pledge for Paris is to cut 30% from 2005 levels by 2030, less ambitious than their previous goal, and far behind the US and EU pledges.

In the oilfields, Saudi Arabia production rose 0.65 million barrels a day, to 10.29 mbd, in March. But this increase is being achieved at a price. The Kingdom burned 5% of its foreign reserves in February & March, offsetting national income lost to the low oil price. For the moment, with the Brent oil price at $63, the desired effect on American shale is being achieved. The US rig count continues to drop, down 53% now, and shale oil production has fallen in North Dakota for the second month running. "The oil industry's 'man camps' are dying," Bloomberg reports, referring to temporary camps for oil workers. "Drillers spent big to house workers in the new boomtowns. No more."

Debts continue to mount on a worrying scale in the oil and gas industry. Bond issuance so far in 2015 is up 10%, $86 billion, on last year. Both independent and state oil companies are involved in this.

American shale champions call for investors to hold their nerve. The price will go up, and shale drilling will start up again, they say. But US shale oil cannot turn off and on like Saudi oil, analysts respond. The response will be "jagged", and "sticky".

There were 61 fracking companies operating in the US at the start of last year, in full-boom mode. Now there are 41. Half of these will be dead or sold this year, an executive from Weatherford International, the 5th largest US driller, tells Bloomberg.

These are the direct economic realities the industry faces. It is also receiving body blows from wholly different directions. The US Geological Survey is now stating clearly that oil and gas drilling triggers man-made earthquakes, in eight states. The waste water injection process is activating long-dormant faults, the government's expert agency says.

Oil industry contaminants have been found in California irrigation water. Tougher fracking regulation is requiring more detailed monitoring, and guess what it is finding. Methylene chloride, for one. Even worse, fracking chemicals have been detected in drinking water for the first time. 2-Butoxyethanol or 2BE, another carcinogen, most likely comes from poor drilling well integrity, scientists suspect.

And so to the inevitable. Accomplished short-sellers are now targeting the shale drillers. David Einhorn, a successful veteran whose winning bets in the run up to the financial crisis included Lehman Brothers, says openly that he has US "frack addicts" in his gun sights. Shares in several of these targets have tumbled as a result.

But the frack addicts, whether in companies or their support base, keep right on justifying their habit, and indeed protecting their sense of entitlement to feed it. A bill stopping cities' right to ban fracking has passed in the Texas House. It now goes to the Senate and after that the Governor. As things stand, it is widely expected to pass into law. If so Texas, will have banned fracking bans.

In the Arctic, Shell barrels on. Their rig arrives in the port of Seattle, en route to the Chukchi Sea, despite the widespread opposition in the city. It is greeted by a fleet of protestors in kayaks. Eni has joined Total in counting Arctic drilling out: it has let its lease in the Chukchi Sea lapse. The Anglo-Dutch oil giant is losing allies in places it once could have expected blanket support. This includes the conservative press. Andrew Critchlow, the Telegraph's Commodities Editor, sees Shell's desperation as a sign that the world is running out of options on oil. This voyage marks the beginning of the peak oil era, he writes.

Now there are a couple of words you are not supposed to use in polite incumbency company. Peak oil is dead, most of them like to say. Killed by the shale boom, right?

Shell, like the Texan frackers, plays dirty in self defence. It emerges that the company has lobbied to undermine the EU renewables target. Documents sought by the Guardian under Freedom of Information show that the union's weak goal of little more than a quarter of renewables in the energy mix by 2030 has much to do with Shell's lobbying for gas.

It seems as though the anti-clean-tech lobbying is going to become more difficult. Tesla has taken $800 million worth of indicative orders for its batteries in the first week since Elon Musk's launch event. "It's like crazy off-the-hook," says Elon Musk. His 5 million square foot Gigafactory will not be big enough if these orders convert: it will be sold out through 2016. Interestingly, $625 million of these orders came from businesses and utilities.

"Can Tesla's battery hit $1 billion faster than the iPhone?" Bloomberg asks.

We also learn how close Tesla came to going under, just two years ago. Google almost bought Tesla in May 2013. Musk was running out of money, and went to his friend Larry Page, Google co-founder, who shook hands on a rescue deal. But sales of the Tesla car then soared in the nick of time, and Musk did not need a rescue. Even megastar entrepreneurs need luck. (Of course, they tend to make their own).

In the solar arena, momentum is not so spectacular as in storage, but hugely impressive nonetheless. SolarCity launches a first-of-kind $1bn solar and storage fund, for commercial and industrial roofs, backed by Credit Suisse. It will fund 300 megawatts in the next 2 years.

China is no doubt watching events in Silicon Valley closely. A new Chinese government report suggests that 85% of Chinese electricity and 60% of energy could be renewable by 2050.

"Renewables ride wave of success as prices fall and spending jumps", the Financial Times enthuses in a headline. Global investment was up 17% on 2013 to $270bn. But renewable energy is still only 9% of global power, if we exclude hydropower.

There are more signs of potential business-model transition in big energy. RWE will explore a split of its businesses like E.ON if the utility sector's crisis intensifies, it says. The German utility has lost nearly €40bn in market value since 2007.

And on 12th May, Statoil announces it is setting up renewable energy division. Statoil New Energy Solutions will grow wind assets initially, and potentially other renewables including solar thereafter.

As the Eurostar rushes through rural France, I reflect on the problems the French face in terms of energy policy in the months ahead. Nuclear must be high on their list of concerns. At the time they volunteered to host the December 2015 climate summit, they must have been content with the role they could claim for nuclear in the zero net-carbon future. Something approaching eighty percent of national electricity comes from the country's nuclear reactors. But now the industry faces disaster. Most of their reactors are aged. A new generation will be needed. The first such has been under construction at Flamanville since 2007. In April, Areva, the government-owned reactor builder, found what it called "very serious anomalies" in the Flamanville reactor: high carbon content in the pressure vessel steel, meaning reduced mechanical strength, and an unacceptable risk that the pressure vessel will split. These "anomalies", currently the subject of an aghast enquiry, raise doubts that the new generation of reactors will ever be completed.

The reverberations of this crisis are international. The French have warned that nuclear reactors being built in China could also have the same high carbon problem. All plans to export a new generation of reactors are on hold. Other nations' nuclear plans have morphed into likely illusions, not least in the UK, where ground was already being prepared for one of the new-generation reactors, at Hinkley Point in Somerset.

The nuclear industry now faces an existential threat, on a global basis. It is most unlikely to feature as the get-out-of-jail card its advocates hoped it would at the Paris climate summit.

I arrive at the OECD's conference centre to hear the indefatigable Al Gore open proceedings. He is tireless in his life's mission. The several hundred invitation-only executives reveal in the discussion that follows his scene setter that they are encouragingly amenable to his message.

A senior executive from an investment bank that would not have talked this way a year ago tells the room that the capital markets have deep pools of capital available to make the transition Mr. Gore describes. These start with $93 trillion of funds under management. Corporate cash in hand is at an all time high: $1.7 trillion. Meanwhile interest rates are at an unprecedented low. Investors are yield hungry. This favours long term agreements with predictable cash offtakes. Mainstream investors once viewed clean energy as a costly asset class. This is changing quickly. There is a virtuous circle: lower costs, increasing confidence, increasing investment, lower costs yet. This means we can deploy capital across a broader toolkit. Hence the recent advent of so-called Yieldcos, for example. In a Yieldco, you take operating clean assets, pool them, and list them in the equity markets. Investors get cash flow dividends. This is a very powerful mechanism in many ways. Yieldcos are driving down the cost of capital. There are other new tools. Renewables have become an investable asset class.

Well I never, I say to myself, as recently as late 2011, people like this were telling the Financial Times that solar, in particular, had become "uninvestable".

In the coffee break, I chat with an executive from E.ON. I can barely contain my fascination with how things are going in the German utility as it tries to execute the board's decree of a 180 degree change of business model.

We are focussing on energy efficiency, storage, rooftop solar and such things, he tells me. Utilities generally have faced a perfect storm over the last few years, and lessons have been learned, believe you me. But let me tell you, we face huge challenges in how to actually execute the transition.

I can well imagine.

I return to the sessions, and multitask, scanning the news of the day. In Berlin, I discover that Chancellor Merkel and President Hollande have both pledged to decarbonisation this century. In a joint statement they say they have "firmly decided" that the Paris summit must produce a binding agreement.

Fantastic.

In the lunch break, Anthony Hobley comes up to me, his face flushed with excitement. I have been invited to lunch with President Hollande and some cabinet ministers on Thursday, he says.

Now we really are in the big league. Very clearly, the French government sees finance as crucial to success in December.

Elsewhere in the news, Shell vows at its AGM to explore the Arctic notwithstanding the Seattle protests. The company has the legal right to use the port, says CEO Ben van Beurden. As for carbon-bubble campaigners, they "ignore reality". He warns that lack of investment in oil production now could result in a 70 million barrel per day shortfall in crude supply by 2040.

Rome, 27th–29th May 2015

I am in a building that began life nearly 2,000 years ago. Today, The Temple of Hadrian is a modern municipal office built on and around what remains of antiquity. It is a place of wonder.

I have been invited here as part of an international group of 12 experts on climate change by the unlikely combination of Michael Gorbachev, the European Space Agency and the Italiani Foundation. The task of these experts is to craft a summary statement on climate change and world development, in consultation with the Vatican, in the shape of the Pontifical Council for Justice and Peace. Our work over the next two days will build on the outcome of a conference organized by the Pontifical Academy of Sciences, "Protect the Earth: Dignify Humanity" on 28th April, and draw on the action proposals in the report by the High Level Task Force on Climate Change convened by President Gorbachev with the support of Green Cross International: "Action to Face the Urgent Realities of Climate Change."

This is going to be a task worth doing, and I am relishing the days ahead. But sadly the former President of the Soviet Union will not be here with us: he is ill in hospital in Moscow.

I was much looking forward to meeting him. I wanted to tell him what an impact he made on me the day he gave a speech to an audience of scientists and others interested in arms control, in the Kremlin, in 1987. The Cold War was still on then. He was in the full flow of his efforts to thaw it. His main thrust was the dysfunctional strategic nuclear overkill maintained by both sides then (and now), and how he wanted to negotiate it away. But even then, in that speech, he showed that he knew where another existential threat to civilisation lay. He was the first world leader I ever heard talk about global warming. I sat there in the Kremlin agog, the token young scientist on the UK board of Pugwash: next to the Archbishop of Smolensk as I recall. (But all that is another story, from another time.)

Alexander Likhotal, a key advisor of Gorbachev's when he ruled the Soviet Union, now president of Gorbachev's Green Cross organisation, reads a speech on behalf of his boss, by way of welcome. He reads of a clear and present danger of existential proportions. The window for strong action is rapidly closing. Paris is our last chance to escape an agonizingly unsustainable path. If we are to succeed, the world will need true leadership.

I wonder what, or who, Gorbachev means by that. I have had e-mails from people who – hearing I was coming here – were keen to tell me how much they hope Pope Francis will emerge as one of the transformative leaders in the Paris endgame.

The Vatican's track record this year indeed offers encouragement for such hopes. Ban Ki-moon visited the Pope in April to talk climate, and emerged saying he expected the forthcoming Papal encyclical on the subject to be strong, and to lay great emphasis on the moral imperative to cut greenhouse-gas emissions.

The emerging rhetoric from the Catholic Church on climate change has prompted a Conservative American think-tank, The Heartland Institute, to send a delegation to Rome seeking to change Pope Francis's mind. Within days of their visit, a senior Vatican official renewed Rome's attack on fossil fuel overdependence.

A recent poll of more than a thousand Catholics shows 76% of them feel moral obligation to help poor people hit by climate change. And two weeks ago, the Pope was very clear on how he sees the spiritual implications of the intersection between climate and world development.

"We must do what we can so that everyone has something to eat, but we must also remind the powerful of the Earth that God will call them to judgment one day and there it will be revealed if they really tried to provide food for Him in every person and if they did what they could to preserve the environment so that it could produce this food."

I sit in the Piazza della Rotunda, nursing a beer solo at the table on the corner nearest by the Pantheon, surveying its amazing columns. A squadron of euphoric swifts swerves between them, emerging to further carve the air above the square, dodging other squadrons. I scan the Roman evening. Japanese students with orderly smiles and selfie sticks, much used. Italians walking home from work, supper in designer carrier bags. A club-foot cripple on a skateboard who all the locals seem to hold affection for. An African, coal black, offering designer handbags for sale to every passing lady save the nuns.

The *passegiata* builds. The locals and their double kisses. The tourists and their justifiable awe.

Calm. Peace. Humanity.

An acoustic guitarist with an amplifier, at just the right volume, offers *Merry Christmas Mr Lawrence* closely followed by *What a Wonderful World.*

It is.

A wonderful and oh so imperilled world. Mark Doherty of the European Space Agency spent half an hour this morning showing us the very latest full-colour time-series graphics of rising atmospheric carbon dioxide concentrations, rising global temperatures, rising sea level, shrinking Arctic ice, shrinking glaciers, and all the rest of the sorry tale.

It's happening right in front of our eyes. And still there are deniers.

My fellow experts are a remarkable group. Let me take just two of them, to qualify that accolade, and give a feeling for the tenor of our discussions. Ian Dunlop is a former senior international oil, gas and coal industry executive who came to see the light on climate change. He is a past Chair of the Australian Coal Association, but now a thorn in the side of the incumbency. He has tried for the last two years to inject himself onto the board of BHP, arguing very publically that they are in process of losing their shareholders a lot of money by essentially ignoring climate change.

Bill Ritter Jr is a former Governor of Colorado, now Director of the Center for the New Energy Economy at Colorado State University. He is an advisor to President Obama on climate change, having chaired an elite committee reporting on "Presidential and Executive Agency Actions to drive Clean Energy in America".

The central thrust of Ian Dunlop's analysis is that a two degree target for capping global warming is too high. He is a man who understands the feedback processes in the climate system, and the considerable scope for significant natural amplifications of warming (those inappropriately named "positive" feedbacks). He follows the science closely, and makes a compelling case that the world would be on course for complete economic and environmental disaster at two degrees of global warming, up just 1.2°C from the 0.8 we have already unleashed since pre-industrial times. In Ian's view, the whole Paris process is aiming to legitimise – as a best possible outcome – something that is guaranteed not to deliver a secure future for civilisation.

Bill takes a different approach, one rooted in the realpolitik of contemporary American society. For whatever combination of reasons, a significant proportion of the Republican party, plus some Democrats, cannot or will not be persuaded that global warming is worth worrying about. In this context, Bill argues, we are lucky to have a President who has decided to make climate change the backbone of what legacy he can craft from his second term. That requires the course of action that Obama is actually on now: doing everything he can to favour a good outcome in Paris that does not involve going to the US Senate in search of consensus. Hence the White House focus on using executive orders, the Environmental Protection Agency's right to regulate American air quality, bilateral agreement with the Chinese and extensive procurement of low-carbon technology for the federal estate. Bill makes a convincing case that his President is trying as hard as he realistically can, that this is as good as we are going to get from the modern United States. We had better do all we can to support the man and not undermine him.

How to marry these two perspectives? One is seemingly too utopian, in the wider geopolitical context, to hit targets. The other appears too pragmatic, viewed through the prism of climate science, to offer hope of ultimate global success.

My argument in the expert group is that there is a way. It hinges on the potential for profound disruptiveness inherent in the survival technologies. If a clear direction of travel can be set towards transition in the Paris process, the disruptive power of solar, storage and all the rest can be awakened.

Readers who have made it this far in the book will be familiar with my line of argument. Ian Dunlop and those who think as he does must make as convincing case as they can, I contend, and not pull punches like so many scientists do when selling the problem. Bill Ritter and his colleagues must continue supporting Obama so that he has the best possible chance of delivering the most he can while surviving the Republican- and (often related) incumbency blowback. This, sadly, is unlikely to involve talk about a two-degrees target being too little too late.

I am tempted to the view that the news of the day, each day since my day in Paris, sits comfortably with this analysis. On May 20th, President Hollande calls for a "miracle" climate agreement in December. Business will be key: there must be a business "revolution", he says, invoking the spirit of the French revolution. President Obama, meanwhile, recasts climate change as a national security threat in a speech to the Coast Guard Academy. This is the kind of thing he has to do, to breathe life into his search for legacy.

On May 21st, Saudi Arabian oil minister Ali Naimi astonishes Paris Climate Week by saying that the Kingdom built on oil can foresee a fossil fuel phase-out this century. Saudi Arabia could phase out fossil fuels, he says, by "I don't know... 2040, 2050, or thereafter".

2040? OK, that's 25 years from now.

GDF Suez (now rebranded as Engie) also unveils a surprise this day. CEO Gerard Mestrallet, he who I saw tell the World Energy Congress not so long ago that gas can solve all problems and that renewables must be suppressed, now sings a different tune. "The choice we have made is very clear", he says, "we have stopped investing in thermal power generation in Europe and we are investing in renewables." Thermal power investment will only happen in the developing world, Mestrallet now says.

Tony Hayward, Glencore chairman, tries to get in on the green-headline-grabbing act. He calls for an end to subsidies for fossil fuels. He still sees a big role for coal though, come what may, as any chair of Glencore would have to. He professes that solar cannot be expected to replace coal in India. Solar executives clash with him, saying that he is defending the past.

On 22nd, insurance giant Axa announces it will divest from higher-risk coal funds and triple its investment in green technology. The company has become motivated to sell €500m of assets by the risks inherent in climate change, it says.

On 26th, the World Health Organisation targets the 8 million deaths per year caused by indoor and outdoor air pollution, and passes a landmark resolution. The co-operation they now intend, aiming to improve human health, will also improve the the prospects of progress on climate change, by dint of default emissions reductions.

On 27th, activist investors win a historic vote at the Chevron AGM. In a breakthrough for corporate governance activists, 55% vote for large investors to nominate a quarter of directors to the board. People like Ian Dunlop will be polishing up their CVs.

But at the ExxonMobil AGM, though, CEO Rex Tillerson stays true to form by mocking renewables. "We choose not to lose money on purpose," he says.

As for the impacts of climate change: "Mankind has this enormous capacity to deal with adversity".

A second day of deliberations. I multi-task, as I am forced to do so often these days if I am to keep up with the simple march of events.

Norway's $900bn sovereign wealth fund is today told by the government to divest from coal.

Nina Jensen and the WWF team I worked with on coal investment in Oslo celebrate all over Twitter. Carbon Tracker colleagues are quick to talk up the significance of this great victory of the Norwegian environment movement. Norway's sovereign wealth fund could trigger a wave of large fossil fuel divestments, Mark Campanale tells the press.

Ambrose Evans-Pritchard is in fine form in the Telegraph today. He tours the carbon war battlefields in masterful form, and reaches an inescapable headline conclusion: "Fossil industry faces a perfect political and technological storm."

The FT's Lex column chases another key dimension of the drama. In Saudi Arabia as much as one million barrels of oil a day, or more than 15 per cent of oil exports, is going up in smoke for electricity production. This is unsustainable. Solar investments beckon, Lex observes.

The FT's Alphaville column picks stranded assets as a theme, in an article by Izabella Kaminski. "The idea of treating climate change as a financial market risk has gained a lot of traction the last few years in no small part due to the efforts of Anthony Hobley and colleagues at the Carbon Tracker Initiative, who understood the issue had to be framed in the language of finance to make progress. That language is now blunt. Trillions of dollars worth of financial assets could be grossly mispriced due to the incorrect valuation of fossil fuel assets – many of which probably can't ever be burned if the world is to limit global warming to two degrees.

And, it's fair to say, investors, asset managers and even central banks and regulators have begun to take note now the concept of a "carbon bubble" has been popularised."

In the Rome expert group deliberations, we are running late. The agenda has long been abandoned. After the tea break, the chairman finally comes to my set-piece ten minutes on stranded assets and all the rest of the story FT Alphaville covered so succinctly. I have barely begun when a clergyman walks in, nods to some Italians he knows, and takes a seat.

I finish my ten minute summary of Carbon Tracker's work. An Italian colleague, Roberto Savio, then introduces the visiting priest as Monseigneur Matteo Zucchi. The Pope, Roberto explains, is the Bishop of Rome. Monseigneur Zucchi is Deputy Bishop of Rome.

The Pope's deputy welcomes us to the city, notes that the Pontifical Council on Justice and Peace will be most interested in the fruits of our deliberations, and takes his leave.

I e-mail the outrageous timing coincidence to my colleagues at Carbon Tracker.

Make of it what you will, I say.

More deliberations on the final morning. Colleagues are debating the draft. My view is that the chairman, Martin Lees, has done a good job, one I can live with. An official of the Pontifical Council on Justice and Peace, Tebaldo Vinciguerra, is with us this morning to observe the conclusion of the document.

More multitasking. I find to my astonishment that the Pope is on Twitter. @Pontifex is his address. One of his tweets catches my eye.

"Better to have a Church that is wounded but out in the streets than a Church that is sick because it is closed in on itself."

I love the honesty there. I retweet it.

President Obama has also recently joined Twitter, as @POTUS. I risk a message aimed at them both.

Someone has posted a comment about *The Winning of The Carbon War*. I repost it with a thought: "@POTUS and @Pontifex on the same side on this one. And I have just learned that a third of the House are Catholics."

Chapter 21

Groundswell dot world

Bonn, 4th–8th June 2015

4th June: The climate negotiations resume. 190 governments sit down in the former German Bundestag for ten working days to slim down ninety pages of draft treaty, full of undecided text in square brackets. The eventual Paris Protocol needs to be only around 20 pages long, with no further need for square brackets. After this session, there are only ten more days of negotiations scheduled before the Paris summit.

As I walk through the corridor that links to the negotiating chamber, I see that the "non-state actors" – cities, companies, regional governments, civil society groups of all kinds – are here in force. Their efforts to encourage the state actors to make progress are clear for all to see. They have been joining forces in what they and the UN call a "groundswell" for action on climate change. Over 20,000 of them have signed up to date, and more are joining every day. An avenue of ten-foot high banners proclaims the commitments they have made. The pledges are diverse, and some of them are huge. The Compact of Mayors, just one city network initiative, commits to eliminate 2.8 gigatonnes of carbon dioxide from the atmosphere by 2020, more than the annual emissions of India. The New York Declaration of Forests, an initiative bringing together governments and private companies, could eliminate 8 gigatonnes by 2020, equivalent to the annual emissions of the United States.

Pictures of the impressive avenue of banners are making their way across the world on Twitter. One tweet, appending a photo of one of the slogans, makes me smile. "Birds do it, bees do it, even cities in South Korea do it."

This rising tide of desire for climate action, in multiple constituencies and multiple countries, must be wracking up pressure on the energy incumbency. A Reuters study shows that just 32 energy giants account for almost a third of greenhouse-gas emissions today. In order of size, the top seven are

Gazprom, Coal India, Glencore, Petrochina, Rosneft, Shell, and ExxonMobil. Increasingly, in the zeitgeist, these companies are discussed more as though they are social pariahs than the responsible fuellers of economic growth they might prefer to be seen as.

Six oil and gas groups – BP, Shell, Total, Eni, Statoil and BG – are now seeking direct climate negotiations with governments. They want to co-operate, they say. They too want climate action. Specifically, they want a carbon tax.

ExxonMobil says it will not be joining this group because it refuses to "fake it" on climate. The implication about the motivation of its peers is clear.

Former Shell chairman Mark Moody-Stuart throws an immediate spoke in this new oil-industry wheel by giving an interview in which he laments "remarkably little progress" by the oil industry on climate change in the last two decades, and as for the companies saying they want a carbon tax: "Being blatantly honest, it is not new. They have been saying it for 15 years. The real question is now what are we going to do about it."

It is clear in the letter that the companies sent to the UN what they want to do: essentially, to transform themselves into gas companies and bolt carbon-capture-and-storage onto the use of gas for power generation.

A new study has cast immediate doubt on the wisdom of their strategy. Research by the Global Commission on the Economy and Climate suggests that methane leakage from gas infrastructure could easily wipe out any benefit gas has over coal in terms of trapping heat in the atmosphere. If policymakers want to use gas as a "bridge", they will need to add what the Commission calls "guardrails": measures to manage and reduce methane leakage, to limit energy demand growth, to direct added gas supplies to the applications that yield the greatest substitution benefit (displacement of coal in the power sector), and to restrict the extent to which lower-carbon technology is locked out as capital flows to gas.

My first meeting is with UN officials. As we gather, I see on a TV screen outside the negotiation chamber that the head of the Australian delegation is speaking in plenary. He intervenes with an air of defiance, trying to pretend that Australia's carbon tax is a major commitment to emissions control. It is a naked bit of political posturing by a man everyone knows is in Bonn primarily to defend the ailing Australian coal industry. My companions shake their heads sadly.

The US, Brazil and China voice doubts over Australia's climate plans. Brazil suggests that Australian emission reduction estimates for 2020 are simply not feasible.

The big coal miners are also in defiant mood today. Glencore, BHP, RioTinto, Anglo American – controlling more than a third of world coal trade – insist that the divestment movement is making no difference to them. Glencore's head of coal sees rising coal demand for decades to come. "Building coal-fired power is still the cheapest way of powering people out of poverty," Peter Freyburg insists.

I hope that growing numbers of people listening to this new mantra of the coal industry see it for the cruel falsehood that it is. SolarAid's million-plus solar light sales, for one, so obviously free up household cash. How could giant coal plants do so in the vast areas areas of Africa where there are no grids, even on what shaky grids there are. How could they ever hope to compete with solar economics?

Carbon Tracker launches its latest report today, and coal is in our gun-sights. "Coal: caught in the EU utility death spiral," we have elected to call this one. The study shows how and why E.ON, RWE, GDF Suez, EDF & Enel – the five biggest European utilities, collectively responsible for 60% of European electricity generation – lost €100 bn, or 37% of their stock market value, between 2008 -2013. In brief, they planned for growth in demand, but demand fell even as GDP grew: this has stranded coal assets, and together with the rise of renewables and increases in the price of carbon, will continue so to do. The report is greeted with the positive press coverage we have become used to. "Investors are taking fright", the Economist observes in a typical article.

Today's unfolding drama mostly involves coal, but in the oil markets matters move apace. The Saudis are still pumping at full speed, trying to force as much of the US oil industry out of the market as they can. Opinions differ as to how this assault is faring. "OPEC is winning the oil war," Bloomberg concludes, though members are still suffering their own casualties. Few are breaking even, and Saudi Arabia continues to burn its currency reserves. The International Energy Agency suggests that the battle has only just started: "it would be premature to say that OPEC has won".

Whatever the state of the market-share fight, warnings over oil debt are growing louder. "Easy Access to Money Keeps U.S. Oil Pumping", a Wall Street Journal headline says of the shale drilling. "Despite poor economics, companies continue to drill because capital flows to them in a zero-interest rate world." The Financial Times points out that the oil industry has about $2.5 trillion of debt outstanding, much of which is considered junk-status by the credit rating agencies.

Petrobras leads the catalogue of oil companies lining up to waste capital. The scandal-mired Brazilian giant returns to the capital markets with a 100 year bond. Yes, that is no misprint: a century-long bond. Having been shunned by investors for a year, Petrobras now sells $2.5bn for ten long decades at 8.45% yield.

Who can believe that the world will be using oil in a hundred years time? Even fifty or less? This one will be a real test of investors' short-termist risk-blindness. James Saft captures the enormity of this spectacle in a Bloomberg column. "Sometimes", he writes, as with a train wreck, "the best thing to do is to stand back and bear witness."

Not all the oil giants try to barrel ahead on all frontiers, of course. Chevron has been forced to delay a flagship oilfield project in the Gulf of Mexico indefinitely. In the latest demonstration of the difficulties drilling in deep water, the loss of a floating platform's connectors has stopped exploration in its tracks. One wonders how Petrobras and its bond investors will process that little setback when contemplating how to "invest" $200 billion plus in the sub-salt.

In a meeting room in the old German Bundestag, my UN companions explain their aspirations for this session of talks to me. Without their ministers to hand, negotiators will struggle to edit the text of the draft treaty down - there are too many big issues yet to resolve. Editing the text down from 90 to 60 pages would be a good result, they say. Their hope is twofold: first, that the closer negotiators get to Paris, the more they will accelerate, and second, that the groundswell for climate action will continue to rise, pumping up the pressure on negotiators to deliver. I can help with this today, they say. They have invited me to give a ten minute keynote speech to kick off a panel convened by We Mean Business, a global coalition of companies determined to see success in Paris, and committed to unilateral actions of their own as a way to try and help achieve it. It would be helpful, they tell me, if my speech could be appropriately uplifting.

I won't let you down on that one, I promise.

I give my speech, and hand over to the panel.

Steve Howard, IKEA's Chief Sustainability Officer, sits on it, visibly raring to speak. IKEA is making a new commitment today, he announces. We will invest €1bn on climate change measures over the next 5 years. €500m will be allocated to wind, €100m to solar and €400m to adaptation, via the IKEA foundation. This comes on top of the €1.5bn that IKEA has already invested in renewables since 2009.

This I find amazing. One billion euros, from just one company! IKEA's turnover last year was €28.7 billion. They must be committing a huge percentage of their anticipated profit for each of the next five years.

Steve spells out the giddy implications. If every business and organisation did what we did, he enthuses, we would flip electricity generation into being renewable-based by 2020 or shortly thereafter.

He goes on to justify IKEA's move in strict business terms. His rhetoric brings happy smiles to many faces in his audience of negotiators, non-state actors, and journalists.

I don't want to hear about burden sharing ever again, he says. That's what camels do. This is all about opportunity. When it comes to targets, we want a one-hundred-percent mindset. We don't have time for anything else. I'm against partial targets. We don't want to be half as bad in 2020 as we are today. It's more fun to commit completely. A hundred companies going 100% renewable would be the same as taking out the emissions of a country the size of Australia.

5th June: Thanks in no small part part to Australia and fellow foot draggers, the Bonn talks make very slow progress. After four days, negotiators have trimmed just five of the 90-page starting draft.

As for collective targets, so far less than 40 governments have submitted their individual pledges to cut emissions. They are not on the same page as IKEA, unfortunately. Analysts are already saying that collectively they will fall far short of preventing global temperatures from rising over two degrees Celsius by 2100. This makes the rise of the groundswell all the more important, in the holistic picture. To achieve a direction of travel significant enough in Paris to keep the winning of the carbon war on course, it is clear that there must be a constant ratcheting up of individual non-state commitments, adding to the pressure on governments, encouraging them to face up to their denier constituencies at home.

This evening I am given an insight into another aspect of the process at work behind the scenes. I am far behind with my e-mails, and repair to a pleasant restaurant on a terrace by the Rhine, there to catch up for a few hours with some excellent German food and wine, and the occasional gaze at my view of the great waterway. It is not to be. A table has been laid for forty across the terrace, and after half an hour the party begins to arrive. They are almost all young activists, clearly fresh from the climate talks. I don't recognise any of them. Then I see a man my own age, who I do recognise from the years of my regular attendance at the climate negotiations in the 1990s. He is Jacob Scherr of the Natural Resources Defense Council.

Jacob and I greet each other as veteran comrades do. He insists I join the party, and I need little persuasion. We spend a pleasant few hours catching up. He introduces me to a host of youngsters keen to hear stories of the early years of the carbon war. I need little persuasion to tell them. My tales go back to 1990. Most of these activists have no recollection or knowledge of those years: the roots of the global problem they are now so passionate about fixing.

Some of them hadn't even been born.

6th June: A side event, organised by the group I was at dinner with last night. Christiana Figueres is the star attraction, speaking on the power of climate action. The French and Peruvian governments have formed what they call a Lima-Paris Action Agenda, in collaboration with the UN, to catalyse climate action at all levels of society. In parallel, the Peruvian government and the UN's climate secretariat has formed an online aggregator that tracks and showcases action by cities, businesses, and non-government actors. They called it the NAZCA Platform: The "Non-State Actor Zone for Climate Action." All this, Christiana explains, is part of the groundswell for climate action.

Five years ago you could pick up the sand of climate action with a few hands, she enthuses. Now you can't, because it is a beach. We are moving into an age of climate action where action on renewable energy, energy efficiency, land use and all the rest are being amalgamated globally the way information was at the dawn of the age of the internet. We are witnessing an exponentially growing phenomenon. If anything, the regulatory system is trying to catch up with where the action is. Your organisation shouldn't be called groundswell dot org but groundswell dot world.

She builds to an upbeat rhetorical ending typical of her leadership of these negotiations, but one that goes further than any I have heard before. All of us here today will witness global peaking of greenhouse-gas emissions, she emphasises. We *will* witness that. That will be a fantastic event in human history. And some of you will witness climate neutrality.

By this she means the point where greenhouse-gas emissions have fallen enough to stabilise atmospheric concentrations, so offering hope of a cap to global warming at levels short of global catastrophe.

At dinner this evening I hear just how far there is to go with that vision. Dr Bill Hare, an old colleague at Greenpeace, is the single most talented technical expert I have met in all my quarter century of climate campaigning. This is because he is both an accomplished atmospheric physicist and is also expert enough in economics to engage the world authorities of that strange discipline, and earn their respect. These days he runs an elite consultancy called Climate Analytics. The name speaks for itself: a mission to keep governments informed and honest on the outcomes of their policies in terms of the science and economics of climate change. Bill and his colleagues have just released a report showing that the Paris climate pledges to date, by 36 countries, will only delay dangerous warming by two years. Two degrees would be locked in by 2038, rather than 2036.

We repair to the party for non-government organisations attending the talks. I went to many of these in the 1990s. Little has changed. Here one can see the full league of nations represented at these talks, at play. Most of the governments negotiating are shadowed by non-government organisation representatives, who attend the negotiations to put what pressure they can on their governments to deliver. The music is appropriately international. There is enough salsa and merengue mixed among the hip-hip, techno and who-knows-what-else-it-is to keep me happy. I watch all the hair flicks and shoulder shimmies for a while, mojito in hand, wondering how a 60 year old might be able to get a dance in this sea of youth without proving the adage that there is no fool like an old fool. I chat to a not-so-junior Kenyan lady for a while, and she turns out to be as keen as me for a dance.

So once again, as at so many NGO parties over the years, it is climate-apocalypse forgotten, at least for the duration of a few cross-body turns.

8th June: Home early, due to other responsibilities. A chance, while travelling, to catch up on the ebb and flow of fortune in the carbon war. At Dusseldorf airport, I discover that today is quite a day in that regard.

"G7 in historic accord to phase out fossil fuel emissions this century." So reads the Financial Times front-page headline. The G-7 leaders have been meeting in a German Alpine village and this unexpected first is part of their communique. Even more impressively, they back emissions cuts of 40 to 70% by 2050. Chancellor Merkel is telling the press "40% is not enough".

The fossil fuel endgame is officially upon us.

This is clearly a diplomatic coup for both Mrs Merkel and President Obama. Interviewees are pleasantly surprised she majored on the issue, and enfolded Canada and Japan – climate laggards – into the compact.

I open an e-mail from Paul Bodnar, an old colleague of Anthony Hobley's who has told me he was persuaded to work on climate policy by reading my book *The Carbon War*. Paul is now President Obama's advisor on climate and energy. The subject of his e-mail is a single word: "Result". There is no message, just an attachment. I open it and see a picture of him holding up the front page of the Financial Times.

The G7 leaders are in tune with the zeitgeist, it seems. A global climate survey released today shows that nearly two thirds of a poll of 10,000 people from 79 countries want governments to do "whatever it takes" for a two-degrees deal in Paris.

There is other encouraging news. Chinese greenhouse gas emissions may peak by 2025 on current trends, an LSE study shows: five years earlier than the current target. If this kind of progress spreads around the world, warming of more than two degrees can still be avoided, the researchers conclude.

The first US state goes 100% renewables in electricity: Hawaii, by 2045. The Governor signs a bill with a 100% renewables portfolio standard.

The proportion of the population viewing fossil fuels as risky investments is soaring, at least in the UK. A poll shows that whereas in 2014 21% in thought fossil fuel investment risky, a year on 66% do. Among 18-34 year olds, the figure is 80%.

80%! Such figures must be very daunting for incumbency marketeers. Unless, that is, they are advocates of transition.

A new YouGov poll looks at the deep climate-deniers around the world: the percentage of the population wanting no climate deal at all. The US tops a league table of 15 countries, with 17%. The UK comes fifth, with 7%. In China, Indonesia and Malaysia the figure is 1%.

How to explain this disparity? Clearly the malign misinformation power of media organisations like the Daily Mail and organs of the Murdoch empire such as Fox News have a lot to do with the depth of minority denial in Britain and America. But for how long will their systematic infusions of poison be able to hold back the rising tide of groundswell dot world? Even their peers in the press are ganging up on them. Global news organisations formed an unprecedented alliance on climate in May. Twenty-five publishers including Le Monde, China Daily, The Guardian, and El País, have made a collective commitment to raising awareness of climate change in the run up to Paris.

11th June: The Bonn talks end with negotiators handing a mandate to their co-chairs to go away and shorten the text. Christiana Figueres explains the state of play at the UN's closing press conference. What is being managed here, she says, is no longer resistance to an agreement but complexity, and enthusiasm. Many negotiators seem to agree with her. There is common talk of a new atmosphere of trust.

But equally, the draft text still stands at 85 pages.

The French meanwhile tee up a backstop of their own. France is ready to step in if the UN climate talks stall, says their climate envoy, Laurence Tubiana. A new text will be produced if nations fail to edit down the current over-long draft. The French also flag the main financial sticking-issue loud and clear. The West must pay the $100 billion a year promised to poor nations by 2020 if they are to secure a deal at the Paris climate change summit, French Foreign Minister Laurent Fabius warns. Poor nations are waiting to see if the climate deal meets expectations, says Ségolène Royal, environment minister. We have to meet their expectations.

Christiana Figueres ups the ante on targets in an interview. The ultimate objective has to be a ceiling on global warming of lower than two degrees, she says. It has to be in the range 1.5 to 1.9, if we are to have reasonable confidence of avoiding climate catastrophe. In this she is drawing on the fears of very many climate scientists.

Eighty British companies call on Prime Minister Cameron for a strong Paris deal, and policies consistent with it at home and abroad. They include two of the Big Six utilities – E.ON and SSE – and huge retailers including Tesco and John Lewis. Truly the groundswell is growing to encompass impressive actors, I reflect.

BP feels moved to concede ground. Company executives talk of a fossil fuel watershed as they launch their annual Statistical Review of World Energy. "In years to come it is possible that 2014 may come to be seen as something of a watershed for the energy industry," says Spencer Dale, chief economist. CEO Bob Dudley joins in. "Something substantial needs to be done. We are conscious of that. We encourage policymakers to move forward on this when they meet in December."

New Orleans, 15th June 2015

The last time I talked to such a huge audience I was at an oil-industry conference. But this is no oil and gas junket: it is a solar junket. There are over a thousand presenters alone here at the annual PV Specialists Conference, from 47 countries, and goodness knows how many more attendees. In the vast ballroom facing me in the opening plenary, most seem to have turned up. I am scanning a sea of faces that is a sign of our times.

These are the people who really know how a solar cell works. They come from university research labs, and research-and-development units in companies, governments, and the military. It is their pioneering work that has allowed the solar manufacturers of the world to bring down the cost of solar via mass manufacturing. The organisers have charged me with offering them a view of the wider context of their work.

I start with the case that energy-incumbency industries face grave threats to their business models, some potentially existential. Your work, and the work of others like you around the world, have played a major role in this state of affairs, I suggest. This is not something you should feel guilty about.

I mention the evidence of the day. "It is time the world kicked its coal habit," reads a lead column in the Financial Times. In Australia, the ABC Four Corners Program tonight is entitled "The End of Coal." The pre-publicity asks: "With the price of coal plummeting and our biggest customers turning to renewable energy, is Australia backing a loser?"

It is like this many a day, these days, I say – for those with the time to check the news.

I run through the rest of the case, and conclude. We are on track, as things stand, for the right signal to be sent from the Paris climate summit. For the incumbency, problems will grow and opportunities will diminish. For the insurgency, opportunities will grow and problems will diminish. You all, and your peers not here today, played a major role in bringing this about. I salute you, and I thank you.

CHAPTER 22

The Churchill of climate change?

Kefalonia, 18th & 19th June 2015

"Love miles", we guilt-stricken climate-campaigner air travellers call them: miles travelled by air to be with loved ones. Getting on for 2,000 of those take me today to an island in Greece, where I and my wife join two dear friends from Australia, for our annual reunion. We think of it as meeting half way. This year we feel less guilty than most: one of our friends has been very ill.

By coincidence, the Greek flag is on the front page of newspapers today. The Guardian's headline reads "Greece: Can't pay, won't pay." This is a crunch day for the Greek economy and the Greek people. The prime minister is preparing a last-ditch offer to the International Monetary Fund and the European Central Bank to avoid default on debts. The stakes could not be higher. Greece's membership of both Eurozone and European Union is in question. The whole European project comes under threat if Greece exits both. As anxious savers withdraw deposits, economists are warning that Greek banks could collapse. That would involve risk of contagion in their global capital markets.

I know where my sympathies lie. The austerity that the IMF, ECB and EU are trying to impose on the Greek economy, as a condition of further lending to the government – money which only goes to banks outside the country – is a cruel delusion, I believe. It plunges Greeks further into poverty with no end in sight. I am with Nobel-prizewinning American economist Paul Krugman on this: austerity won't work, for any economy. The case for cuts was all wrong, he argues: this you can see clearly today in their failure to stimulate economies, including Greece's. Why do the powers that be still believe in it?

The whole tragedy is a stark reminder for me that my cautiously growing optimism about the prospects for the ultimate winning of the carbon war is heavily caveated by the foibles of modern capitalism. The global economy still dances to the tune of a deeply flawed operating manual, as it is currently drafted: one capable of crashing the system at very short notice. And we are not spoilt for choice as to potential triggers: just ask the Bank of England about their multiple fears on that front. If a crash erupts in the markets, and we enter a global depression deep enough to threaten social cohesion – potentially in the process handing power in multiple countries to ruthless libertarian rightist governments or worse – then all bets are off for the winning of the carbon war.

A good day, therefore, to reflect on the ethics of both climate change and power in society. Pope Francis publishes his encyclical in Rome.

This much-heralded and already-leaked missive, a book in its own right, talks not just about climate change but the wider global inequality and abuse of power that allows carbon emissions to keep growing, notwithstanding the threat they pose to creation. It tells rich nations to pay back their debt to the poor. It casts blame for the entire global ecological crisis on the indifference of the powerful, and the extreme consumerism on the part of some.

The message of the encyclical is fast echoing around the world in media coverage, both mainstream and digital. There is every chance, commentators are saying, that many people will come to view climate change very differently under this Pope's guidance.

The message on the energy transition could not be clearer. "There is an urgent need to develop policies so that, in the next few years, the emission of carbon dioxide and other highly polluting gases can be drastically reduced, for example, substituting for fossil fuels and developing sources of renewable energy." That reduction should go all the way to phase-out. "We know that technology based on the use of highly polluting fossil fuels – especially coal, but also oil and, to a lesser degree, gas – needs to be progressively replaced without delay."

The encyclical has a name, "Laudato Sii" (Praised Be), after a line in St. Francis of Assisi's prayer of praise, "Canticle of Creatures." Another line of that prayer reads: "Praised be you, my Lord, with all your creatures, especially Sir Brother Sun, who is the day, and through whom you give us light."

In the encyclical itself, the Pope is more prosaic about Sir Brother Sun's role. "Taking advantage of abundant solar energy will require the establishment of mechanisms and subsidies", the wording goes. "The costs of this would be low, compared to the risks of climate change."

In the preparation of the document, the Vatican has taken great care to consult many the world's foremost experts on climate change. The Pontifical Academy of Sciences drafted in top climate scientists to advise. I can see their footprint in parts of the encyclical. On the scientific underpinnings, and the treatment of uncertainty, for example: "If objective information suggests that serious and irreversible damage may result, a project should be halted or modified, even in the absence of indisputable proof."

American conservatives have launched pre-emptive attacks on the Pope in recent days. They fear mass conversion of currently climate-denier Catholics who have been core to political conservatism. Jeb Bush has joined this backlash: the second of five Catholic presidential contenders so to do.

But the leadership of America's nearly 80 million Catholics is diverging from the incautious early responses of Catholic Republican politicians. Joseph Kurtz, president of the US Conference of Catholic Bishops and the Archbishop of Louisville, is clear on how he views the encyclical. "It is our marching orders for advocacy. It really brings about a new urgency for us." He talks of a vast and highly organised process of outreach across the United States.

Across the political divide from Bush and other Republican defenders of fossil fuels, another US Presidential election candidate backs 100% clean energy, citing the encyclical. Martin O'Malley, a Catholic Democrat, has served as the governor of Maryland. He sets a target of decarbonisation by 2050. An outsider in the presidential race he may be, but it now seems clear that this issue will be a defining one in the forthcoming primaries, and subsequent national election. And you wouldn't want to bet on the diehards now, I tell myself. One in four of the population that is Catholic. I sense the scope for U-turns by the Republicans. Even grovelling appeals for forgiveness.

The resonance of the encyclical extends far outside the United States. Other religions are beginning to echo the Pope's plea. Scholars from Hinduism, Islam, Judaism and leading evangelicals are offering comment to the media today on why the environment and the climate matters to their religion.

I am not alone in my rising sense excitement. "The Pope is the climate change Churchill humanity desperately needs", veteran climate expert and campaigner Joe Romm writes. "He has just elevated (climate change) to its rightful place as the transcendent moral issue of our time."

Damian Carrington, another influential commentator, asks: "Will Pope Francis's encyclical become his 'miracle' that saved the planet?"

London, 23rd June 2015

The aspiring drillers of British shale, and their support base, convene for a morning conference in a hall in the City. Outside the building, a small group of protestors gathers, led by the fashion designer Vivienne Westwood. I stop to talk to them, wearing an unfashionable suit, before going in.

Inside the conference, the geology of British shale deposits is first item on the agenda. I sit listening to Professor Mike Stephenson, Director of Science at the British Geological Survey, talking about the Upper Palaeozoic shale of northern England. I once wrote a research paper entitled "British Lower Palaeozoic black shales." Shell and BP both funded part of this work. I google the title of my paper. It is still there, downloadable today from the web pages of the Geological Society, whose journal it appeared in back in 1980, eleven years before the internet became available for use.

Stephenson tells the conference that the Bowland Shale, a deposit of Carboniferous age stretching across much of the north of England, is by far the biggest UK shale resource. His team has a median estimate of fully 1,300 trillion cubic feet of gas for the resource. Of course, resources are not reserves, he reminds us. For calculating reserves, you need to know how much is recoverable. We don't really know that, he admits. We don't know the gas content of the shale because we haven't drilled much: nowhere near as much as the Americans. How much resource can be turned into reserves is therefore impossible to say. As for the Weald, in southern England, we know there is no shale gas there. Shale oil perhaps, maybe 4 billion barrels. But there is no gas.

This hardly seems like the resounding basis for a gold rush to me. But then I have a different belief system to most of the people in this hall.

Tony Grayling of the Environment Agency – the regulator of the aspiring industry and everything else with environmental implications – speaks next. It is not our role to decide if shale gas is extracted, he begins, that is for society to decide. Our role is to weigh the environmental risks. We feel the risk of water pollution from fractures in fracking is low at the depths targeted by the drillers. Key risks are at the surface and in borehole construction and operation. In that regard, we won't allow drilling in protected groundwater source areas. The drillers must publish the chemicals they use. Waste liquids produced from fracking must be stored on site in sealed containers and taken to wastewater treatment sites by lorry: drillers won't be allowed to store them in open pools and pump them back underground like they are in America. We will not allow venting of gas at the surface, like the Americans do. We will require "green completion" at production stage – in other words, the use of gas waste.

And so he goes on, telling the drillers in the room how much additional cost the British government's regulator is going to load onto their operations should their fracking ever produce gas or oil.

Then a strange thing happens. The lights go out. It is a power cut. Sitting in the back row, I laugh out loud at the irony, in the deep gloom.

The conference goes on in the dark.

Nigel Mills, Member of Parliament, Chair of the All-Party Parliamentary Group for Unconventional Oil and Gas, gives a speech on energy security. He is a strong supporter of fracking. Gas will have to form part of our electricity supply for a very long time, he asserts. The Chancellor wants shale gas, and will provide local incentives. But, he adds, "any more cock ups and there will be a big problem." We can't have any of those horrific accidents the US has had. We must show compliance or political backing will evaporate.

I sit in the dark, marvelling how he doesn't mention the economics of shale drilling. Neither did Mike Stephenson or Tony Grayling.

Four days ago Bloomberg ran another article on fears that the US shale train may soon crash. "The shale industry could be swallowed by its own debt", the headline read. "The debt that fuelled the U.S. shale boom now threatens to be its undoing." That debt pile stood at fully $235 billion at the end of March. Incumbency hype focusses on how US oil production has been pushed to the highest level in more than 30 years. It ignores the cost of achieving this. The drillers have consistently spent money faster than they have made it. This was true even when oil was $100 a barrel. The 62 companies Bloomberg follows in its shale index have spent $4.15 for every dollar earned selling oil and gas in the first quarter, up from $2.25 a year earlier.

The cost-cutting by distressed drillers is creating severe hardship in the shale-drilling regions. Williston, a shale hub in North Dakota, has clocked up over $300 million in debts of its own. Like many a municipality in the shale regions, it faces a crime wave. Nobody mentions this kind of thing in the London meeting. Nor the growing frustrations for shale drillers outside the USA. Conoco has become the last global oil firm to quit Polish shale gas. Seven wells were drilled, $220m invested, no gas found.

The presentations continue. A representative of the water industry opines that if shale drilling is going to happen, his industry can help it go ahead safely. There are risks, but we are confident we can meet the challenges, he says.

A lawyer, a QC no less, suggests there are three reasons why there has been no UK shale boom like the American one. First, ownership by the Crown of subterranean rights, hence application are needed for licences to drill: a serious impediment to exploration. Second, no right to trespass without landowner's consent.

At this controversial point – is he advocating a right to trespass? – the lights come back on.

Third, multiple agencies have to be consulted or give consent: fourteen of them. This means planning applications 800 pages long. They take a week to read, and a week to critique. The problem in the US was an absence of regulation. The problem in the UK is too much regulation.

Our system needs changing, booms the QC. If we don't we will miss out on the economic success in the US.

Ah-ha, a mention of economics!

Now we hear from the Right Honourable the Lord Smith of Finsbury, Chairman of the Task Force on Shale Gas. This is a body funded by six oil and gas companies to report to the public. We are completely independent, says Lord Smith, with a straight face. A lot of the early concern about shale drilling comes from the early days in the US. There are risks with shale. Those risks can and must be mitigated. Well integrity has to be properly ensured. Proper baseline monitoring is needed.

Ken Cronin, chief executive of UKOOG (UK Onshore Oil and Gas – trade body for the industry) keeps the drumbeat of support beating. Like everyone else so far he doesn't mention economics.

Ed Heartney, Councillor at the Embassy of the United States does the same. Heartney injects a nakedly tribal element to the praise book. If you think favourably of the USA, he observes, you tend to be in favour of exploiting shale drilling. If you don't think favourably, you are likely to oppose shale.

He has a few lessons to share. We didn't do baseline studies before we started, and it has been a real problem. The UK should. You should also put sheds over drill sites to cut down noise, and reduce the size of well pads.

John Blaymires, Chief Operating Officer of iGas Energy, says that the UK needs to have an evidence-based discussion. He goes on to offer his version of evidence. I hate the word unconventional, he says. There is nothing unconventional in what we do. I'm not keen on the word fracking either. The most important thing is that we construct wells with integrity. In the UK onshore today we are producing 20,000 to 25,000 barrels of oil equivalent, operating smoothly. Time is an issue for us. From beginning planning, we can expect to start fracking in month 16-17. The first full scale production cannot be before 2020. Our costs are going to be higher than they are in the US, we're not going to be able to get them down for quite a time.

Finally, in the last session of the morning, I get a turn. I am only one of two open critics invited to speak.

I am suffering from cognitive dissonance, I begin. For nobody has given economics more than a passing mention this morning. The American drillers have clocked up nearly a quarter of a trillion dollars in debt, most of it rated junk. Speculators have started shorting these companies, just as they did debt-mired mortgage companies in the run up to the credit crunch in 2007. The Bank of England is worried about a threat to the global capital markets. Is this what we want to import into the UK?

Even if you get drilling underway, you risk stranding assets. The industrialised nations have a compact involving cuts in greenhouse-gas emissions of up to 70% 35 years from now. You would face the plunging costs of renewables and storage. You would need to load the unknown but undoubtedly high cost of carbon capture and storage on to the holistic economics. You would face the high cost of compliance with British regulation.

Then there is the politics. Approaching half the population say they haven't made their minds up about fracking yet. How is the constant drip of negative stories coming out of the US shale likely to impact them? Nobody has mentioned Denton, where the Republican population of the first Texan town to be fracked voted to stop the drilling. If that can happen, how can you hope to foist fracking on the conservative rural population of England?

There is one more speaker, after me: John Beswick, Director of Marriott Drilling Group. The oil and gas industry can win back the hearts and minds of people, he suggests. We have lost their support because of a few people spreading fear.

Bupa, a British private healthcare company, wants to do something about climate change. Their chief executive, Stuart Fletcher, convenes a dinner party discussion. Hugh Montgomery, Professor of Intensive Care Medicine at University College, is invited. So am I. Hugh is a member of the Lancet Commission, an eminent committee of medical practitioners studying the medical impacts of climate change.

Five years ago, the first Lancet Commission report called climate change "the biggest global health threat of the 21st century". The second report of the

Commission, published today, raises the stakes. Current greenhouse-gas emissions-projections, and consequent climate change, pose an unacceptably high and potentially catastrophic risk to human health, the Commission reports.

A report for the IMF, published earlier in the week, allows such warnings to be translated into estimates for economic losses. If we count both the direct cost of subsidies to fossil fuels, and the cost of damage created by burning them, the total impact is $5.3 trillion per year, 6.5% of global GDP, a sum exceeding the total annual health spending of all governments. Just over half involves money governments are forced to spend treating the victims of air pollution, plus income lost because of ill health and premature deaths. Ending the subsidies would also halve premature deaths from outdoor air pollution, saving some 1.6 million lives a year. The authors emphasise that all their estimates are conservative.

How does the Lancet Commission propose to approach this enormous collage of loss? Hugh Montgomery explains at the Bupa dinner. Climate change is a medical emergency, he says, and there are five main solutions to it.

First, we need to think of it in terms of opportunity. Tackling climate change could be the greatest global health opportunity of the 21st Century. Apart from the human misery escaped if we cut some of those lost trillions of dollars, imagine the net benefits if we invest the deferred losses in social good!

Second, achieving a decarbonized global economy is no longer primarily a technical, economic or financial question, it is political. As Al Gore loves to say, our capacity to survive climate change is all down to political will. And political will is a renewable resource.

Third, we need to proceed as though global health equity, sustainable development and the international policy response to climate change are inseparable. We cannot hope to cut emissions deeply enough on a global basis if the sum of our national greenhouse policies leaves billions with no prospect of poverty alleviation, and the poor health that attends poverty.

Fourth, we must make the most of the vital role the health community can play in tackling climate change. People have a justified tendency to respect their health professionals, as the response to the recent Ebola crisis showed. If the medics speak out, and act, positive change on the scale needed to defeat the climate threat becomes more likely.

Fifth, framing climate change as a health issue will help counter opposition from vested interests, accelerating progress towards meaningful action. Recently, for example, the British Medical Association joined the fast-growing divestment movement. Imagine you work for a fossil fuel company, and you read in the papers that doctors now consider that you work for an industry on an ethical par with the tobacco industry. Is that likely to make you more or less likely to argue within your company for a change of business model to one that embraces energy transition?

Lisbon, 26th & 27th June 2015

Lord John Browne reclines in a sofa in the bar of a five star hotel, enjoying a bottle of fine red with his chef de cabinet. I see a vacant armchair next to him, and wander over. He greets me cordially, and we are soon deep in conversation. This is the kind of thing that happens at The Performance Theatre, an annual two-day retreat of business leaders worried about the state of society in general, and climate change in particular.

John Browne was CEO of BP between 1995 and 2007. He led a company with which I have spent much of my professional life in a state of polite but acute confrontation. He remains a BP loyalist, but with caveats. At this gathering, he describes BP and the other major oil companies as having dysfunctional cultures wherein, to use one memorable phrase of his, failure is routinely dressed up as just another form of success. Why do they have such cultures? They are creatures of an Establishment that itself has got much wrong about the world. BP, in another memorable phrase, was born as a standard bearer for the British Empire, and an arm of the Foreign Office if not the Secret Service.

Disarmingly, he takes his own share of blame for the dysfunctional culture. I myself, he admits, made my own crop of mistakes along the way.

But his reaction to climate change, early in his tenure as Chief Executive, was not one of these mistakes, according to people like me. Quite the reverse.

We talk about those years, and the importance of the turnaround he led within BP on climate change. I explain that in my view the Kyoto Protocol could not have been negotiated without BP's U-turn. The company broke ranks with the rest of the industry in 1996, admitted there was a problem, and said they intended to do something about it by accelerating clean-energy investments. That split in the corporate world, I believe, gave Al Gore and the progressives of the political world the air cover to deliver a treaty at the Kyoto climate summit. So I argue in my first book, *The Carbon War*.

Why, I'm so pleased you think that, Jeremy, John Browne says.

Yes, I say. But.

I emphasise the "but" and we both laugh.

You have built the biggest private-equity fund investing in renewables since leaving BP – another historic first – but now you are running a multi-billion dollar oil and gas investment fund with Russian capital. How you invest that – how much green energy you have in it, and how you position the oil and gas investments you make with respect to the carbon bubble and the degree of global warming – will define your place in the history books. You could really help the world on climate change, and make more history. You could maybe even be the man who turns the tide completely, if you wish. You have it in your power.

London, 28th June 2015

The cars streak past with a reverberant whine, flashes of dragonfly colour behind the wire mesh guarding the race track. I recall a sunny day at Silverstone, another Grand Prix, and the earplugs I needed to shut out the Formula One roar. This is definitely very different.

The Formula E racing teams are in their time trial to decide placings on the start grid ahead of the race later this afternoon. The track for the British Grand Prix in Battersea Park is narrow, with limited opportunities for overtaking, so the time they achieve is even more vital than usual.

I am shown into the Virgin team's command centre in the pits. At the back is a narrow space for observers to stand, out of the way of the mechanics and their kit.

One of the two drivers, Sam Bird, is on the course, striving for the best speed his batteries can give him. I am handed earphones to hear the comms between pits and driver. The team boss gives terse information to Sam on his lap times, those of others, and track conditions. Sam doesn't say much. I imagine what he is doing requires a degree of concentration.

He finishes, and the team boss seems pleased. The next group of cars goes out to be timed. As they zing round the track, it begins to rain. A change of mood descends on the pit. Nobody says much, but this is clearly a development. I ask ignorant questions. No, it has never rained before in the other Formula E races around the world – this is a first. No, they won't change the tyres on the cars.

On the big TV screens, I see a car wobble as it traverses a chicane. Clearly grip is becoming more difficult. The lap times drop. A car slides sideways, slowing right down for a moment. There goes the championship leader, someone says. Another slips just a little, but scrapes a wall hard. Its front left wheel collapses sideways.

I have seen my first Formula E crash, and I have only been here a few minutes.

The rain falls harder.

Taking pity on my ignorance, a member of the team explains to me that this is good for Sam. He will be near the front of the grid for sure now.

But I am hoping the rain won't be keeping the public away. The Formula E organisation joined with the Low Carbon Vehicle Partnership to stage a conference earlier in the week. At that, Alejandro Agag, the Formula E Chief Executive, explained that 99% of people who come to Formula E races leave saying they are more likely to buy an electric vehicle as a result of seeing the racing. He finds that enormously encouraging, and so do I. But then it needs to be. I also heard the President of the Automobile Association explain that a poll of AA members who plan to change car in the next three years shows only one percent currently favouring an electric vehicle. Richard Branson likes to say that he hopes 10 years from now the smell of exhaust from cars will be a thing of the past, much like the smell of cigarettes in restaurants. That one percent had better expand fast then.

I look around the pit area at the technology on show. Most of the cutting-edge stuff is below the shiny low bonnet of the car. At the conference, I listened to Pascal Couasnon, Director of Michelin motorsport, speaking with Gallic passion about how electric vehicle racing is driving innovation and doing so faster than mainstream Formula 1. Vincent Geslin of Renault was even more enthusiastic. You cannot imagine the kinds of things we discover every day on batteries, he said. We cannot predict where we will be in 5 years, except in a much better place.

Alejandro Agag thinks Formula E will have doubled the power of the batteries within 5 years: a time frame – he foresees – in which solar will increasingly be charging the batteries.

You won't be doing this alone, I suggest to him. What about, say, Apple and Tesla, and their plans? What might the world look like ten years from now, five years after Apple says it wants to be mass producing electric vehicles; nine years after Tesla completes its giant battery factor in Nevada? How might those developments feed back into Formula E?

The more electric vehicles there are in the world, the more there will be, Alejandro replies. The better they do, the better we do. Maybe in 5 years we will see an Apple Formula E team and a Tesla Formula E team.

I'm sure you will, I respond.

But today, in Battersea, with the soon-to-be out-of-date technology of 2015, it is that ever-adventurous brand Virgin that is mopping up. Sam Bird flies with minimum noise to glory in the inaugural British Formula E Grand Prix.

Chapter 23

All that gas

London, 7th July 2015

Launch day for Carbon Tracker's gas report, the third and final in the series of global fossil fuel market analyses. After the impact of the first two, on oil and coal, this one has been much anticipated. We have found the compilation of it more complex than the other two, but we have kept our debates behind closed doors, resolved them, and the result – I can say this as chairman, I think, not one of the authors – is another clearly drafted and well designed headache for the energy incumbency. The launch is held in a lecture theatre with floor to ceiling views of the Thames and the City of London, and is full of inhabitants of that square mile. The Carbon Tracker team's goal, as ever, is to provide another boost for the prospects of success at the Paris climate summit, and beyond.

It has been a busy few weeks for Paris watchers. China has submitted its treaty commitment: to cut its greenhouse-gas emissions per unit of GDP by 60-65% from 2005 levels, and to "work hard" to peak emissions before 2030. That wouldn't be enough, from the number one emitter, for the world to stay below the two-degrees global-warming danger-ceiling: but that goal, and even a lower ceiling, would still be achievable in principle, provided nations commit in Paris to ratchet up emissions-cuts meaningfully in the years after the summit.

President Obama continues to pursue his hopes of a legacy rooted in climate action. He visits the Brazilian President, who uses the occasion to pledge zero deforestation by 2030. Both leaders commit to 20% renewable electricity by that year.

New Zealand tables its commitment: emissions cuts of 11% on 1990 levels by 2030. Analysts immediately condemn it as an inadequate "slap in the face" to Pacific island nations.

It must be becoming increasingly clear to governments that simply tabling commitments that they judge feasible in terms of realpolitik, as opposed to imperative in terms of climate science, is not going to mean an end to all the pressure building on them. If they were tempted to hope otherwise, Dutch citizen activists achieve a world first. They take their government to court, alleging that the politicians are essentially betraying their duty to the citizens by agreeing a climate target that falls well short of ensuring a safe future. On June 26th, a Dutch court orders the government to cut carbon emissions by more: 25% within five years.

Environmental campaigners erupt in joy on Twitter. The climate-liability significance of this landmark ruling is immediately clear, notwithstanding the prospect of an appeal by the government. International lawyers working with environment groups will be sharpening pencils and putting on thinking caps, wondering which corporations and governments might be most vulnerable in the courts. "The Hague climate change judgement could inspire a global civil movement", writes Emma Howard, co-ordinator the Guardian's divestment campaign.

The Calderon Commission hands more ammunition to campaigners. Halting global warming without denting global economic growth is now "within reach", the star-chamber of former political leaders and economists argues in a new report. Indeed, 96% of the cuts needed to abate global-warming danger are feasible as soon as 2030.

Carbon Tracker adds further ammunition, joining with Client Earth and others in a letter to the Financial Reporting Council which alleges that fossil fuel companies are failing in their legal duty to report climate risk.

Meanwhile, in America, Catholics and non-Catholics alike seem to be listening to the message in the Pope's encyclical. A major US Protestant denomination, the Episcopalian Church, votes in Salt Lake City to divest from fossil fuels. This is a moral issue, the church leaders say.

Other luminaries speak out. The Dalai Lama supports the Pope's message during an appearance at the Glastonbury pop festival. Prince Charles gives a speech in London in which he backs divestment from fossil fuels, and the Guardian's "Keep It In The Ground" campaign.

All this is looking ever more likely to be an existential threat to coal. "It may be too late to save the coal industry from looming financial disaster", John Dizard writes in the Financial Times. The biggest US coal miner, Peabody Coal, is in deep trouble, its shares down 70% this year. Arch Coal bonds are selling for as little as 14 cents on the dollar.

And so to gas today.

Anthony Hobley opens the batting, setting the scene, teeing up Mark Fulton, James Leaton, and an analyst making his debut – Andrew Grant – to summarise Carbon Tracker's conclusions. We have elected to focus on the potential for wastage of capital invested, or scheduled to be invested, in Liquefied Natural Gas (LNG) projects. Fully $283 billion of actual and intended projects would be uneconomic if governments pursue policies consistent with a two-degree world, we calculate. Sixteen of the twenty biggest LNG companies are considering major projects that probably won't be needed in the next decade.

There is still limited room for growth in the gas market while keeping global warming within two degrees, we conclude, but not anything like as much as the industry projects, and certainly nowhere near at the levels of the new "golden age of gas" that many gas enthusiasts envisage. 97% of the LNG required in the next ten years can be met by projects already committed to, and 82% of LNG required to 2035. $221 billion of the $283 billion now not needed globally for LNG over the next ten years are in Canada ($82 billion), in the US ($71 billion), and in Australia ($68 billion).

As for shale gas, in Europe as a whole there is limited scope for the growth, we conclude, because it requires higher prices and is vulnerable to competition from cheaper Russian gas. Over the last decade there has been no significant increase in EU gas consumption, and Europe's emissions targets leave little room for new large gas power plants. In the UK, we conclude that shale is likely to supply less than 1% of gas demand in the next decade.

That is the consensus view of the Carbon Tracker analysts. The chairman's personal view is that I will be very surprised if UK shale gas production contributes a single molecule to UK demand, ever.

The overall conclusion of the Carbon Tracker gas report is cautionary. Unlike coal and oil, there is some scope for growth in gas use in a low carbon scenario, but not much. Investors should scrutinise, in particular, the true potential for growth of LNG businesses over the next decade. Shareholders need to question whether the strategy presentations of the oil and gas companies add up. They can't all expand LNG as fast as they can build it.

As for the climate implications, projects to convert shale gas into LNG for export do not fit in a low carbon future, we conclude, categorically. LNG is less carbon efficient than piped gas because gas equivalent to one fifth of the amount delivered can be needed to liquefy and transport the gas. On top of this comes the vexed question of fugitive emissions. Both conventional and unconventional gas operations (as well as oil and coal) have "fugitive emissions" of methane, a greenhouse gas much more powerful than carbon dioxide. Studies show that gas needs fugitive emissions of less than 3% if it is to provide a climate benefit over a typical coal plant. There is currently no consensus on levels of fugitive methane emission across the industry, which vary by project and geography. The Carbon Tracker gas report assumes a level of 1.4% leakage of methane for both conventional and unconventional gas. This may well be conservative. Some studies suggest that unconventional gas has higher levels of leakage, but the data are insufficient to enable a firm conclusion to be reached. Either way, efforts by industry to minimise leakage and make processes as efficient as possible will be important in determining gas's contribution to the energy transition.

With this significant caveat about the uncertainties around leakage, the presentations end and panel discussions begin. I look across the Thames at the City of London and wish more of its financial legions could hear what I suspect is coming.

The first panel is asked to reflect on the conclusions of the Carbon Tracker gas report. Gerard Moutet, Vice President for climate and energy at Total, has the floor. Clearly not all the oil and gas reserves will be burned, he says. It is vital that the Paris agreement refers to a price of carbon, and this must favour gas over coal. Total sees a 2% per annum growth in gas consumption, much bigger than the 1.4% Carbon Tracker uses. In Total's scenario, LNG will grow at 4% per year. Of course not all LNG projects will be built, but we need much more than in Carbon Tracker's scenario, he says. Solar will grow and Total is big in solar. But it will take time. So we need gas.

Gerard has given a perfect summary of the gulf that still exists between the relatively progressive oil and gas companies, and advocates of ensuring a two degree future at all costs.

Paul Spedding, former Oil and Gas Financial Analyst at HSBC, now Senior Advisor to Carbon Tracker, is also on the panel. History teaches us that even small drops in demand for oil result in big drops in oil price, he says. The same is true for gas. I don't like LNG, he says. LNG projects are the gas industry's equivalent of the oil sands. They don't give shareholders a good rate of return.

In the back row, I smile, or probably more exactly smirk. Paul has a quietly spoken manner, but rarely minces words. His comment about LNG being akin to tar sands leaps onto my computer screen as participants tweet it.

Others agree with him. We do not see significant increase in global demand for gas, says Richard Chatterton, European Gas and Carbon Markets Analyst at Bloomberg New Energy Finance. it will become increasingly uncompetitive. The potential for renewables to take market share from gas is material, says Ashim Paun, Director Climate Change Strategy at HSBC Global Banking and Markets. And it is increasing as the economics improve.

I look at Gerard Moutet and wonder how embattled he feels, on a scale of 0 to 10, listening to these senior finance people cast public doubt on Total's version of the future. It was good of him to come and talk at the Carbon Tracker event. At least the French oil and gas giant is engaging positively. The same cannot be said of some of their peers.

An evening soiree: the summer party of the London Speaker Bureau, at the glitzy Royal Thames Yacht Club in Knightsbridge. To qualify for their cocktails, the bureau's clients have to listen to four 15-minutes speeches by speakers the bureau view as being among their best offerings. The first is Paul Mason, Channel Four economics correspondent, dissecting the tragedy unfolding in Greece. The second is a PR guru, arguing that PR is dead. I am the third. The fourth is a mystery speaker.

My subject is of course the energy revolution, and the renaissance beyond. I weave Formula E into my talk as one of my pointers to the road ahead. The head of world motorsport, Former Ferrari chief Jean Todt, has said recently that Formula E is the "right way to go". Renault's boss has gone further, suggesting that his company may leave Formula One for Formula E. "The only thing that is certain today is that we are going to be bigger in Formula E", he now says. "Everything is open."

As it is in the solar revolution, I argue. En route to winning the carbon war.

And so on.

I finish on time, sure that the audience will be impatient to see who the fourth speaker is. As am I.

The mystery speaker takes the stage. He is David Coulthard, ace Formula One driver.

Online, 9th July 2015

The American Petroleum Institute does not like the pattern of play these days. It launches a new campaign to roll back regulation of all things oily and gassy in America. "Vote 4 Energy" aims to win support for Presidential candidates who favour increased access and decreased regulation for the oil and gas industry. They have been campaigning for such "freedoms" all the time that I have been working on climate change. I have witnessed a quarter-century rearguard action by the freedom-loving peoples of American oil.

Actually, it has been longer. An e-mail from one Leonard Bernstein, climate expert for Exxon then Mobil before the two giants merged, makes headlines. In the note, sent to an academic researching the history of efforts to combat climate change, Bernstein admits that Exxon knew about the dangers of climate change way back in 1981. But it nonetheless elected to fund climate denial, obfuscation, and worse, pretty much ever since.

That little insight might have considerable significance, once the climate-liability lawyers get going in earnest.

I knew Bernstein in the 1990s. He was Mobil's representative at the climate negotiations. I describe his antics *The Carbon War*. I portrayed him as a "master of propaganda": an architect of intended sabotage at the climate negotiations so malign as to amount to "a new form of crime against humanity". He took great exception to that depiction, and sent me a series of long e-mails in 2000 and 2001 seeking to justify his views, his actions, and to admonish me for misrepresenting him. (I responded to them in full). His key argument then was that "as a scientist, I look into my data, not my heart." Now, 15 years later, he says this of the Global Climate Coalition, the now-disbanded lobbying super-group for American fossil fuel energy companies in the 1990s, including Exxon and Mobil.

"I was involved in the GCC for a while, unsuccessfully trying to get them to recognize scientific reality."

Conscience-stricken retired oil-industry hitmen can try to rewrite history as much as they wish, but the catalogue of capital being wasted by their successors goes on building by the week, climate change or no climate change. And dollars by the multiple-hundred-billion will echo longer in history than words by the thousand. The Wall Street Journal reports on July 2nd that US regulators are now warning banks about the extent of their loans to oil and gas producers. They fear a growing inability to service the loans, it seems. Loss-exemption insurance is expiring for US shale drillers. This way of delaying the day of reckoning will no longer be available to them going forward: the drillers are about to become "zero hedged." As for the alternative to debt, US equity sales by shale drillers are also slowing as the low-oil-price financial-noose closes. Only $3.7bn has been raised in the second quarter.

All this nuance is lost on the typical motorist. Sub-hundred-dollar oil is encouraging American consumers to trade up to gas guzzlers. Given the low gasoline prices, old appetites are reviving fast. Ford factories are responding, cranking up output of fuel-inefficient vehicles.

How short term and blind it all is. How ignorant of history. Even basic statistics. As though to make the point, it becomes clear that China's oil production is set to peak in 2015 and then decline. China is both a massive oil importer and among the top five oil producers in the world: behind only the US, Russia, and Saudi Arabia, and about equal to Canada. Falling Chinese production will now add to the list of future supply constraints junk-debt-laden American shale drillers will have to compensate for if they can.

Shell are consistent on supply constraints: they major on the need to drill on, so as to sure of being able to keep meeting demand in the future. (But not in US shale: they have written off billions there already.) This is one of the ways they justify their gamble on Arctic drilling. And what a gamble it is. They intend to spend another $3bn on up to four wells in the Chukchi Sea this summer and next, using two giant rigs and 28 other vessels. This makes $8.4bn of cumulative Arctic oil capex and counting.

Enter the uninventable, once again, in our saga. Shell's Arctic icebreaker develops a 39 inch crack in its hull. The suspicion is that in its hurry to maximise drilling time on site, the fleet plotted a course across shallow water, and the ship's hull hit an obstacle. In any event, it has had to turn back to Oregon for repairs. It will be gone from the fleet for weeks. And it carries equipment vital to containment of oil spills.

The Archbishop of Canterbury, a former Shell employee, now voices fears for the Arctic. Justin Welby says he doubts an oil spill could be contained.

No, Archbishop, not without emergency well-containment it couldn't.

Back on land, it turns out that drought-stricken Californian farms are using oil-drilling wastewater to grow crops. Chevron has piped almost 8 billion gallons of treated wastewater to almond and pistachio farmers last year. Others are looking to copy them.

Nibble on, Chevron shareholders, as you enjoy that fine Californian wine while you can. Before repeats of the current searingly unprecedented Californian drought do for its vines.

On fugitive emissions, a new report estimates that enough US natural gas is leaking to negate all the climate benefits over coal. The report, by ICF International, concludes that 2.2% is leaking from drilling alone. The same day the report is released – uninventable, again – a gas pipeline leak in China raises fears about the whole national network. It is a brand new pipeline operated by China National Petroleum Corporation. Modern infrastructure like this should really not be leaking at all. I wonder if Carbon Tracker's gas report has been too lenient with the industry on the subject of fugitive emissions.

In the UK, another disaster hits intended shale gas drilling plans. Cuadrilla's application for a fracking licence in the Bowland Shale, the deposit underlying much of the north of the country, is rejected by Lancashire county council. Campaigners are overjoyed. "In the future, this may well be seen as the day the fracking dream died", the Guardian's Damian Carrington writes.

And so to another inevitability. The UKOG Chairman David Lenigas, he of the Saudi Arabia below Gatwick Airport based on one well with no flow measurements, "steps down" from his post. He launched a "reserves" media frenzy with his ludicrous and mostly uncontested hype. But the London Stock Exchange then asked some inconvenient questions about the information he was providing for investors. Unlike the BBC on the day. And, needless to say, Lenigas's "stepping down" attracted somewhat fewer column inches of media coverage than the "Saudi Gatwick" myth did.

The UK Conservatives are undaunted in their enthusiasm for shale, however. Chancellor George Osborne presents a budget on June 8th in which he proposes a "sovereign wealth fund" for communities that support shale gas. He also has multiple spanners to throw into the works of the competition. More on that tomorrow.

London, 10th July 2015

A launch event for a battery storage product, UK-style. Simon Daniel, CEO of Moixa Energy, a luminously smart and infectiously enthusiastic inventor-entrepreneur, is our host. He risks the observation in his introduction that the BBC has compared his company to Tesla. The audience of entrepreneurs, investors, solar executives, and journalists mull this one in silence. No whooping and cheering, Tesla style, for the British.

Simon has a vision to set out for us. Most if not all the people in the room share it, but it is still powerful to see the picture painted so compellingly by someone who really knows what he is talking about. It is a vision of a revolution in direct-current (DC) electricity use, emerging hand-in-glove now with the solar revolution. The "internet of things", Simon says, is seeing exponential growth in electronic and battery operated devices that use DC power. Moixa is redesigning the whole architecture of local power, distribution and storage, so to better manage global energy demand. The product they are launching today, called Maslow, will help them do this, to the benefit of we the customers.

The picture of a Moixa Maslow on the screen looks not unlike a Tesla Powerwall.

Maslows can be aggregated and managed across communities to provide storage services to gigawatt-hour scale, Simon enthuses, so providing a range of solutions for grid balancing, network capital deferral, load shedding or community renewables. Add on a few other DC technologies and you can provide a local smart DC network to efficiently power and control LED lighting, electronic devices, and act as back-up. And you can control it all remotely from your mobile phone!

He gets deeper into detail, enthusiasm rising, delivering at machine-gun pace. I begin to lose his thread. I am hit by the bizarre but sadly accurate thought that I would find it easier to follow at an oil and gas industry seminar on the geology of shale.

But the basic message is crystal clear. All the problems of supposed renewable-energy intermittency and grid imbalance that the energy incumbency falls back on in its self-defence arguments are evaporating with every technical advance and product test that we are hearing about today, and in all the other storage discussion events, not least Tesla's.

Out in the wider world, the march of events encourages belief in Simon Daniel's vision. In the USA, the Obama administration sets a new goal of installing 300 megawatts of solar and other renewable energy in affordable housing by 2020, tripling a goal set in 2013 that has already been achieved. The White House's aim is to create solar jobs for the poor, and there can be little doubt that this kind of ambition will soon extend to storage. American firms are already using energy storage for savings. Companies in California, New York, and Hawaii are installing battery banks as costs fall fast. By 2020, on current course, paybacks will be down to 5 years across 46 states & 455 utilities. Meanwhile, battery second-use is being primed to offset electric-vehicle expenses and improve grid stability: the US National Renewable Energy Laboratory is researching 10 year extensions of operational life for hybrid-car batteries so that they can be used in utility storage. New uses for technology are not always a result of high-tech research breakthroughs.

The link between accelerated solarisation and the rise of electric vehicles is becoming clearer all the time. Why did Warren Buffett bet a billion on solar, a pundit asks on the website oil.price.com? The answer: Because the per-mile cost to operate an EV is already well below that of a standard internal-combustion-engine model. That advantage is set to widen. Watch out, oil industry.

Here in the UK, sunny weather sees Britain break its own solar power record on June 30th. 16% of the UK's electricity demand is comes from the sun during the afternoon. Renewables supplied a record 22.3% of electricity in the first quarter of this year, up from 19.6% last year. This is 7% of all energy.

And yet, just as the shape of the solar-storage revolution becomes clearer in the crystal ball, UK Chancellor George Osborne has made clear that he is going to lead an assault by the new Conservative government to put clean energy back in its box and try to screw on a lid. With his budget he has completed a blitzkrieg transformation of the government from cautious green advocate to confirmed green saboteur. They are now applying the Climate Change Levy not just to coal, oil and gas, but to renewables. Electric vehicles pay tax at the same rate as four-by-fours. The government plans a road-building spree. They have ended their zero-carbon homes policy. As for solar, they intend to end both current forms of subsidy by April 2016.

If they do this, the UK Energy Research Centre says, the domestic solar market will drop off a cliff.

Osborne seeks to justify his actions by saying he is acting to protect hard-working British families from rising energy bills. This is the stuff of black arts: one of the long-running smokescreens that fossil fuel defenders pedal to the public, echoing the myths in the conservative tabloid press. UK energy bills are not rising due to clean energy charges. Some 80% of a typical bill is wholesale power and network costs. Energy efficiency is around 3%, and low-carbon power 3%. Subsidies specifically for solar are a mere 0.7% of bills.

But if you repeat a fib often enough, busy people not paying attention will come to believe it.

As George Osborne repeats his mantra about helping the hard-working Brits with their energy bills, he avoids mentioning the savings and gains to the economy of a low-carbon energy path. He does not mention the income tax paid by, and consumer spending of, green-energy workers. Yet such workers are far more numerous per unit of national energy produced and saved than the incumbency equivalents. He does not mention savings on soaring fossil fuel bills. The first energy-positive house in the UK has just opened in Wales. It produces and sells more energy than it uses. For every £100 spent on electricity imported to the house, £175 is generated from electricity exported. Neither does he like to mention other countries' successes with renewables. Denmark's wind power is now generating up to 140% of the nation's electricity demand. At those times, there is plenty to spare for Germany, Sweden and Norway.

And of the multiple millions the Chancellor intends to throw at shale-gas drillers to encourage them to drill the rural Tory heartland, and bung to local communities to try and bribe them into accepting the upheavals of fracking, we hear nothing in his budget. Instead, he promises a sovereign wealth fund based on the profits of the shale gas they allow to be drilled in their backyards.

Chapter 24

Dung of the devil

London, 21st July 2015

A Solarcentury day. In the morning, an update meeting with the innovation team, in their lab in Bermondsey. In the afternoon, an all-company quarterly review in a conference suite in the Barbican. There are too many Solarcenturians to gather all at the same time in the headquarters in Clerkenwell.

There are eight in the innovation team. They are keen to walk me through the story of their latest product, a slate the size of a table top that they call Sunstation. This is BIPV perfected, they tell me proudly. (BIPV is building-integrated photovoltaics, meaning solar PV that makes a weatherproof roof itself, rather than on-roof solar, the standard method of installation: solar modules bolted onto a roof in a frame structure).

The challenge that the sales and marketing teams gave the innovation team was this. They needed a product to sell for use on residential roofs that would be attractive, quick and easy to install, and crucially, the same cost as on-roof solar. They also needed it to be flexible enough, when being fitted, to deal with the multiplicity of real-world roof tiles and rooftop shapes.

No pressure there then.

The innovation team have met this challenge.

They show me on a mock roof of wooden battens how they did it. I look around the group as they take turns telling me the story. You can see in the body language how proud they are. Especially now that, as we speak, the fruit of their labours is being mass produced in China, and receiving rave reviews from the first installers using it in Europe.

The product has no sub-frame, a Lego-like click-in interlocking design that can be adjusted easily for different roof and tile shapes, and a small number of components, all of low cost.

They install a tile. All the clicking, adjusting and screwing to the roof battens takes just a few minutes.

It is clever, even a non-techie like me can see that. And it looks really good; much more stylish than on-roof modules. I tell them as much.

A representative of the marketing department is permanently attached to the team. Photos of me being educated are taken, and tweeted. I am requested to retweet them.

I do so. How different the marketing game was, I remark, when we first started working on our own products. You know, around the turn of the century.

The marketeer tells me about the market research that led to this product. 91% of UK adults want lower energy bills, and 58% of homeowners find on-roof solar unattractive, she explains. Critically, 86% of them want new additions to their homes to be stylish. With that in mind, she gives me a selection of the feedback in the field. It is, to say the least, encouraging.

We sit over coffee, talking about prospects for the product. Jon Sturgeon, the dynamic team leader, tells me that the team is worried that the UK government now seems set on axing the small but still-vital subsidies for rooftop solar, and that this will mess up the residential roof market, and Sunstation's prospects in the UK with it. What do I think is going on?

I think there is real hope in the world outside the UK, I say, but at home our new Conservative majority government seems determined to position itself on the wrong side of history. It doggedly favours shale gas and nuclear, still, despite all the evidence that both are not just uneconomic, but coming off the rails operationally.

There is a whiff of madness about all this, I tell the Solarcentury innovation team. Osborne is suppressing renewables in order to promote a strategy that is not working in America. In the latest pages of this catalogue of fast-brewing ruin, BHP has written down $2 billion of US shale assets. The tally of drillers that have bankrupted themselves by drilling at costs above sales price now stands at four: Sabine Oil & Gas, American Eagle Energy, Dune Energy and Quicksilver Resources. Others are in visible distress. A bond offering by Swift Energy has just failed: a $650m deal dropped because of low investor appetite. The shale train is coming off the tracks, for those with the eyes to see it. There don't seem to be any in the UK Treasury or Downing Street.

As for the tight environmental regulation that Whitehall has promised for UK shale drilling, the gas industry's own taskforce, the Task Force on Shale Gas, has reported that the government is failing to implement a key fracking safety recommendation: the monitoring of wells to ensure no leakage. Meanwhile, the latest government initiative on shale is to renege on a fracking ban in the Sites of Special Scientific Interest that cover 8% of England.

The nuclear train is also heading nowhere fast. The French nuclear regulator is warning of more faults at the Flamanville reactor, the fore-runner of the reactor that the UK government wants installed at Hinkley Point. The supposed French flagship project is now a minimum of 5 years late and €5 bn over its €3.3 bn original budget. This appalling track record can only get worse. Remy Catteau, head of pressurized equipment at the nuclear safety regulator, tells a hearing in Paris that "there is no doubt" that the carbon anomalies found in the steel of the pressure vessel fail French regulations and international standards.

Seeing all this, Austria loses patience with UK nuclearphilia. The EU member state files a lawsuit against the EU Commission for giving permission for the putative new UK nuclear plant at Hinkley Point. The project goes against the EU's aim to support renewable energy, the Austrian government argues.

I can only imagine how George Osborne's advisors explain all this away to him, a man with a ruthlessly hawkish approach to economics whenever he senses the state erring to profligacy – as he sees it – in sectors of the economy other than energy. Yet evidently they do. To the extent that Osborne wants to suppress the alternatives.

I have to remind myself to tone down my frustration in exchanges like this. I have to mix the squalid realpolitik with the seeding of hope. And that must include the hope of a U-turn by the UK government on their doomed policies. People and institutions do change their individual and collective choice. Sometimes they have no choice.

The innovation team are scientists. They ask me how it can be that a government ignores scientists so readily when rushing through policies that sabotage the climate objectives they claim to champion at the climate negotiations heading towards Paris.

I talk about neuroscience. About the tribalisation of belief systems that is so commonplace in the human condition. But it is indeed so very difficult to understand, when the facts seem so very obvious.

As ever, the recent additions to the library of evidence on climate change are impressively depressing. Warming of the oceans is now unstoppable for centuries, US scientists report in their *State of the Climate in 2014* report. Even if we stop emissions now, the heating will continue. The seas could rise 6 metres even if governments curb warming, a US-led international team reports. Huge tracts of Greenland and Antarctica will melt at average temperatures not much higher than today's. As the temperatures rise, global risk of wildfires is already soaring. A recent report shows that the wildfire season lengthened between 1979 and 2013 by an average of some 19% over more than a quarter of the Earth's vegetated surface.

An unprecedented coalition of 24 UK scientific and medical bodies has urged the government to take effective climate action. Mitigation will bring direct economic and health benefits beyond the abatement of climate-change risk, they say. The Former chief of the UK Navy, Rear-Admiral Neil Morisetti, joins them, lambasting the UK government for its climate disconnect. Recent policy retreats ignore "one of the greatest 21st-century challenges to our prosperity and wellbeing", he insists. Even the International Energy Agency is sounding shrill alarms now on a regular basis. Fossil fuel firms risk wasting billions by ignoring climate change, says new boss Fatih Birol in a speech. If they think they are immune, they are simply wrong.

The rising tide of evidence is undoubtedly fuelling the divestment movement. Warwick becomes the latest university to divest. After a two year student campaign, all fossil fuel investments are now being dropped. The Church of England synod votes for divestment from coal and tar sands.

Dissident voices are appearing in the most senior levels of the investment industry itself. "Fund managers need to stop chasing returns and start tackling global warming," writes Saker Nusseibeh, CEO of Hermes Asset Management, the man who accompanied Mark Campanale and me to the Bank of England to sound a carbon-bubble warning back in August 2013. "The financial system has failed."

A financial system now summing to $87 trillion.

As it prepares for the Paris summit, France continues to offer encouragement that it will lead the endgame well. The government seeks a short, concise, but long-lasting legal core for the Paris climate-change deal. A draft of such has been being prepared for climate ministers attending preparatory talks in Paris.

Japan submits its Paris plan: 26% cuts from 2013 levels by 2030: steeper than US (18-21%) and EU (24%) over the same period. But if all countries copied it, average global warming would be more than 3 to 4°C.

Meanwhile governments demonstrate that they are still capable of negotiating complex treaties. Iran agrees a breakthrough nuclear deal with the US and other states.

City governments continue their impressive work in the UN's climate groundswell. At a two day conference in Rome aiming to keep pressure on national governments, 60 mayors of the world's major cities urge climate action, standing full square alongside the Vatican.

I head off to the Barbican with the Solarcentury innovation team for the quarterly review. I love these. I sit in the back row listening to the speakers update their hundred-plus colleagues, and dozens of others around the world receiving live feeds to their computers.

Today the multinational aspect of the Solarcentury team is on full display. A map of the world showing the countries where we are now operating, plan to operate, or are contemplating operating in, is covered with different coloured dots.

A Chilean reports on Chile, a Panamanian on Panama, a German on Germany. We are one of the few remaining large-scale developers in Europe, says the German. Our €300m turnover is impressive. The cash we have in the bank is even better. Absence of that was what killed so many large downstream companies, he says.

He knows, from bitter experience.

In the tea break, I circulate, hearing as much Spanish being spoken as English. I chat to longer-serving colleagues, old friends without exception. One is Kirsty Berry, these days filling the vital role of head of health-and-safety.

I can't believe we're finally in Germany, she says, beaming. Jeremy, do you realise we have worked together for 16 years! We talked about going to Germany 16 years ago!

I remember clearly the first time she spoke to me, at an amateurish Solarcentury solar display in an even more amateurish show of green technology. She was a recently-graduated engineer working for BP Solar at the time. I said just enough to encourage her to defect to the upstart competitor. Now she has two young children, and a wonderful husband. She met him when he was working at Solarcentury.

After the break we return to hear about the policy state-of-play. Seb Berry (no relation to Kirsty) is head of external affairs. Like Kirsty I have been working with him for well over a decade. Like Kirsty he is ex-oil-industry: he used to write speeches for Jeroen van der Veer in Shell.

Seb tells the sorry tale of the government's about-turn on solar subsidies. They are intent on breaking pledges to investors retrospectively, he says. This is the worse thing you can do in terms of investor confidence.

What will we do? We are mulling our options. We might take them to court again, like we did in 2011, after their previous effort to destroy the solar industry in favour of the gas lobby.

Pope Francis gives a speech on the messages in his encyclical during a visit to Bolivia. He refers to unbridled modern capitalism as "the dung of the devil". He is given a standing ovation.

The Pope is ploughing fertile ground. In the capital markets, dysfunctional behaviour is everywhere to be seen in daily events, well beyond the mountain of junk debt piling up so dangerously in American shale. A mergers-and-acquisitions boom surges on. "2015 feels like the last days of Pompeii", one banker tells the Financial Times. "Everyone is wondering when will the volcano erupt."

Meanwhile the Greek economy continues to melt down. Nobel Prize-winning economist Paul Krugman writes in the New York Times that German "vindictiveness" and Greek "fecklessness" are "killing the European project". The Eurozone hangs in the balance.

By vindictiveness, Krugman means the doctrinaire insistence that ever more austerity can solve the problems, despite abundant evidence that it doesn't work.

The former Greek Finance Minister likens the bailout proposed by the EU and IMF to the Treaty of Versailles. It requires the state to sell state $50 billion of assets: a cruel twist seen by many as the imposition of a total surrender of sovereignty. I wonder if the mostly conservative European governments would treat a fellow conservative regime the same way.

At the same time, Europe's leaders are saying they worry about violence and revolution if Greece goes wrong. European Council leader Donald Tusk talks of the similarities to 1968.

"It is a complete car crash," the Guardian's Larry Elliot says of the crisis. "Any suggestion that Europe provides an alternative to the nastiness of Anglo-Saxon capitalism has disappeared."

Which is the Pope's point, is it not, about the root causes of climate change. The system, governmental and corporate, is in danger of having become simply too nasty to do anything meaningful about climate change or poverty alleviation. Too blinded by collective nastiness to do anything but march on over a cliff.

This is a question that adds a certain urgency to SolarAid's microcosm of one form of alternative: a social venture with a mission to rid a continent of a whole category of fossil fuel use using a different method of capitalisation and hence ownership to the norm in companies. An organisation funded by philanthropy and people-power debt, not the equity of conventional capital-markets players. An organisation wholly owned by a charity completely focussed on climate-change risk-abatement and poverty alleviation, with none of the restraints imposed by enrichment of short-term investors.

An organisation called SunnyMoney.

Nairobi, 28th July 2015

I tear myself from mangled sleep, shower, and climb into the car that will take me to the SunnyMoney office. Africa has long since woken up, and is on the move, the suited and the ragged alike, walking the dusty roadsides, or crammed in dilapidated minibuses.

As I expected, I sit breathing diesel fumes in near gridlock as my driver inches along. He is listening to a radio station that is broadcasting a lesson in the Chinese language.

I watch the street scenes crawl by. The entrepreneurship of extreme poverty is all around. Women tend plants in regimented rows of pots by the highway, arrayed by species the easier to sell to people in cars in the jams. A troop of young men has set up a crude carwash by a grey stream, ferrying the bilge water in huge plastic buckets to pour on the grimy battered cars of their customer base.

Not a good job, I say to my driver, to check if he speaks anything other than Swahili and Chinese.

He laughs good naturedly.

Meanwhile, I add, in the rich nations the one percent get richer every single day.

He guffaws, and turns to look at me with a smile.

I watch the compounds, residential and commercial, go by. Is it my imagination, or are their high walls more festooned with barbed wire and razor wire than last time I visited? Are the security guards on the gates more numerous?

Many of the vehicles in the traffic are four-by-fours. I have visited the suburbs of the one-percent in Nairobi, where the people being driven in these four-by-fours live. The close juxtaposition of ostentatious wealth with extreme poverty must have something to do with the need for high security everywhere, I reason. I haven't seen the same security consciousness in Dar es Salaam, or Lusaka, or Lilongwe, where wealth disparity is much less.

We make it to the main highway. A huge banner shows the face of the US President. "Welcome Home Obama", it reads.

He visited a few days ago. How amazing it must have been for Kenyans to listen to a man of mixed African and American blood who, as one of them proudly told me, was riding public transport around these streets not long so ago. People at SolarAid and SunnyMoney know how easy it is to inspire

Kenyan school kids with a bit of extra light, of better quality, to do homework by. I can only imagine how additionally inspiring it must be to see evidence of how quickly it is possible, with a bit of such homework, to get on in our world.

We edge past a Total filling station. I wonder how many solar lights they will be selling today from that forecourt. Continent wide, how close will the company come to allowing their solar sales to replace their kerosene sales?

Probably the oil giant's sales are approaching the same level as SunnyMoney's now, because ours have fallen fast in Kenya and Tanzania in recent months. The crisis has morphed since it became clear earlier this year. At first our cash crunch came from too much stock, but we have liquidated most of that, repaid debts to manufacturers, and sales continue to fall. We are depleting cash fast. Painful cost-cutting and restructuring has become an inevitability.

SolarAid has long delayed a big multi-year ask of foundations for unrestricted funds for SunnyMoney to expand into additional countries, preferring to use small asks and donations until we had our proposition just right. This seemed like the right thing to do when sales were soaring, but with the benefit of hindsight the delay has been a huge mistake. And now making our case will be greatly hindered by our falling sales.

There may well be a further disincentive to potential funders, one that wouldn't have been in play as SunnyMoney's sales soared. Middleweight solar companies like Solarcentury are being acquired at healthy valuations by bigger companies these days. My 9% ownership of the company may look like a barrier to grant-giving for SunnyMoney, from this perspective. Foundations are likely to think twice before extending significant grant funding to a social entrepreneur if he is potentially going to bank much more money than he needs to live on.

I have to find a way to begin to solve these problems in the week that I am in Nairobi, and in the month beyond. I will have to come up with something completely new, I feel sure. Traditional proposals are not going to work.

Some have advised me to declare the venture over. SolarAid's SunnyMoney has done something amazing, they say: catalysed the first two mass markets for solar lighting in Africa, improving the lives of more than ten million people with the 1.7 million solar lights we have sold, using our unique donor-plus-debt model, preparing the way for other more-conventionally-capitalised players. So let the conventional companies get on with it now, they say. Step back, declare the job done, transfer SunnyMoney staff to the companies moving in, and move on yourself.

But 200 million kerosene lanterns remain in use in 52 African nations, fifty of them not yet catalysed. And SunnyMoney's early success in kick-starting markets in Kenya and Tanzania strongly suggests that traditional models of capitalisation, and hence ownership, may not be the best way to catalyse frontier markets in poor nations. The African solar lighting market was resistant to takeoff for years until SunnyMoney, a brand wholly owned by a charity, came along.

There are other disadvantages to traditional finance- and ownership models. Problems of short-termism and greed, personal and institutional, can easily arise. The Pope's encyclical shines a stark light on this, in terms of the root causes of climate change. In this respect the SunnyMoney ownership model – which exists solely to benefit poor nations, beyond the fair remuneration of its staff – shows the way to a possible alternative for the 21st century. It is not a competitor to traditional models, but a second option, operating alongside them. And of course potentially softening them in the process.

That is why I feel I must try to find a way out of the current trap, not step sideways and move on. To keep SolarAid and SunnyMoney alive is to keep hope of a "nicer capitalism" alive: one that would pass Pope Francis's Dung-of-the-Devil test.

I have an idea how to do it. It came to me when running with my dog. It is that I plug the holes in SunnyMoney's current finances by promising to deploy some of the cash I would get if I were to sell Solarcentury shares – which I would have to, if the majority shareholders decided to sell the company to an acquirer. I ask for grants in that context: help me now, I say, and I will pay you back immediately upon a "liquidity event" at Solarcentury. What I am thinking of, in essence, is a new type of philanthropic capital. It isn't a grant, in all probability. It isn't a loan, that any bank would recognise. It is a hybrid. We can think of it as a "groan".

Somehow the name seems suitable.

Of course, organisations and people would still have to give straight donations and grants to SolarAid and SunnyMoney, if our basic proposition – that we can catalyse solar lighting markets in multiple new countries quicker than conventionally-capitalised companies could – is to work. But funders will surely be more minded to do this, I tell myself, if they see the founder deploying his own money alongside theirs. Especially if they agree with me that I have made mistakes that account for a good deal of the current financial crisis. What better way to pay penance?

In August, I shall be asking people for groans.

The mood in the Kenya office is sombre. Caesar Mwangi, the CEO we finally found for SunnyMoney, began work in May. He has the retail experience and track record we were looking for in order to keep a modest market share in Kenya and Tanzania, grow our three other existing national operations in Zambia, Uganda and Malawi, and establish beachheads in new countries. That way we create space to achieve our stated mission: to play a lead role in eradicating the kerosene lantern from Africa by 2020. But we have left his appointment very late, maybe too late.

Caesar greets me warmly as I enter the office. Over coffee, he runs me through the state of play. He is cutting costs everywhere he can. He has found that the field operations selling solar lights to schools in far-flung places are operating on increasingly shaky economics. Transport costs are wiping out gross profits to too high a degree. He is reshaping the sales force, asking people to stay for longer in towns out of Nairobi, so as to develop networks they can sell into. There are other problems: an amalgam of competition from fake and substandard products, distribution and channel conflicts with manufacturers, inadequate business systems to monitor performance and financial controls, and fundraising shortfalls, all of which deplete operating cash. The situation will be tough to turn around.

He wants me to talk to the entire team, and tell them how the situation looks from the SolarAid board's perspective in London.

Just give them hope, please, he asks me.

Last week I went through a similar sad exercise with the SolarAid team in London.

We walk into the main office areas. Worried Kenyan faces line the desks.

Caesar introduces me briefly. I take a deep breath, and begin.

Dinner with our main manufacturing partner, d.light, in a Chinese restaurant. We join founders Sam Goldman and Ned Tozun and two of their senior sales executives. Caesar and I know we must find a way for SunnyMoney and d.light to collaborate better if our sales are to rise again. SunnyMoney sales teams are reporting that d.light dealers are going direct to our customers in the field, cutting them out. This is contributing to our reduced sales, they say. The reports are complicated by the fact that we know people who have nothing to do with us are pretending to be both SunnyMoney and d.light representatives. Sometimes they tout fake d.light and other products that do not work as well as the real thing, or do not work at all, so eroding customer confidence in solar.

Notwithstanding these problems of a "mature" market, I know Sam and Ned, and know that they would not have sanctioned any of what may have been going on in the field.

d.light was born in 2004, when Sam was a US Peace Corps volunteer in Benin, Nigeria. On that tour of duty, he witnessed a neighbour's son badly burned by a toppled kerosene lamp. He knew that this would never have happened with a solar light. His experiences in the field also showed him close-up the many other benefits that solar could bring to the two billion people in the poor nations who have no access to reliable energy. He returned to America, and enrolled in a class at Stanford University with an interesting theme: Entrepreneurial Design for Extreme Affordability. Ned Tozun also enrolled. Or rather he tried to, and was rejected. He turned up for lectures anyway.

Sam and Ned struck up a partnership and friendship in that class that gave birth to d.light in 2006, the same year I started SolarAid. Now d.light is a social venture selling over 5 million solar lights a year from ten field offices and four hubs in Africa, China, South Asia and the United States. They have designed and sold over ten million solar light and power products in all, in 62 countries, improving the lives of over 50 million people in the process. Their mission is to reach 100 million people by 2020. SunnyMoney is one of their biggest distributors. Most of our 0.6 million light sales last year were d.light products.

We drink beer, we eat chow mein, we fix problems, we plot ways to synergise more effectively going forward. We agree that each social venture is stronger if the other is strong.

We talk excitedly about d.light's brand new product, the lowest-cost solar light on the market. It will retail for around $5, 40% less than the current cheapest product, also a d.light design. We strategise how this product can help lift SunnyMoney's sales, in which countries, in which market sectors.

Sam and Ned show us pictures of President Obama inspecting the light last week, on his trip to Nairobi. They have created something amazing, in this and their other products, and in the organisation they dreamed up and breathed life into.

As I am driven back to my hotel afterwards, the car passes a Shell filling station and a thought strikes me as I look at it. d.light is the most successful manufacturer of solar lights in the world, so far. SunnyMoney is the most successful distributor of solar lights in Africa, so far. These organisations were set up by two students and a maverick environmental campaigner. Where were the big organisations? Both Shell and BP had solar companies in 2006.

The Shell Foundation started life in 2000, six years before we did, with a $250 million endowment from Shell. The World Bank and the International Finance Corporation have been flying consultants around the world writing reports on how to set up solar lighting markets since at least 1995, the first year I went to one of their meetings. I could go on at some length.

What were they all doing? How hard were they all really trying?

I feel the anger rising in me from deep down, once again, and one more time I battle to suppress it. I repeat my mantras. Focus on the snowballing positive news. The zeitgeist is changing. History is not destiny. The foot draggers can yet come to the party. And when they do, they can do so quickly, bringing with them hundreds of billions of dollars currently flowing to fossil fuels.

And meanwhile I have maybe a month to find enough funding to get SolarAid and SunnyMoney, my candle for hope that "nicer capitalism" can accelerate climate-risk abatement and poverty alleviation, back on course.

Chapter 25

Guilt and disruption

London, 7th August 2015

In a community garden project in north London a British politician is about to join the long list of those purporting to hold the solution to Britain's energy policy. Today's aspirant is Jeremy Corbyn, contender for the vacant leadership of the UK Labour party. He is doing rather well in the polls, against expectations, including his own. The launch of his energy policy platform is attended by an exotic mix of young people, and – given his ascendancy in the polls – the full British political media pack. I sit in the back row of seats with a festoon of TV cameras and paparazzi lenses pointing over my head.

Corbyn begins his speech with a recent quote of the Pope's. "We are faced not with two separate crises, one environmental and the other social, but rather one complex crisis which is both social and environmental. Strategies for a solution demand an integrated approach to combating poverty, restoring dignity to the excluded, and at the same time protecting nature."

This is beautifully put, says Jeremy Corbyn, and were he to be elected, this might as well serve as the underlying principle for his energy policy.

Nothing to disagree with there then, I think.

Corbyn's speech paints a picture of a Britain providing international leadership on climate change, at the same time as socialising and decentralising its own energy supply. The country would work its way to a future economy one hundred percent renewable powered. Over the next few decades, a host of cities, regions and countries around the world are aiming to have 100% renewable electricity, heating/cooling and/or transport systems. Over 180 German cities and towns are taking over their local electricity grids, selling themselves cheaper and cleaner electricity that they increasingly produce themselves with renewables. It would currently be illegal to do this in the UK. All this would change under a government led by Jeremy Corbyn. Britain needs an energy policy for the Big 60 million, not the Big 6, he says.

As for subsidies, why are we putting more subsidies into fossil fuels than renewables? We should be looking at feed-in tariffs as an investment in green jobs.

After he has finished, the rows of young listeners clap enthusiastically.

I ask a question. If you are elected, you suggest you would have five years to confer within the Labour party to make sure all opinions on energy are heard. But is it not more likely that the current government's catastrophic miscalculations on energy will come home to roost within those five years? That you'll be having to add specifics to your energy agenda under fire, in a national emergency? In particular, the government is fixated on fracking, yet the US shale boom they seek to import looks as though it is about to go bust. They are in favour of nuclear at all costs, yet the French nuclear industry they seek to import is imploding before our eyes. They are actively suppressing renewables at the same time much of the rest of the world is enthusiastically deploying them. There is every chance of an energy crisis on their watch. And yours, were you to be elected leader of the opposition.

Let me be clear, says Jeremy Corbyn. I am against fracking and shale gas exploitation. The water use and pollution alone would make it impossible. Nuclear is enormously expensive and has obvious security issues. I'm in favour of a German-style phase out. The solution lies in job-rich green infrastructure building.

How he would fund all this climate-friendly refiring of job creation has already been the subject of discussion in the national media. Richard Murphy, a colleague of mine on the Green New Deal group of economic and environmental thinkers, has been influential in the Corbyn camp. Richard advocates what he calls "people's quantitative easing", a strategy Jeremy Corbyn has enthusiastically embraced. In the traditional form of quantitative easing, used by governments in the financial crisis, central banks bought liquid assets such as bonds as a way to inject money into the economy in an effort to rev it back up. Much of that ended up in the pockets of the bankers who created the financial crisis in the first place, adding inflated value of the mansions they elect to live in, pumping up house prices to increasingly unaffordable levels for the rest of the population. In people's quantitative easing, debt would be deliberately created and issued by agencies with a specific interest in new green investment in the economy, such as local authorities and health trusts. The Bank of England would be mandated by the government to buy those debts. The decisions on how the money would be used would rest solely with the government, with the Bank of England simply acting as a bank. Governor Mark Carney has already said it is technically and legally possible for the BoE to act in this way.

That doesn't sound too scarily unfeasible, does it?

Not to the Financial Times's Alphaville column. "Corbyn's 'People's QE' could actually be a decent idea," it concludes. "We don't understand the negativity."

And there is a lot of that negativity. Predictably, it extends across the full spectrum of rightist political thinking. Somewhat less predictably, it spans members of Corbyn's own Labour party, including fellow contenders for the leadership. Corbyn and his exploding support base would say that includes rightist political thinking, of course.

I make my way to my next appointment, in the West End of London, reflecting on the politics at work here. I have always said that climate change in the UK is not a party-political issue. It looks as though I might have to modify that opinion. I know generalisations are dangerous when it comes to politics. But it is beginning to look, in the UK, as though the political left is lining up to defend what I would have to say are the correct policies for our times. The political right – on the whole, at the collective level – is lining up to defend, well – what other way can I put it, really – the indefensible.

I suspect Pope Francis might agree with that thought, given the tenor of his encyclical.

I say this not as a card-carrying member of the political left, or as a Catholic. I say it as a citizen of the world desperate for humankind to avoid the fate that unconstrained global warming holds for us. I am convinced that Pope Francis is right that we will need not just to count emissions and cut them, but refashion society's operating manual to address the root causes for our headlong rush to self-destruction.

Tea with a senior executive who must remain nameless from a major resource company that must also remain nameless. He is guilt stricken, and in a swank London hotel he tells me why.

His story is not surprising to me. The only thing that surprises me – still, after all I have experienced this past quarter century – is that I don't come across more people like him.

The climate clock ticks on, louder than ever as a result of reports these last few weeks. The Earth is now halfway to the UN's intended global warming ceiling. All but one of the main trackers of global surface temperature are now above 1°C. Glaciers are retreating worldwide at a historically unprecedented rate. The World Glacier Monitoring Service reports annual ice loss now at 2–3 times the 20th century average.

President Obama continues to hear the ticking. He has imposed the most far-reaching carbon restrictions on the US power sector yet. His latest announcement targets a 32% cut for power-sector emissions-cuts by 2030, and encourages bypassing of gas, favouring renewables.

US oil and gas drillers are furious that Obama's plan gives solar and wind a chance to compete. Yet the inventor of the device they commonly use for measuring industrial methane leakage reports that a systemic fault in his invention means leaks may greatly exceed estimates, cancelling out natural gas's supposed greenhouse-gas heat-trapping benefit over coal.

My tea companion's company has a coal business. It is not doing at all well. Hardly surprising: the second biggest US coal producer, Alpha Resources, went bankrupt recently. It was $3.3bn in debt. With "clean coal" ever more clearly an expensive and thus-far impractical pipe dream, analysts expect that other giants, such as Peabody and Arch Coal, will soon go bankrupt too. Investors are fleeing. "Do you really want to be the last coal investor?" asks Henry de Castries, chair of giant insurer Axa. He reports no negative shareholder reaction to Axa's sell off of coal shares.

As for the coal industry's last ditch argument that coal can cure poverty, even the World Bank has rejected that notion. The World Bank climate change envoy has said that continued use of coal is exacting a heavy cost on the world's poorest countries.

Things are little better for oil. Since the oil price peaked above $100 in June 2014, 157 energy companies have lost market capitalisation amounting to $1.3 trillion. The amount producers are now set to lose in the next 3 years, relative to figures a year ago, is fully $4.4 trillion. Deflation is spreading through the global oil industry. Oil groups have shelved $200bn in new projects. 46 big oil and gas projects have been deferred. Only a handful of major projects have been fully approved. Deepwater drill ships worth $300 million apiece sit idling at anchor, burning $70,000 a day waiting for low oil prices to abate, turning the Caribbean into what Bloomberg calls an expensive parking lot. The oil price collapse is hammering the big energy groups.

So my question for my guilt-stricken energy-incumbency companion is simple. Why don't his colleagues change course? Why do they keep defending the status quo so trenchantly, when those of them who are Catholic have been told, in not so many words, that their souls are in mortal danger if they maintain course, and those who aren't Catholic are no doubt often given a barrel of grief over the breakfast table by the kind of sons, daughters and grandchildren flocking to the new, fresh, politics offered by Jeremy Corbyn's cause?

I think I know what his answer will be. It is about an enculturated belief system involving short termism and resistance to change, insulated from the possibility of individual or collective self doubt by dysfunctional corporate culture, cemented in place by machismo. It is about power, money, and the ease with which politicians can be bought, indirectly or directly, assuming they don't hold the same belief system as energy-incumbency diehards in the first place.

London, 18th August 2015

How the Imperial College of Science and Technology, has changed, since I researched and taught here, in the 1980s. Then, fossil fuels ruled the roost unopposed. Now one of the buildings facing the austere quad is entirely devoted to just one category of environmental science: that involving the greenhouse gases. Inside the Grantham Institute – funded by and named after the legendary investor, Jeremy Grantham – I meet the Carbon Tracker team and a group of climate scientists and economists on the faculty. The aim of our afternoon is to explore the flip side of divestment from fossil fuels: investment in clean energy. Our particular interest is to skill-share on the role of disruptive innovation in the creation of companies to invest in.

One of the scientists kicks us off with a reminder of what passes for common understanding. He projects on a screen the Intergovernmental Panel for Climate Change's global energy scenario for a world with a fair chance of staying below the two degrees danger ceiling. The transition from the fossil fuels that today dominate the energy mix to the low-carbon future is smooth over the decades to come. The mix of low-carbon and zero-carbon energy technologies that dominate that future occupy roughly equal sized shares. This was the best that the hundreds of world authorities who cooked up this scenario for the IPCC could envision: a future where no one or few technologies dominate in a mix of biomass, carbon capture-and-storage, energy efficiency, wind, solar and the rest.

Why are we so scared of modelling disruption, the scientist asks. We know that the future isn't going to to be like this. We know that disruption, when it happens, can unfold at breathtaking pace. Why is the energy transition going to be any different to the digital and internet revolutions?

Indeed. And so the discussion begins.

The developments of recent weeks give only small hints as to what is likely to come. A highlight of energy efficiency, which generally receives so little media coverage, is in the news. On US electric grids, LED lightbulbs are outshining solar and wind. Efficiency accounts for more emissions reductions in the USA than renewables, and LEDs replacing compact fluorescent lightbulbs constitutes most of that. You would never guess this, from so much of what there is to read on energy. Yet it is profoundly important: it points to a potential future where renewables might have to do much less heavy-lifting when meeting demand, because efficiency is lowering the demand for them.

Hints of the dramatic developments unfolding in energy storage are there for those following the detail of play. One giant German utility, RWE, invests in a storage company from Silicon Valley. Another, E.ON, begins construction of a 5MW modular large-scale battery system at a German university.

Most of the business news focuses on a fresh phase of perturbation in the capital markets. China has triggered turmoil by devaluing the Yuan. Stocks are coming under pressure amid fears of a global currency war. Solar panel manufacturers are among those suffering: Chinese panel makers in particular. Expansion of the solar industry is coming under threat as investors begin to perceive that solar companies face too much production capacity, too low profit margins and potentially crushing debt. The oil, gas and coal industries are not alone in their problems with conventional capital.

The big downstream solar companies press ahead with the rollout of their expansion plans as best they can. SolarCity is making good progress with construction of its its own gigawatt-a-year module factory in New York: the biggest such facility in the western hemisphere. It will produce its first modules early in 2016, and be in full production by 2017. SunEdison, meanwhile, has been replaced as the most valuable US quoted solar company by First Solar. Its stock has fallen 55% since investors began to fret about its speed of expansion in late July.

I read about developments like this frustrated that the discussion is always framed by perceptions of market reality voiced only in terms of the short-term value of money. Discussion of solar energy by analysts and journalists is generally devoid of any sense that we are talking about vital strategic assets in a mass mobilisation for war. Yet that is what the science of climate change, and increasingly the politics, demand. A research report just published by the

Economist Intelligence Unit makes the point. Even if warming holds at two degrees, investors could lose $4.2 trillion due to climate change impacts.

The preparations for the Paris climate summit continue to look cautiously promising, offering hope that the endless grind of short-termist money-centricity might change thereafter, if a strong enough signal can be sent. The French government is shaping to lead from the front. It has passed a sweeping energy bill cutting fossil fuel consumption 30% by 2030, boosting renewables, raising a €100 per tonne domestic carbon tax, and more. In a speech at a "Summit of Conscience for the Climate" in Paris, President Hollande says that 80% of fossil fuels must remain in the ground. Clearly, Carbon Tracker CEO Anthony Hobley's briefing in the Elysée Palace found fertile ground.

A Paris deal might be taking shape. The co-chairs of the negotiations have edited the 88 page treaty draft agreed in Bonn draft down to 19. This document is now being mulled over by governments.

Much will depend on President Obama holding the line. Fifteen states have now lined up in opposition to his climate plan. But attorneys general for 15 others states have issued a statement supporting the Environmental Protection Agency's rules to cut emissions, saying they would oppose legal efforts to block them.

American firefighters, meanwhile, struggle to contain the largest ever US wildfire, in Idaho. Spanning more than a quarter of a billion acres, the fire has been fed by drought. It should be so clear to the fossil fuel diehards and their political support-base that this is the kind of thing that awaits us in spades, should they win the carbon war. It is clear enough to young climate campaigners around the world. Activists descend on RWE's German lignite mines. 1,500 march under the banner Ende Gelände: "here and no further", putting their bodies on the line to stop the giant digging machines. Says one, to a camera, but speaking to governments: "if you are not going to solve it for us, we are going to solve it for you."

Australia maintains its shocking record, setting an emissions reduction target by 2030 that leaves the nation trailing most other developed countries in efforts to combat climate change. Meanwhile, a giant coal mine planned at Carmichael looks like ending up stranded. Indian coal imports will drop to zero by 2021, a new analysis shows, as renewables and grid upgrades progress. The Australian coal won't be needed. Investors will have poured billions uselessly into a hole in the ground, never mind about climate change.

Standard Chartered sees the danger and abandons the mine.

Coal companies continue to suffer on a global basis. "Glencore's investors want it to stop digging," a headline reads. The resource conglomerate's shares are down two-thirds since its May 2011 flotation. It has a debt mountain of $48 billion.

The biggest problem for climate campaigners targeting coal remains Turkey's incomprehensible rush for the black stuff even as most of the rest of the world turns its back on it. The Turkish government plans 80 new coal plants, the third biggest programme after China and India. In that sunny country, they envisage only 5% electricity from solar by 2023.

Both Saudi Arabia and American shale drillers keep pumping oil with gusto. The IEA suggests the glut will extend into 2016 now. Supply is outstripping consumption by 3 million barrels a day. Canadian tar sands crude has halved in price in the last six weeks, down to just $23, less than half the price of Brent: a level not seen for 12 years.

But still Shell presses on in the Arctic. The capping stack is now back with the fleet and they have told the US regulators they are good to drill. I watch TV pictures of Shell's Arctic icebreaker threading a human barricade of Greenpeace climbers dangling on ropes from a bridge as it departs from Oregon. Police arrest protestors both on the ropes and in kayaks on the water below. I feel proud of my old colleagues. There are increasingly few opportunities for their classic non-violent direct actions to reach global news bulletins. The Arctic frontier is one of them.

Notwithstanding all the great work the US President is doing on climate, the Obama administration grants Shell a final permit for Arctic oil drilling. Some complex calculus of political expediency has been made, no doubt, somewhere deep in the White House. But the concession causes Hillary Clinton to break with Obama on climate for the first time in her Presidential bid. "The Arctic is a unique treasure," she says. "Given what we know, it's not worth the risk of drilling."

Shell disagrees, and once again attempts to reassure an increasingly incredulous world. "We remain committed to operating in a safe, environmentally responsible manner", says a spokesperson.

It would have to be difficult to say such a thing if you were a devout Catholic, I reflect. I wonder how the Pope's message is playing among those of Shell's employees who are.

In Istanbul, a large group of Islamic leaders gather to issue a clarion call for rapid phase out of fossil fuels. A declaration issued calls for phasing out of greenhouse gas emissions by 2050, and 100% renewable energy as a means for doing so.

I wonder how that is going to play in Riyadh.

Online, 27th August 2015

The week begins with the biggest stock market wobble for a long time. In China, stocks plunge 8.5%, triggering a global rout. On "Black Monday", as Beijing's official mouthpiece is calling it – just as a good capitalist would – spooked investors are selling off virtually every asset class in virtually every market. Global stock markets have lost more than $5 trillion since the People's Bank of China devalued the renminbi on August 11th.

This new twist in the drama compounds Big Oil's problems. "Oil Industry Crash Leaves Wall Street Playing Catchup", a Bloomberg headline shouts. Bad news is raining down down on the American oil industry so quickly that analysts are having difficulty keeping up. The US oil price slides below $30 for first time since the financial crisis. The price of Brent oil falls to $45.

In the UK, a 100-organisation coalition writes to the Prime Minister urging protection for small-scale renewables. Among them are IKEA and Solarcentury. We fear that the government is about to launch another assault on solar in the misguided effort to clear the way for shale gas. A thousand square miles of England are to be opened up for fracking, we now know. The government is saying it will fast-track planning applications by drillers. If councils delay beyond 16 weeks, government may remove their decision-making powers.

A retired bishop summarises a good deal of (conservative) local reaction. "The government is saying you can have local democracy as long as it agrees with us", he complains.

Even the Conservative press is unsettled by all this. "Britain's shale fracking revolution comes with big risks," argues the Telegraph's Andrew Critchlow: risks that may outweigh the gains.

Indeed. The latest news from the US is another report that methane leaks in the gas supply chain far exceed estimates. A study shows that gathering facilities, serving multiple wells, leak some 100 billion cubic feet of gas a year, 8 times the estimate by the Environmental Protection Agency. Often this leakage can literally be smelled. In New Mexico, a plume of leakage is visible in satellite imagery around gas wells. Press reports quote locals recounting how foul the air smells, and how worried they are for their health.

In the face of reports like this, the Obama administration is reacting. Just two weeks after its new power sector regulation, it moves to require cuts in methane leaks.

SunEdison makes a timely announcement about the economics of solar versus gas, never mind the leaks. One of its solar farms is beating gas on dollars per unit of electricity in Colorado. This is the first time utility-scale solar PV resources have come out cost-effective head to head with natural-gas fired generation.

And so from Black Monday to Grey Thursday. Today the UK government exceeds the UK solar industry's worst fears. It announces its intention to slash solar feed-in-tariff rates by 87%, from January 2016.

Industry leaders describe the move as "alarming", "damaging" and "absurd". "Today's proposed solar FIT cuts add to the calculated turmoil that the new Government has unleashed on the solar market since the election", Solarcentury CEO Frans van den Heuvel fumes. My e-mail inbox fills with e-mails from aghast colleagues, in Solarcentury and other companies. "They are trying to kill us", says one.

Indeed they are. And the government's own impact assessment of its proposed policy assault reveals UK cuts would wipe out 6 gigawatts of solar by 2020. And doing that they would only save around one percent on average household energy bills by 2020. It is now completely clear that their assault on renewables is doctrinaire, driven by desire to eliminate competition for their collective fixation on shale gas. The only reason they didn't abandon the solar feed-in-tariffs completely was presumably to leave a fig leaf of "support" behind which to obfuscate.

Elsewhere in the press today, a former Shell man who is now trying to make a career advising on the energy transition, Adrian Kamp, has a thought on the gas-versus-solar carnage. "Managing a wind farm or solar project is nothing a good oil and gas man who has built or organised facilities cannot manage," he observes.

CHAPTER 26

Things are coming together just as things are falling apart

Paris, 1st September 2015

The divestment movement is gathering in Paris to review progress with three months to go to the summit.

I take the Eurostar and sit marooned for three hours just outside Calais because there are migrants on the track. I take the opportunity for a major e-mail catch-up. I am interested and encouraged to hear nothing but sympathetic comments from the trapped passengers for the desperate people causing the delay.

At the conference next day, many of the movement's stars perform. 350.org's dynamo May Boeve is an early speaker. We have a new saying these days, she recounts. Things are coming together just as things are falling apart. In Copenhagen we thought climate change would be something in the future. Today we see all around us that it is happening now. But we also see a vibrant and growing divestment movement, and many other active campaigns to stop climate change.

Bill McKibben gives a rousing speech. Another difference between Copenhagen and now is that the alternatives are clearly within grasp, he says. The price of solar has come down eightfold since then. Then there is the impact of the carbon bubble. The work the divestment movement has done with the numbers provided by Carbon Tracker has been so important in creating the hope that many of us now have of success in Paris, and the fears that fossil fuel companies now feel for their future.

These are not normal companies we are talking about any more, he continues, a quiet anger in his voice. These are rogue companies. Think of Shell. They watched the Arctic melt just like the scientists said it would. Then instead of doing the obvious thing and saying we should stop drilling and go for the new alternatives to fossil fuels, they go ahead and drill for yet more oil where once the ice would not have allowed them to. What kind of people are they, that agree to invest in such madness, and to execute it?

Marjan Minnesma, architect of the court case launched by Dutch citizens against their government over the inadequacy of its carbon-emissions targets, tells the story of how the campaign was won. It began with a letter to the government in 2012. The government replied that they agreed the Dutch emissions-reductions targets were insufficient, but they did not want to become a front runner in climate policy. There was and is no danger of that, Marjan says. Holland is 24th out of 25th on renewables deployment in the EU today. So she and her colleagues appealed to ordinary Dutch citizens to become co-plaintiffs in a legal action. Hundreds handed in the summons in November 2013.

We used civil law, claiming an unlawful act, and we won, says Marjan with a beam. We think you can do this in almost every country now.

In the evening, a private dinner convenes in a historic dining room. Funders of the divestment movement, both wealthy individuals and philanthropic foundations, gather to compare notes and plot future action. They have much to be pleased about. More than 400 institutions have now divested from fossil fuels: 28% of them foundations, 23% faith-based groups, 15% pension funds, and 10% educational institutions. New commitments are being made daily.

We begin dinner by going round the table introducing ourselves, in the manner of such gatherings. One of the participants is Stephen Heintz, President of the Rockefeller Brothers Fund. He raises smiles with a frank statement about how the Rockefeller family made its money from the founding of Standard Oil, later to become ExxonMobil, and has been trying to make amends ever since.

Mark Lewis and I are asked to talk for ten minutes each to kick off the discussion: him addressing the impacts of divestment in the capital markets, me the scope for reinvestment in the clean energy transition. Mark, who has just told me he is to become head of European utilities research at Barclays, explains that in his view the world's major oil-and-gas companies now face the same challenges that the European power-industry incumbents faced ten years ago: first, increasing policy focus on the impact of their industry on the environment, and second a growing commercial threat from the dramatic and ongoing drop in the cost of renewable-energy technologies. Will the oil-and-gas companies make the same mistakes that most of the European utilities made a decade ago and persist with an unsustainable business model, he asks. Or will they learn from these mistakes and embrace the opportunity of diversifying into renewable energy and energy storage?

I give a highly personal account of how life looks and feels on the renewable energy and energy storage frontlines. The impact of divestment is indeed impressive, I say, and will certainly continue to grow. The impact of divested capital in the clean-energy revolution I have yet to see much evidence of at all. I am sure that will change.

Oslo, 10th September 2015

A small group of energy experts assemble to advise three utilities how to respond to the energy transition. We are in the HQ of one of them, Statkraft. The day has been organised by a Norwegian consultancy specialising in helping big corporations in all matters involving potential major transitions, on an international basis, Xynteo. Statkraft's chief executive and his senior management attend all day. I contribute but also learn. It turns out to be one of the most interesting and constructive business events I have ever participated in. If all utilities were to approach the strategic challenges of the great energy transition the way I am seeing today, the task of climate negotiators would become much easier.

It needs to. In Paris, President Hollande issues a gloomy warning. The summit in December could fail, he warns. Another session of talks in Bonn has become bogged down. The vital agreement on financing for developing countries is still missing. The co-chairs have been given a mandate to prepare another draft agreement, by the first week of October, which parties will then discuss in Bonn on October 19th. But time is running out to clear the way to a treaty now.

With a migration crisis playing out live in Europe – Syrian and other refugees are now fleeing war zones in their hundreds of thousands – Hollande suggests that failure in Paris would entail millions of refugees over the next 20 years simply because of the impacts of unmitigated global warming. Which of course could well involve fresh wars, as liveable land areas and food and drinking-water supplies shrink.

The carbon diehards launch their latest advocacy of unmitigated global warming and refugee creation. The Minerals Council of Australia runs a TV advertisement with the punchline "coal is amazing". Their case is immediately labelled "ludicrous" and "desperate" by environmentalists.

A comprehensive survey of coal burning in EU underlines the environmentalists' claims in graphic air-pollution terms, never mind climate change and its impacts. Europe as a whole had equivalent mortality costs in 2013 of between €21 and €60 billion, an estimate which includes deaths from coal-related respiratory and cardiovascular illnesses, such as heart disease and lung cancer. As for the economics, famous solar advocates David Hochschild and Danny Kennedy write gleefully in the San Francisco Chronicle that "coal is fast becoming the telegraph to renewable energy's internet". The top four US coal stocks have dropped 98% in value since 2011.

Utilities continue to wrestle with the implications of coal's descent. A coal-dominated Texas utility, Luminant – the largest power generator in the state – takes its first solar step. It signs a power purchase deal, with SunEdison, for 116 megawatts of photovoltaics.

As for Solarcentury, we prefer a statement more long-lived than a TV advertisement. We elect to solarise the Tate Modern, donating a roof installation that will be installed on the iconic art gallery next month.

I love the inventive irony of our marketeers' idea. The world-famous Tate is housed in an abandoned oil-fired power station.

It occurs to me that there is another irony. We join other corporate sponsors that include BP.

Bucharest, 23rd September 2015

Another capital city, another company seeking advice, another roomful of senior executives. This company is itself in the advisory business, on a global basis. Their clients include many of the world's biggest companies. They are contemplating just how far the great global energy transition is going to transform their clients' business models, and hence their own. Crucially, they are wondering how proactive and far-reaching to be, in the advice they give.

My case to them is simple. With respect, you and your competitors in the space you occupy did not cover yourselves in glory in the run-up to the financial crisis. You did not advise anyone to change course, that I know of. Yet with the benefit of hindsight the warning signs were pretty clear. Now, with the unfolding global energy transition, the warning signs are even clearer, and many people are vocal about them already. The conclusion is pretty obvious, I contend. The advisory firm that gives the most insightful proactive advice on transition, when the history is told, is the one that will do best in the brave new world. Let it be you.

The executives have flown in from many different countries. They are sobered by emerging news of one of the biggest scandals in corporate history.

Volkswagen has been caught by the US Environmental Protection Agency cheating on emissions tests for diesel cars on a massive scale. The EPA alleges that the carmaker fitted nearly half a million VW and Audi vehicles with devices designed to bypass environmental standards. VW is issuing no denials. Today we read that the scandal has caused nearly a million tonnes of extra pollution from the 11 million cars apparently involved.

The talk among the executives in my seminar group, before we start, is about whether we have heard all there is to hear about the scandal yet. Have other carmakers been involved in this kind of fraud? The suspicion is that they have.

The stakes are existential, in more ways than one. A recent study from the Max Planck Institute shows that more people die prematurely from air pollution than from malaria and HIV/Aids combined: 3 million people each year. Most of these are from wood and coal burning, but a significant minority derive from auto emissions. It is easy to imagine the class actions that must be on the drawing boards of law firms today. Can VW hope to survive, when all the accounting is done?

The wider implications are clear. VW, or some gang of suitably senior executives therein, has somehow been willing to engage in criminal acts in order to maintain a lie about diesel vehicles being better in terms of emissions, including greenhouse-gas emissions.

Another report points to a bigger context: 48 of the world's hundred largest industrial companies are actively obstructing climate legislation. They may be doing so via lobbying tactics that are currently legal. But will such tactics remain legal? And if not, what future retrospective legal action might this group of companies, and their directors, potentially face?

My audience today will be weighing such considerations very carefully in the times ahead.

BP tops the list of the points table of corporate climate-legislation saboteurs. In Europe, for example, it joined with Shell to craft a winning formula for the scrapping of the EU's renewable energy and energy efficiency goals, in favour of a single greenhouse gas target for 2030 that could be met by an increased use of natural gas. The two companies are taking heavy flak already over this kind of behaviour. Shell has just quit the Prince of Wales's Corporate Leaders Group of Climate Change. It is not clear whether they jumped, or were pushed, or jumped ahead of the growing probability of being pushed.

ExxonMobil needs to worry too: perhaps more so than BP and Shell. New evidence has emerged that they knew about the problems associated with greenhouse-gas emissions decades ago, and have actively suppressed it ever since. A past employee admits that he wrote an advisory note warning about emissions causing global warming as long ago as 1982.

The impacts of the story the oil giant has been trying to suppress for decades become ever clearer. Burning all fossil fuels would melt Antarctica completely, ultimately raising sea levels 50 metres, wiping out land where 1 billion live, so a new study led by the Potsdam Institute shows. 2015 and 2016 are set to break global heat records, the Met Office announces. Natural climate cycles in the Pacific and Atlantic oceans are reversing and will amplify man-made global warming. As for the worrying rise of methane levels in the atmosphere, Russian scientists publish photos of giant craters caused by escape of this potent greenhouse gas from melting bogs in Siberia.

In my presentation to the executives in Bucharest, I avoid climate impacts and potential liabilities, and major on the opportunities. More than half of my presentation focusses on the evidence for an emerging clean-energy revolution. I hold back on the downside part of my story: the cost-increase problems that their clients in the energy incumbency are suffering. Everyone here reads the financial pages of newspapers.

The emerging pattern is clear for anyone who regularly spends a few minutes a day on those pages. The Wall Street Journal reports that credit-line reviews due in October mean that oil and gas drillers are "expected to finally face a financial reckoning", with "carnage occurring as early as this month." Drillers both small and large are facing that carnage. Goldman Sachs is now predicting 15 years of low oil prices. Expect $20 a barrel when refineries shut in October or March for maintenance, their head of commodities professes. The investment bank admits it has been wrong in the past. But its opinion shifts markets, rightly or wrongly.

The International Energy Agency expects US oil production to fall sharply next year: 80% of a 500,000 barrel per day drop in the non-OPEC countries. Total global oil supply was 96.3 million barrels a day in August. Bloomberg, reporting "an Oklahoma of oil at risk as debt shackles U.S. shale drillers", foresees as much as 400,000 barrels per day of oil production disappearing.

The surprisingly large market for sand in America has just collapsed. Huge amounts of sand are used in fracked wells, wherein the sand grains help to hold fractures open in the target strata. "When the shale boom went bust", Bloomberg reports, "it took down the sand industry with it."

Note the use of the past tense here.

Another major shale driller, Sansom Resources, joins the list filing for bankruptcy. Its debts had piled up to >$4bn, most with no collateral behind it. That makes eight bankruptcies now, and counting.

Around the world, $1.5tn of potential investment in oil is now uneconomic, Wood McKenzie reports. That includes much of the US shale. In a dissection of the Energy Information Administration's Annual Energy Outlook, David Hughes of the Post Carbon Institute – author of the seminal 2014 report exposing hype by shale drillers – criticises the US government's "rosy" shale predictions.

Whole areas of the North Sea oil are at "serious and urgent risk" of shutdown, the new head of the UK Oil and Gas Authority reports. The more companies leave, the less others can share infrastructure costs. The fall in North Sea rig orders threatens thousands of jobs. Fabricators are facing the worst downturn in orders in 25 years.

Total slashes its capex further to defend dividends: as much as 15 per cent next year, to $20bn-$21bn, and by a further $3bn in 2017. These cuts are the most aggressive yet by any oil group.

Drilling for Arctic oil is also not viable unless the price is higher, says the International Energy Agency chief, Fatih Birol. But he stops short of advocating a drilling ban.

Shell boss Ben van Beurden heads to the interview studios once again to make his case. World energy demand is rising, he insists, and renewables provide electricity, which is only 20% of energy. Solar will eventually dominate global energy, he tells the BBC, but will only replace fossil fuels over multiple decades.

I wonder if Shell conducts follow-up opinion surveys to check what percentage of the public still believes their boss. And if they do, whether it gets reported back to him.

Coal remains in a worse plight than oil and gas. Goldman Sachs called the peak yesterday: before 2020. The resource is in "terminal decline", the bank concludes in a note to clients. BHP is warning that coal miners are losing the battle for investor & public support. "In the court of public opinion", an executive admits, "the 'no coal' camp has been more effective."

Carbon capture and storage appears now to be a dying horse the industry can flog no longer. "Carbon storage will be expensive at best", the Economist concludes. "At worst, it may not work". There are only three real projects around the world, each keeping just a million tonnes of carbon dioxide a year out of the atmosphere. Offsetting US electricity-sector emissions alone would require 1,500 such sites.

Funds worth $2.6 trillion have now pledged divest from coal. The value of funds shifting away from coal, oil or gas has increased fifty-fold over the past year. The total value of the holdings that these investors are selling is as-yet uncompiled, but one research firm uses a sample of just 7 per cent of the investors involved to estimate the whole group having sold, or promising to sell, some $6 billion worth of fossil fuel investments.

The impact on coal firms is clear everywhere. "Mines in America's Coal Country Just Sold for a Total of Nothing," Bloomberg reports. A group of Appalachian surface and underground mines has had to be given away for free.

The company I am engaged with today also numbers ministries and entire governments among its clients. This compounds the difficulty of their own strategic positioning. It is quite clear that front-running governments on climate change are far ahead of where they were just a few years ago. Right now, as I speak, the Pope is visiting President Obama in Washington. President Xi Jinping will follow him in a few days time. What will come from these meetings? It is a safe bet that the US President won't be contemplating backward steps in his intended legacy issue.

The US and Chinese have just announced joint action on climate by cities, states and provinces. Eleven Chinese cities, with collective emissions equal to Japan's, have set out targets for faster cuts in carbon emissions than the national target. The California Assembly has just passed a 50% renewable portfolio standard by 2030, up from the present 33% by end 2020. In a landmark Senate bill the voting splits 51-26. A 50% energy-efficiency target in buildings by 2030 also passes.

Should the Democrats win the next national election, backsliding from all this seems unlikely. Hillary Clinton has come out in opposition to the Keystone Pipeline. She now says it is "a distraction from the important work we have to do to combat climate change." At the same time, Republicans seem to be losing support at the margins. Seventeen of them break ranks with party leaders in a call for climate change action. At least 10 House Republicans sign on to a resolution endeavouring to put pressure on presidential candidates.

The European Union, meanwhile, professes itself united for an ambitious, binding agreement at the Paris talks. Ministers agree that 5 year reviews and a target of climate neutrality in the second half of the century should be in the treaty.

In Australia, arch climate-denying prime minister Tony Abbot has been ousted by a climate-action seeking ex-lawyer. In interviews and blogs, Turnbull has historically described Abbott's climate strategy as "bullshit".

The British are staking out increasingly lonely ground in the foot-dragging stakes. In Ernst & Young's annual Renewable Energy Country Attractiveness Index, the UK falls to eleventh place. Panasonic is among many companies criticising the UK government's intended cuts to solar subsidies. "Do not push the bird out of the nest before it can fly", their UK boss says.

But the new Conservative government seems perfectly happy to do just that, ignoring the potential for more than 20,000 job losses, meanwhile backing a bid by fossil fuel firms to kill new EU fracking controls. Leaked letters reveal that Europe's biggest oil and gas firms are trying to block environmental controls on shale exploitations.

As for UK electricity supply, mandated conventional power-plant closures now pose a danger to the UK electricity grid. Energy experts are warning of a "breaking point within months", the FT reports. Nuclear clearly cannot help with this. EDF has just admitted a new delay to its intended first new UK reactor at Hinkley Point. The planned £24.5bn plant in Somerset will now not come online until after 2023.

George Osborne's response is to announce £2bn in new government loan guarantees for the project. His goal is to entice EDF and two Chinese nuclear companies to finalise investment in the plant. China could then develop and own a nuclear plant of its own in Britain, the Chancellor announces on a trip to China. Help with financing Hinkley Point could open up their use of other sites, potentially Bradwell.

Few seem to support the government, including in their own usual constituencies of support. City analysts and industrialists join Greenpeace and Friends of the Earth in saying Hinkley Point is a white elephant. The Financial Times concludes that the reactor should be killed off. "The economics of EDF's project looks less and less desirable," an editorial concludes.

The Confederation of British Industry chief slams the UK government more generally, for losing credibility on climate, and harming the economy. UK companies stand to lose hundreds of billions in export opportunities because of the renewables roll-back being led by the Treasury, he says.

Online, 24th September 2015

NBC is covering the Joint Meeting of Congress live. A motorcade of tank-sized SUVs pulls up on Capitol Hill. It includes a small black Fiat containing Pope Francis.

He has made his first point.

The chamber is packed, abuzz. The Pope walks in, to protracted clapping, cheering and a standing ovation. Fully a third of Congress identify themselves as Catholics. Outside, a vast crowd watches his speech on big screens.

He begins reading, his speech slow, his manner gentle.

I am most grateful for your invitation to address this Joint Session of Congress in the land of the free and the home of the brave.

Another standing ovation, one sentence in.

He reminds the assembled politicians that they have a duty to defend and preserve the dignity of their fellow citizens in the tireless and demanding pursuit of the common good. He says he wants to address the entire people of the United States, through his audience, their elected leaders. He invokes great Americans, and alludes to their achievements. He builds his platform carefully and respectfully.

He speaks of the human tendency to form into polarised camps. All religions have been subject to extremism, he says. This is why we must guard against it, in all its forms. The challenges facing us today call for a renewal of the spirit of cooperation which has accomplished so much good throughout the history of the United States.

He recalls Martin Luther King's march for civil rights fifty years ago. He is pleased that America is a land of dreams.

Second standing ovation.

Dreams which lead to commitments, and action. We are not afraid of foreigners. Most of us are descended from immigrants. In our treatment of migrants today we should not repeat the sins of the past. Our world is facing a refugee crisis of a magnitude not seen since that Second World War. Let us remember the golden rule. Do unto others as you will have them do unto you.

Third standing ovation.

He talks of common good. Common good includes the Earth. Now he mentions the encyclical, that he recently wrote in order to enter into dialogue with all people about our common home.

The camera focuses on the stony face of Senator James Inhofe, an extraordinary climate-change denier. It is difficult to read, in those hard eyes, how this is going down.

The Pope elects to talk about climate change without mentioning the words that trigger so many Republicans. In Laudato Si, he says, I call for a courageous and responsible effort to redirect our steps, and to avert the most serious effects of the environmental deterioration caused by human activity. I am convinced that we can make a difference and I have no doubt that the United States – and this Congress – have an important role to play. Now is the time for courageous actions and strategies, aimed at implementing a culture of care and an integrated approach to combating poverty, restoring dignity to the excluded, and at the same time protecting nature. We have the freedom needed to limit and direct technology, to devise intelligent ways of developing and limiting our power, and to put technology at the service of another type of progress, one which is healthier, more human, more social, more integral.

On a visit to President Obama in the White House yesterday Pope Francis was more forthright. Climate change is a problem which can no longer be left to future generations, he said then. But today in Congress he has said it all without using any of the language that would bring the resistance shutters down in the minds of ideologues like Senator Inhofe. This Pope is not just a mighty religious leader, but a wonderful diplomat. It will be fascinating to see the extent to which his obvious popularity carries his perspective on climate change into American climate politics from here on. The outcome will be a key factor in the outcome of the carbon war, of that I have little doubt.

Another critical factor in the struggle brings good news today: the US-China bilateral accord on climate. The New York Times carries a story that President Xi Jinping will make a landmark announcement when he visits President Obama in the White House tomorrow: a national cap-and-trade programme to limit carbon emissions, starting in 2017.

This is a substantial step forward. A cap on emissions in China, enshrined in national legislation, will send a substantial signal into global markets, favouring clean energy, and adding to the difficulties of trying to to cling to the incumbency.

London, 28th September 2015

An investment bank, GMP Securities, funder of much fossil fuel historically, has discovered solar. They host a discussion afternoon for their stakeholders in the Royal Society of Chemistry, and call it "The African Energy Revolution". There is standing room only. Attendees span the financial sector and companies of all kinds operating in Africa.

I am asked to kick proceedings off with my view of the world. I have fresh ammunition for the part of my case that involves soaring incumbency costs and increasing business-model questionability. Shell has announced this morning that it is pulling out of oil and gas exploration in the Arctic, for the foreseeable future, after disappointing results with its latest drilling. Having wasted $7 billion finding nothing, the company remains defiant. "Shell continues to see important exploration potential in the basin", the director responsible says.

By coincidence, we must presume, the company also launches a new gambit on climate change today. They have convened a group of like-minded companies and other players to advise governments on the energy transition, funding it together with BHP Billiton, General Electric (GE) and others to the tune of $6 million. They call it the Energy Transitions Commission. Sixteen commissioners are to sit on it, seven from oil companies, nine from outside the industry. These include some excellent people, with well-established concerns about climate change. But Carbon Tracker has elected to lead what has fast become a chorus of criticism of the move. "We question the credibility and independence of any entity set up by energy incumbents with an interest in maintaining the status quo", CEO Anthony Hobley tells the Financial Times.

Shell are not here today, but GE are. After my opening presentation their boss in Africa offers a different view of the future to mine. We have nothing against solar, he says, but we see a gas-to-power revolution coming in Africa. Gas turbines will proliferate all across the continent. GE already sells a lot of gas turbines, and wants to sell more.

He does not address the dismal implications for climate change, should GE, Shell and others achieve their current aim of spreading profligate use of gas across Africa and the rest of the developing world.

Acwa Power's CEO, Paddy Padmanathan, gives a presentation. Acwa, Saudi Arabia's biggest power-plant developer, has built many a fossil fuel plant, but is increasingly deploying renewables, including solar. They have developed the cheapest solar power in the world: 5.8 cents per kilowatt hour from a 200 megawatt plant sold to the Dubai Government in 2014. Paddy forecasts the next 800MW will be cheaper still: competitive with any generation technology, including coal, on price. With eight Saudi conglomerates as investors, and many billions of dollars to invest in the years ahead, Acwa Power is an icon of the energy transition.

We are expanding from the Gulf countries into Africa, Paddy says. We look for reasonable returns, not excessive profits. To get returns over the long-lifetime power-plant projects we are locked into, we need to take an interest in the long-term prospects of the economies where we are working. So it is essential to partner with local communities on projects that build social cohesion.

This is where solar can come in, I think to myself, at all levels of the energy ladder from solar lights up to solar farms.

Another presenter, an investor, wonders how Sub-Saharan Africa, with all the resources it has, over such a vast area, can today muster less collective GDP than Germany's. Think of the growth potential, he exhorts.

James Cameron, chair of the Overseas Development Institute, and my long-time friend and collaborator, elaborates on this theme. Distributed power *distributes power*, he enthuses. You have to look where the players with the big balance sheets are investing effort in the world today. You have to have to have faith in the new technologies that, crucially, can get you power quicker than the old incumbents. Roughly $100bn has gone into African infrastructure this last year, much of it into renewables.

And so the day goes on, as presentations and discussions come and go around the two end member visions: a revolution focussing on insurgency disruptor industries, or an explosion of the old technologies and industries. What nobody disagrees on is the need for Africa to be lifted out of general energy poverty as soon as possible. Several speakers show the famous cloud-free-collage image-from-space showing Europe and Africa at night: Europe mostly lit up, Africa mostly dark. This has to change. If economies are to develop, and poverty is to be eradicated, African energy supply must grow fast.

Simon Catt, the GMP executive who convened this fascinating and useful meeting of minds, visions and business models, is a true believer in the solar revolution story, and very much wants it to extend all the way to the bottom of the pyramid. He is motivated by both the thought of the prosperity the revolution will spawn, and the social good it can do. He is a keen supporter of SolarAid, accordingly.

My SolarAid colleagues man a stand at Simon's event all through the day, doing brisk business explaining our mission to the executives attending.

Simon has managed to persuade GMP to donate the proceeds of a day's share trading to the charity.

I and the surviving members of the team wrestle on with SolarAid's cash-flow crisis. I have raised £280,000 of the £500,000 target in my "plug the gap" chairman's appeal so far. Finding a few more supporters like GMP will help enormously.

Elsewhere in the City today, unknown to me, another new development strengthens the case for believers in a fast transition away from fossil fuels. In a speech and a BBC interview, Mark Carney, Bank of England Governor and Chair of the Financial Stability Board, describes climate change as the biggest future threat to economies, concludes that the majority of fossil fuel reserves must be deemed unburnable, and voices concerns that the risk of stranded assets might destabilise markets. It is imperative that investors be provided with information that allows them to "invest accordingly", he says, clearly meaning that they should stop further inflating the carbon bubble, so as to increase the chances of an orderly retreat from fossil fuels.

This is the first time a regulator has defined climate change as a major financial risk. It is a landmark in the carbon war.

Carney's intervention shines a whole new light on the UK Treasury's version of energy policy. Their obsession with exploiting UK shale, and suppressing clean energy to make that exploitation easier, is now by corollary a threat to the stability of capital markets. This is what their own financial regulator, by clear implication, is now telling them.

Chapter 27

We are backed by billions

Birmingham, 13th October 2015

The UK solar industry's main annual trade show, at the National Exhibition Centre. Walking the avenues between the exhibition stands, it is not obvious that this is an industry in crisis. Elaborate displays manned by earnest people sprawl into the distance, brands to the fore and slogans shouting positivity. The crowds seem no thinner than previous years, no less a mix of nationalities. But I am stopped half a dozen times before I have progressed fifty yards into the great hall: each time by people worried for their companies, or their jobs, or supply prospects. Wondering if there is anything I can tell them that they don't already know. There isn't.

"Wind and solar keep getting cheaper and cheaper", a Washington Post headline announced a few days ago. President Obama himself tweeted that one. But in the UK? In recent days more than a thousand jobs have been lost as three of the bigger companies fold. The solar trade association says tens of thousands could follow. The government's own estimate for the impact of its proposed subsidy cuts show that up to 22,000 jobs could be lost in the solar sector. This to achieve a cost saving of only £0.04-0.10 a month on the average electricity bill: around £1 a year. Meanwhile they pile multiple billions into a single nuclear plant, and they intend £100,000 bungs for each and every shale fracking pad they can get away with.

I have written to the new Secretary of State for Energy and Climate, Amber Rudd, asking for a meeting. She professes herself too busy.

I do what I can to comfort anguished industry colleagues. I make my way slowly to the Solarcentury stand. It is completely dedicated to our new Sunstation product this year. I talk with the team running it. They report brisk business and multiple accolades for the product. But most people enquiring make no commitments to buy.

I counsel the team to hold their nerve. Solarcentury is in twelve countries on four continents now. Our prospects are much better than those of all the businesses, and investors, that made the mistake of trusting that the Conservatives meant what they once said about clean energy.

Evening. Dinner in a casino on the vast site. I walk in to find that it is a black tie event. I had forgotten that. Or maybe I never knew. I borrow a black tie, but not one of the bow variety, and my crumpled suit makes a poor substitute for a tuxedo. I hope the lights will be low.

Six hundred people sit at round tables around a dance floor and a stage. The occasion is quickly raucous. For many, this will be their last gala dinner in the British solar industry.

After the meal, awards are handed out. Each individual or team winner walks to the stage accompanied by blasts of pop music.

I have won the night's final award, it turns out. I am asked to give a short speech.

I feel embarrassed, I say. I still don't know how to wire a solar array.

I was expecting a small laugh at that, but there is no change in the buzz of a hundred whispered conversations.

It is the teams at Solarcentury and SolarAid that really won this award, I suggest, waving my glass plinth.

That gets a muffled cheer. Probably from the Solarcentury team.

I hasten off the stage.

The band starts to play.

Here I am, Signed, Sealed, Delivered, I'm Yours.

The black ties and long dresses pile onto the floor. I see the young Solarcentury crew dancing together. For a moment I am tempted. In the old days, I wouldn't have hesitated. But I pick up my piece of glass and slip out of a side door back to the hotel.

I make the mistake of logging on to check my e-mails. Sue Lloyd-Roberts, an amazing investigative journalist, a friend, has died.

I look at the minibar, and reluctantly decide against.

Next morning, in the hotel lobby, I sit with a coffee and my substantial e-mail backlog. Sales conversations are going on all around me.

We can do this in sunset yellow, someone with a German accent says loudly at the next table.

Bay Area, California, 23rd–25th October 2015

I am killing time, waiting to visit America's most successful solar installation company, sitting in a Starbucks in San Mateo. The air conditioning is on sub zero and Joe Cocker is oh lording on the sound system.

Any*waaaay*.

I attack my jet lag with a supercharged flat white and check the internet for the news of the day.

A draft climate treaty has been completed in Bonn and forwarded to Paris. "We now have a Party-owned text that is balanced and complete", Christiana Figueres tells the press. Negotiators seem to have put a fractious opening day of the final pre-summit negotiations, just a few days ago, behind them.

France is playing a solid game. President Hollande has been engaging every French ambassador in a global drive for a climate deal, it turns out. Every one of them has been educated in climate change, so they can speak with authority to all comers. And no doubt twist arms where necessary.

In the USA, as many as 25 states are set to challenge President Obama's emissions plan, which is to be published today. The conflict will likely end up in the Supreme Court, the New York Times reports. But the incumbency can expect little succour from their lawyers. The brawling in court will inevitably be protracted. Meanwhile, many states and companies, including some of those that are suing the administration, have already started drafting plans to comply with the rules.

Writing seems to be appearing on walls in the utility world on both sides of the Atlantic. In Europe, Italian utility Enel has joined E.ON and GDF Suez in U-turning away from fossil fuel dependency. The fossil fuel phase-out must be by 2050, it now says. It has even joined forces with former enemy Greenpeace to make the case. Enel will retire 13 gigawatts of fossil fuel power by 2020 for starters.

Abysmal results can be expected in the tar sands this quarter, analysts report. In the shale belt, Bloomberg is marvelling at the early write-downs in the current earnings reporting season: $6.5 billion.

In the UK, the solar redundancies continue to roll out. The company I am visiting today, Solarcity, has closed its UK subsidiary, blaming the British government's assault on subsidies.

Meanwhile the Hinkley Point tragedy forges on. The new nuclear station gets a go-ahead during a visit to the UK by Chinese President Xi Jinping.

Construction could start "within weeks" EDF says. They also report that the reactor isn't coming online until 2025 now.

By then, Apple will have been mass producing solar-charged electric vehicles for five years. Goodness knows what Google will have come up with. Will the hideously expensive nuclear electricity from Hinkley Point even be needed?

Investec Securities, a broker, immediately tells investors to sell EDF shares because of the potential for nuclear liabilities.

The UK government has finally admitted that it is subsidising nuclear power, and will drop its previous pretence that no public funds are involved. Where this leaves the logic of their argument that they can't support solar because it needs subsidies is not clear.

SolarCity CEO Lyndon Rive sends me an e-mail as I scan the dramas on the news sites. He has issues today. He can give me only 20 minutes.

Never mind, I e-mail back. We'll just have to talk fast.

I do know what it is like to have issues, of course. So does everybody who has founded a solar company.

I walk the few hundred yards from the coffee stop to SolarCity's HQ building, in a typical Californian business park by a typical Californian freeway. But I have forgotten that there is no such thing as a typical Californian pavement. Half way there the sidewalk runs out, leaving me walking with my wheelie luggage along a six-line highway thickly populated by speeding cars and trucks.

In the SolarCity building, the foyer is decked out for Halloween, with spiders, webs, and pumpkins. I sit waiting, watching a video of staff partying. It seems they like to dress up for their parties. This they have in common with the staff at Solarcentury. Maybe it is a solar thing.

SolarCity has other things in common with us. But not capitalisation. We are a middleweight. They are very much a heavyweight. "We're backed by billions", their website explains. The market capitalisation of SolarCity today is more than $3 billion.

There are a lot of staff. They file through the lobby, casually dressed without exception, some in running clothes. The company is expanding so fast that it tried to hire 500 people around America in a single day last week.

After my machine-gun-fast conversation with Lyndon, I repair to the Pacific Ocean, half an hour away across the San Andreas Fault and some hills clad with redwoods. I figure I deserve to treat myself to a decent lunch. On Half Moon Bay, I find a crab restaurant, watch surfers wait in vain for surf, and contemplate the hazy curve of the Californian coast. Right now, 2,000 miles to the south, the strongest hurricane ever recorded is heading for Mexico. Patricia is expected to hit the coast at Category 5, with 200 mph winds. Hurricane Katrina, which killed more than 1,800 people, was a Category 3 storm when it reached landfall.

A record 22 storms of Category 4 and 5 have formed across the Northern Hemisphere this year. The previous record was 18, set in 1997 and equalled in 2004.

A weekend living in luxury where the venture capitalists live. I stay in the wonderful hillside home of Stephan Dolezalek, looking out across a wooded valley with occasional neighbouring luxury homes tucked in the trees. Stephan and I spend hours talking about the state of the world. He is moving on from Vantage Point Capital Partners, forming a new fund that he is calling Resourcient. Bruising as his experience has been in the first phase of clean-energy investing, he feels confident that the next phase is going to be different. The fundamentals, as the analysts like to say, are stronger than they have ever been. Stated another way, what other choice is the world going to have now but to invest trillions of dollars in clean energy, instead of fossil fuels?

Stephan has a metaphor for the state of play. The captains of the energy incumbency, and their investors, sit in deckchairs on the luxury ocean liner, smoking big cigars, looking at the tiny lifeboats bobbing in the freezing waters below them. Why would anyone jump into one of those, they ask each other, observing that some people have taken the leap.

But that nasty noise they heard a while ago should be persuading them to do otherwise, if they would but appreciate what it meant. Their ship has hit an iceberg, and it is sinking.

Dinner in another luxury home, this time looking out across the ocean at Pacifica, not with an investor but the other side of the equation: a successful entrepreneur. Dan Shugar and I go back to the mid 1990s. He co-founded a solar installation company, Powerlight, the world leader of its day in solar installation. I was on the board of the first private equity fund investing solely in renewables. We invested in Powerlight.

Powerlight was different from Solarcentury. They were plastering flat roofs of commercial buildings in sunny California with hundreds of kilowatts of solar while we had barely got off the ground with our smaller and more varied installations on homes and commercial buildings under cloudy British skies.

Powerlight were acquired by American solar manufacturing giant Sunpower in 2006 for $332 million. Dan has since built another solar company, NEXTracker. This company was acquired by electronics firm Flextronics in early September for £330 million. He has a golden touch.

We drink fine Californian wine, reminiscing, looking across a bay where whales breach and dolphins play, so he tells me, though not today. Just up the coast twenty great white sharks were sighted a few days ago. Incomprehensibly to me, dozens of surfers are now in the water below us. But not to Dan. He was in there himself a few hours ago.

The water is far warmer than it normally is, he tells me. He didn't need a wetsuit.

We agonise about this and other emerging footprints of global warming. Thankfully, the Mexican hurricane lost strength before landfall and did not produce the damage expected. But as for the long running Californian drought, none of the news is good. The rains have not fallen for four years. The agricultural region inland, in the Central Valley of California, is suffering terribly.

With us is a trio from As You Sow: CEO Andy Behar, President and Chief Counsel Danielle Fugere, and Dan's son David, an Associate. As You Sow targets foot-dragging corporations with shareholder actions and what they carefully call "innovative legal strategies". They filed a resolution on carbon asset risk at ExxonMobil's AGM in 2014 based on the Carbon Tracker report, and the company responded with a report stating "we are confident that none of our hydrocarbon reserves are now or will become 'stranded'."

This evening, it is the global-warming blame question that particularly interests the As You Sow team. ExxonMobil tops the blame list by head and shoulders. In recent weeks two separate teams of investigative journalists have published further evidence that the oil-and-gas giant knew exactly how dangerous global warming was, decades ago, before intergovernmental negotiations started in 1990, as a result of their own detailed in house-research. The journalists drew on evidence from archives, leaked documents, and ex-employees willing to speak out.

ExxonMobil knew about the horrors of global warming, did nothing to alert the world, and everything they could torpedo climate policymaking in all its forms.

"No corporation has ever done anything this big or bad", Bill McKibben says of the revelation. He marvels at "the sheer, profound, and I think unparalleled evil" of the company.

The media's spotlight is firmly on the biggest oil company in the world, in consequence. Increasingly it is falling on the Koch Brothers too, fellow travellers in the suppression of climate truth in perverse defence of the fossil fuels that are the source of their massive wealth.

Two brave Congressmen have said that the ExxonMobil scandal warrants a federal inquiry, and that the "sustained deception campaign" by the company could be prosecuted through truth-in-advertising and racketeering laws.

Racketeering indeed.

Hello innovative legal strategies.

Sacramento, 27th October 2015

I am in the state capital for a day with my host on this trip, old friend Danny Kennedy, recently appointed boss of the California Clean Energy Fund, CalCEF. Like me, Danny once worked for Greenpeace, and left to set up a solar company. His, Sungevity, competes head on with SolarCity. When I open my internet browser, it is the ads of these two companies that leap out at me.

We are here so that I can give a presentation at the California Energy Commission, and talk to senior state officials, who had better remain nameless, about the implications of Carbon Tracker's work for theirs.

Everyone is fascinated by the position the Bank of England has taken. They all want to know what this might mean for other regulators, not least in the USA, both at federal and state level. After all, the state of California is the eighth biggest economy in the world.

The pushback against Mark Carney from the financial services sector has been surprisingly weak, I report. The Financial Times published an article based on only two anonymous sources and a few named executives with very tepid statements. None of them was a CEO. Their argument was that the business of the Governor should be ensuring the stability of the financial markets, not worrying about climate change. In saying this, they illustrate the very problem Mark Carney has chosen to speak out about. Financial institutions are tending not to see that unmitigated global warming is a clear and present danger to both fossil fuel investments and stable capital markets.

I suggest to the California state officials that it is difficult to imagine that the Bank of England is not talking to other regulators around the world about joining them on the barricades. Were they to do so, an already dire new problem for defenders of the energy status quo becomes altogether more difficult.

The news from Europe this morning adds to this case. My Carbon Tracker colleagues organised a gathering of key players in the Guild Hall in London yesterday. At it, Christiana Figueres said that the Paris process is "unstoppable". "No amount of lobbying is going to change the direction," she professes.

Coverage of the event in the Telegraph spells out the implications loud and clear. "Fossil fuel companies risk plague of 'asbestos' lawsuits as tide turns on climate change", the headline reads. Ambrose Evans-Pritchard writes with both barrels. "The old energy order is living on borrowed time", he concludes. "You can, in a sense, compare what is happening to the decline of Britain's canals in the mid-19th century when railways burst onto the scene and drove down cargo tolls, destroying the business model."

I struggle to suppress Schadenfreude, relaying the state of play in my Sacramento meetings. Incumbency defeats are breaking out all around, it seems. Shell halted a tar sands project yesterday. It quit the Carmon Creek facility in Alberta, adding another $2bn to the write-off pile already clocked up in the Arctic and the US shale. Its retreat brings the total of project cancellations in the tar sands this year, industry wide, to eighteen.

Such cutbacks are causing pain right across the oil and gas industry. "Crunch clouds future for oil firms", the Wall Street Journal reports. Oil companies are struggling to generate enough cash to cover spending and dividends, despite all their efforts to cut billions of dollars from operating budgets. Spending on new projects, share buybacks and dividends at the "supermajors" – Shell, BP, Exxon Mobil and Chevron – outstripped cash flow by more than a combined $20 billion in the first half of 2015.

Meanwhile, asset managers are suffering as sovereign wealth funds rooted in oil withdraw billions from funds in order to support their own economies. They are also seeking a cheaper investment approach, the Financial Times reports.

None of this is lost on Sacramento's regulators. If I were on the board at CalPERS and CalSTRS, the two giant state pension funds, I would be increasingly concerned about what I was investing in, and why, in the energy space. Especially if I professed to be worried about climate change.

San Francisco and Oakland, 28th October 2015

CalCEF's office in the heart of the financial district. I am giving a webinar for more than a hundred financiers, foundation representatives, and others interested in the threat of stranded assets. I have so much to talk about that it becomes increasingly difficult to organise it coherently, and hold back from talking too fast. As well as the daily recent pattern of play, I am armed with a new Carbon Tracker report, not much more than a week old. Entitled "Lost in Transition", it takes issue with big energy demand scenarios that routinely assume fossil fuel use will continue to expand in the years ahead. Not so, our analysts argue. Rapid advances in technology, increasingly cheap renewable energy, slower economic growth and lower than expected population rise are among nine themes that could dampen fossil fuel demand significantly by 2040. Typical industry scenarios see coal, oil and gas use growing by 30%–50% and still making up 75% of the energy supply mix in 2040. These scenarios do not reflect the huge potential for reducing fossil fuel demand.

I show a particularly sobering chart from the paper. It plots the IEA's estimates of the forward growth of solar energy globally in 2000, 2002, 2005 and 2007. None exceeded 20 gigawatts by 2014. The reality was nearly 180 gigawatts.

I am thrilled by this report not just because it seamlessly maintains the standards of the earlier catalogue of Carbon Tracker reports, but because younger analysts, Luke Sussam and Tom Drew, co-wrote it with James Leaton. The Carbon Tracker team is developing both resilience and strength in depth.

The webinar is an hour long: 20 minutes for my thoughts and the rest questions and answers. The time evaporates.

I make my way by subway to Oakland for a lunchtime meeting with entrepreneurs in solar and related businesses at an incubator, co-founded by Danny Kennedy, called PowerHouse. Twenty entrepreneurs sit around a room for a weekly session they describe as Open House. In it, they ask for and offer each other help in areas where they can usefully skill share: personnel, finance, contacts, IT, and more. I am asked to say a few words as a nominally successful solar entrepreneur, and answer their questions.

The questions are sharp. The entrepreneurs are mostly young. The occasion has a campus feel to it.

I tell them the obvious: that they are the future. I ask them to spare a thought for executives in their fifties at the top of utility and oil-and-gas companies, and the chances those culture-bound people have of leading their companies across the energy transition successfully. Not high. Many of these

"Men of a Certain Age", I say – for they are almost all men – will not even have rudimentary understanding of, much less expertise in, the tools of the information age that California's young clean-energy entrepreneurs are using routinely in their start ups.

I do not elaborate, for fear of revealing myself as a man of a certain age. But all through this trip I have been increasingly struck by the impact of Silicon Valley innovation in the fabric of Californian life, and how much it fuels the thesis that civilisation is in danger of winning the carbon war.

Most of the cars I have been driven around in are electric vehicles or hybrids. Stephan Dolezalek owns a Tesla Model S. Danny Kennedy has a neat little Fiat 500e. The people driving me search for traffic problems ahead, and how to avoid them, using real-time data processed by Waze, a company acquired by Google for an unknown but undoubtedly huge sum in 2013. I am staying in peoples' homes, in rooms ordered up through Airbnb, a company that has built an empire valued at $25 billion in June simply by leveraging other peoples' assets on the internet. I sit in these rooms communicating on the web for free via the software of a company called Skype, acquired for $8.5 billion by Microsoft. As I do so, hyper-personalised advertising flashes up on my screen, from the likes of Amazon, which began life as an online book company and is now valued at more than $290 billion. I ferry between meetings, when not driven by hosts, in cars laid on at a few minutes' notice by Uber, a company that leverages both other peoples' assets and creates a whole workforce in so doing. And so it goes on. Most of these companies did not exist a decade ago. Now they are worth multiple billions.

How will such corporate manifestations of the information revolution overlay on the energy transition? There is no doubt that they will, of course. And when they do, how fast can things then move?

Stand back and watch evolution at the speed of light, I suspect.

Pick a relevant news story of the day to help make the point. One can do so with ease, most days, these days. Today, a group of chemists at the University of Cambridge announces a super-battery breakthrough. A discovery in lithium-air batteries could allow five times the energy storage of today's best batteries, they tell the world.

Oakland, 29th October 2015

A lunchtime update in the 350.org office. I talk divestment, not engagement, today. I am in favour of both, as readers know well by now. The activists attending the discussion are a wonderful mix of the very young and surprisingly senior.

When we are finished, I listen in to SolarCity's third quarter earnings call. I want to see what Lyndon Rive's issues were a few days ago.

He tells the listening analysts that the company is effecting a major change of strategy, favouring cost reduction and a positive cash flow over installation growth. Customer acquisition costs will be cut in a variety of ways, allowing positive cash flow by the end of 2016, he says. Rooftop installations in 2015 will now be 878 to 898 megawatts, down from previous guidance of 920 to 1,000 megawatts.

The SolarCity share price falls by 18% on this news. It had already fallen 38% over the last six months, wiping away all of its gains over the past two years.

How short term these investors are, I reflect for the umpteenth time. SolarCity will still be growing rooftop installations at 40% a year, in this plan: just not at the breakneck rate in recent years of 80 to 90%, with its associated high costs. The company has well over a quarter of a million customers on its books. It has installed 1.67 gigawatts of solar in all, and is on track for installation of over a gigawatt a year. There is no other solar company like this in the world.

The change in strategy is manifestly sensible because it positions the company better for a key event in Californian solar politics, scheduled for the end of 2016: a cut in the Investment Tax Credit that solar customers receive, from 30% to 10%. This tax credit has been a vital element of forcing growth of the solar industry in California. Politicians may reinstate it, because of the popularity of solar, but that cannot be relied on.

How glad I am, listening to it all, that Solarcentury is not quoted on a stock exchange. How on Earth would our CEO, Frans van den Heuvel, explain to "the market" the ravages that the UK government is inflicting on the British solar industry? We would be a penny stock before he could even begin talking about our overseas pipeline.

Former Secretary of State for Energy and Climate Change Ed Davey has joined the choir of people speaking out. "Rarely has so much damage been done in such short a time", he tells Recharge magazine. "The problem is Osborne and the motley crew backing him."

A survey by the UK's National Association of Professional Inspectors and Testers shows that 70% of UK solar businesses are so under threat from feed-in tariff cuts that they are planning to leave the industry. The wrecking-ball intent of the British government is clear in two new developments. The first is a U-turn on energy tax relief, in amendments to the latest finance bill, for communities planning solar projects. Conservatives once talked with pride of how their notion of "The Big Society" required proliferating community ownership of solar projects. They also talked, in the time of the coalition government, about wholesale installation of solar roofs on the government estate. Solar Power Portal has now used a freedom-of-information request to show the total of solar installed across all the vast acreage of buildings in every government ministry, eighteen months on from the announcement of the Government's "solar strategy". It is precisely zero.

There can be no bigger example of the Conservatives' hypocrisy on climate change.

Meanwhile, their favourite newspaper, the Telegraph, has an interesting headline this morning. "Paris climate deal to ignite a $90 trillion energy revolution." "The old fossil order is on borrowed time as China and even India join the drive for dramatic cuts in carbon dioxide emissions", the paper reports.

Just not in the UK, if Prime Minister David Cameron, Chancellor George Osborne, and Secretary of State for Energy and Climate Change Amber Rudd have their way.

Silicon Valley, 30th October 2015

Sand Hill Road, the most famous road in the Valley. The place where entrepreneurs come to seek money. Many venture capital firms have their spacious offices along this Menlo Park road. I have pitched here, in the early days of Solarcentury, experiencing what it feels like to have your business plan eviscerated by super-confident people with razor-sharp minds, dressed in chinos and tasselled loafers.

Today, though, I am talking to another kind of money nestling among the cacti and palms: philanthropic foundations. I am wearing two hats: the Carbon Tracker one, and the SolarAid one. My hosts permit me a thick glass wall between the two.

My foundation meetings have been going well, when wearing the Carbon Tracker hat. The story is now so very compelling. My main role is to let key funders see the face of the board, and give them the confidence that there is solidity, hopefully, behind the doers in the full-time team.

SolarAid remains a worry. A proposal for our Phase 2 of African solar-lighting market-catalysis has gone to key targets, and we are apprehensively awaiting the outcome. This will probably take a few months to come.

Meanwhile, we have learned that we are short listed for the Zayed Future Energy Prize, the winner-takes-all prize of $1.5 million, for the second year running. We are on a short list of four this time: the only organisation short-listed last year to make it to this year's list.

So, I have another sweaty palmed day to look forward to in Abu Dhabi in January. Meanwhile, I am hoping that the possibility of winning the prize doesn't detract foundations from what I am hoping are the attractions of our proposal, notwithstanding our continuing problems on the solar-lighting frontier in Africa.

The latest setback is that a cadre of our Uganda team have been caught with hands in the till. The police are involved.

Try not to worry, I have told Caesar, SunnyMoney CEO. The same happened at Solarcentury. An employee we trusted totally stole a pallet of solar modules and sold them on. The police were involved there too.

I talk with the foundations doing my level best to understand the world from their perspective: just how many worthy projects there are, and how very difficult it must be to decide priorities, and make choices. But then I think of all those endowments languishing in investments of every shape and size – so many billions, that will become valueless if global warming runs out of control. It is difficult not to become a little frustrated at that thought. We need so much more of what SolarAid has done to date. We need it done so very quickly, on so many fronts. And here I am grubbing around for a few million dollars in a world where any holder of a fortune made in Silicon Valley, or for that matter any holder of a fortune based on a historically high oil price, could write a cheque from small change that would in a heartbeat free SolarAid and SunnyMoney up to catalyse more African solar lighting markets, and so speed up the solar revolution at the bottom of the pyramid.

The latest report on global warming impacts, published earlier this week, shows that the Middle East may suffer heat beyond human tolerance towards the end of this century, on a regular basis, if greenhouse-gas emissions are not cut deeply, quickly. I cannot read such developments without feeling like finding a soundproofed room and having a good scream.

I catch the CalTrain back to San Francisco. There is no sound proofing on this train, and California is partying. I had not realised that Halloween is such a big deal here. The train is heaving with witches, wizards, zombies, and people with enormous spiders on their clothing. Alcohol is being taken. Boys are being boisterous. Some girls too.

But I am catching up on my world on the web, as ever.

"The world is ready for change", says Christiana Figueres in an op-ed. "We should remember that international negotiations don't cause change, they mark it.... We will look back at this moment as a moment of remarkable transformation, as the indisputable turning point of this century. Let us open our eyes now and see it as it happens."

How I have come to love this woman. She should get a Nobel Prize, win or lose in Paris.

Buddhists leaders are joining the Pope now, calling for strong Paris climate deal. It is the first time that a critical mass of leaders of the world's billion Buddhists have spoken out on a global issue. They are arguing for a 1.5°C limit to global warming.

A billion Catholics here, a billion Buddhists there, plus manifest interest from Muslim, Protestant, and Jewish leaders, and pretty soon you have a pan-religions movement.

Perhaps the Middle East won't boil after all.

Chapter 28

Lies, organised crime, and doomed policy resets

Colorado Springs, 5th November 2015

Mark Campanale and I are required for a double-act presentation on Carbon Tracker's work for American investors at an annual conference of those endeavouring to invest responsibly. There are many more such investors in America than there used to be a few years ago, the organisers tell me proudly.

The conference is in a luxury resort capable of housing thousands, in parkland tight against the foothills of the Rocky Mountains. I stare up at the colours of fall splashed across the steep rocky slopes. I feel like taking a walk up there. A long walk. The first snow is on the top peaks. But I will have no time as ever.

The good investors are not alone in the resort. Signs point the way to a Tight Oil Symposium. I follow them, and hold ajar the door to the venue, a ballroom, wondering whether to sit and have a listen. I hear geotechnical data drawled from the stage within. I close the door and walk away. That's enough cognitive dissonance for today.

US shale drillers have dodged a bullet for another three months. Only $450 million of credit lines have been cut by banks in October, a tiny fraction of the total lending extended. The day of reckoning for the US oil and gas industry has been postponed, analysts are reporting.

Bloomberg reports that ExxonMobil and Chevron are cutting their spending on conventional-oil megaprojects in order to focus on shale, where the smaller operations are deemed safer ways to generate cash. I sense an air of desperation as I read between the lines on the business pages. Profits have plunged at both companies. Chevron is in the process of cutting its global workforce of more than 64,000 by 11%. And ExxonMobil is a much diminished giant: its cash pile has shrunk from $28.7bn to $4.3bn since 2006.

Today we hear dramatic news of further scope for diminishment. New York Attorney General Eric T. Schneiderman announces he will be conducting a sweeping investigation into whether ExxonMobil has lied about climate risk to its investors, and the public, over many years. He has issued a subpoena to the company. His investigation will extend right back to the 1970s.

The New York Times reports that other state attorneys general may follow suit. Professors of law are saying that this move could open up years of litigation, similar to action against the tobacco industry. They believe that more oil giants could be joining ExxonMobil in the courts, trying to persuade sceptical lawyers steeped in forensics that they haven't lied to their investors, or worse.

That the American oil and gas giants have distanced themselves from other global giants adds to their problem. On October 16th, ten oil and gas companies pledged their support to the Paris climate summit. The group included BP, Shell, and Saudi Aramco. Global warming should stay below 2°C, the ten CEOs said, without prescribing how to do this, much beyond their well-aired mantra of favouring gas over coal, and pledging that they will invest in carbon capture and storage.

Governments, and attorneys general, may or may not believe all that. But they will surely attach significance to the absence of ExxonMobil and Chevron from the attempt to look supportive of climate-change action by ten of their peers.

All this has to be viewed now in the potentially game-changing rear-view mirror of Volkswagen's global fraud. The latest news here, the day before news of ExxonMobil's subpoena, is that the emissions-dodging now extends to 800,000 cars that may have false carbon dioxide levels, now including petrol cars as well as diesels. Skoda, Audi and Seat cars are also affected. Volkswagen is setting aside £1.4 billion on top of the £4.7bn earmarked for the original 11 million diesels.

I wonder if £6 billion will be enough, in these times, to shake off criminal accountability for the organised crime that some of their executives seem to have been party to.

I know a few lawyers who are betting not.

Balcombe, 11th November 2015

Evening in the church in the village where the oil and gas industry first wanted to frack in the south of England. I am giving a talk to the churchgoers, at the invitation of the vicar. His church is lit with the solar lamps SolarAid sells in Africa, and it looks enchanting. I will be talking more about the light these lamps bring than about fracking tonight.

The reason is that the parish is split, between those who oppose fracking and those who welcome it. Both constituencies are represented here. The anti-frackers won the battle to expel the frackers from Balcombe, with a little help from outside protestors and unhelpful geology thousands of feet under the Weald. But the issue remains divisive, and my instructions are to accentuate the positive.

The main reason for focussing on solar is that we fixed the date for this talk soon after the community had won permission to build a solar farm, on October 1st. Going solar was something both anti- and pro-fracking parishioners felt they could support. They knew a solar farm would make economic sense, with the subsidies available, and that it could produce nearly all the electricity used by the village. They thought it would be a positive, unifying, thing to do: a means of helping lead the way forward.

Then George Osborne cut the tax credits available to communities backing renewable-energy projects.

A few days ago, the villagers reluctantly decided to shelve their plans.

The government is engaged in a scorched earth assault on solar now, it seems. They want no opposition to gas and oil, fracked from British shale or otherwise produced at home and abroad. They seem unembarrassable by their willingness to shovel large subsides to shale and nuclear while torpedoing solar subsidies.

Two days ago somebody in either the civil service or the Tory Party leaked a letter from Amber Rudd to ministerial colleagues. It shows that in June, when she insisted that the UK was on track for its legally-binding European commitment for renewables in the energy mix, she misled Parliament. Now she admits that the government is on course to miss the target by some distance. She suggests some shameful ways of wriggling out of the commitment, like somehow creating renewable energy "credits" abroad so as to claim the targets have been met.

Today, facing calls for her resignation, she has given an interview insisting that she still has the confidence of the renewables sector. I know of no leader in the renewables sector whose view deviates much from derision. The whole spectacle has descended beyond farce.

But all this unpleasant politics I must eschew tonight, for fear of antagonising the Conservatives in the congregation.

In my upbeat proposition that the global energy transition is underway, unfolding right in front of us, I do risk a reference to the Pope. His call for action on climate change has shifted US views, I recount. A survey just completed shows that after after Pope Francis's visit to Washington, the number of Catholics "very worried" about climate change has doubled. The Protestant vicar, sitting in the audience, lifts his thumbs, turns to his parishioners, and memorably announces that he thoroughly approves.

I quickly add that the Church of England is playing a solid game as well.

Gothenburg, 19th & 20th November 2015

The City of Gothenburg presents me with one of its Sustainable Development Awards. City officials hope such awards will one day come to be recognised as the green equivalent of the Nobel Prize. It is a wonderful occasion, but quite surreally uncomfortable for me at the same time. I won the award for my solar work with SolarAid and Solarcentury, yet previous winners include Al Gore, Kofi Annan, Gro Harlem Brundtland, and Paul Polman. I can't stop wondering how they could justify an award to me, in that company, when Solarcentury is a middleweight solar company, not a rock-star heavyweight, and SolarAid may have catalysed the first two African solar lighting markets, but is very much unfinished business. Solarcentury is in fair health – fingers crossed – notwithstanding the UK government's wrecking assault on solar and much else low-carbon in order to make room for shale and nuclear. We are genuinely international now. But SolarAid remains in a highly vulnerable position, with an ebb and flow of encouraging news and bad news. There is no rule that says it is easy helping the poor, it seems.

Brussels, 23rd November 2015

On a Eurostar to Belgium, I catch up on my backlog. The storm that is breaking on the fossil fuel industries immediately before Paris is not yet what users of metaphor would call a perfect one. The US shale story is still widely viewed as a tale of success. But the Bakken shale belt in North Dakota is pumping less oil than it did a year ago. September production is down 1.1%. "The shale boom in North Dakota has softened to a whisper", Bloomberg declares. The future of shale lies in the Permian Basin, the Financial Times tells us. Production is falling less steeply there than in the Bakken and Eagleford shale regions.

Meanwhile the warnings of widespread debt default among shale drillers continue. Chief investment officer of USAA Investments R. Matthew Freund is the latest to speak out. He tells CNNMoney to expect many. The collective global debts of the oil and gas industry are an increasingly worrying component of a general tendency to indebtedness and default. 99 companies have defaulted globally since the beginning of the year, 66% of them American, mostly in oil and gas, mostly in shale.

In conventional oil, the retreat from frontier projects continues. Statoil follows Shell in pulling out of Alaska, exiting 16 operating leases in the process, plus its stake in 50 leases operated by ConocoPhillips.

The City of North Vancouver marks the retreat from oil in a different way. Fuel pumps are to be labelled with cigarette packet-style climate change warnings, the mayor announces.

The first city in the world to label climate risks at the pump will be Canadian.

US Presidential hopeful Bernie Sanders announces unashamedly that in his view ExxonMobil lied on climate. He is not waiting for the courts to take a view. They knew the truth and "lied to protect their business model at the expense of the planet", he tells anyone who will listen.

President Obama seizes the zeitgeist to make a huge move. He rejects the Keystone XL tar-sands pipeline and hails the US as leader on climate change. "Frankly, approving that project would have undercut that global leadership", he announces.

Mark Fulton, Carbon Tracker advisor and analyst extraordinaire, is ecstatic. He considers the work he did for us on the economics of Keystone – or rather, the lack of workable economics – as among his finest. As do many others. And that is saying a lot, given everything he has produced over the years for Deutsche Bank, Citi, and now us.

In Belgium, I am talking to 300 people involved in the nation's clean-tech industries. Tesla has recently opened its first European factory not far from here.

I am going to have a fun day.

City of London, 25th November 2015

The Carbon Tracker team launches its synthesis report, the one where we join up the threads of previous individual reports, since May 2014, on oil, gas and coal. CEO Anthony Hobley kicks off proceedings as ever. It is 8.30 in the morning, but most people on the carefully-constructed invitation list of investors have turned up.

Business history is littered with examples of incumbents who failed to see the transition coming, Anthony says. Kodak, Olivetti, Blockbuster, The American Locomotive Company to name a few. Today, fossil fuel incumbents seem intent on wasting capital trying to hold onto growth by doing what they have always done rather than embracing the energy transition and preserving value by adopting an ex-growth strategy. Our report offers these companies both a warning and a strategy for avoiding significant value destruction.

Anthony is in full rhetorical mood this morning. He speaks of Henry 5th, and King Canute.

Next comes Mark Fulton, beamed from Sydney on video, talking about how the team went about the research. We took the International Energy Agency's scenario for keeping atmospheric carbon dioxide concentrations at 450 parts per million – which the IEA reckons to be enough to offer a 50% chance of limiting global warming to two degrees – and looked at the difference between that and the industry databases we use for working out the capex and carbon-emissions inherent in business as usual.

Mark airs his favourite conclusion. Our work shows thermal coal has the most significant overhang of unneeded supply of all fossil fuels on any scenario, he says. No new mines are needed globally in a 2°C world.

James Leaton then runs through the top takeaways of the research. Fossil fuel companies risk wasting up to $2.2 trillion in the next decade, he says, threatening substantially lower investor returns, by pursuing projects that could be uneconomic in the face of a collage of factors including international action to limit climate change to 2°C and rapid advances and reducing costs in clean technologies. Not only will no new coal mines will be needed, but oil demand will peak around 2020, and growth in gas will disappoint industry expectations. There is a danger zone between industry business-as-usual strategies and action that would be needed to meet the UN commitment to limit climate change to 2°C. If the industry misreads future demand by underestimating technology and policy advances, this can lead to an excess of supply and create stranded assets. This is where shareholders should be concerned. They should question whether companies are committing to future production which may never generate the returns expected. They should insist that companies they are invested in do a two-degrees stress-test.

James dives into the granular detail: the risk by country, by company. The US has the greatest financial exposure, he says: $412 billion of unneeded fossil fuel projects to 2025, at risk of becoming stranded assets, followed by Canada ($220 billion), China ($179 billion), Russia ($147 billion) and Australia ($103 billion).

The companies that represent the biggest risk in a demand misread to the climate and shareholders alike in the next decade are a mix of state and listed companies, including oil majors Royal Dutch Shell, Pemex, Exxon Mobil, and coal miners Peabody, Coal India, and Glencore. Around 20 to 25% of oil and gas majors' potential investment is on projects that will not be needed in a 2°C scenario, and cancelling them would mean forgoing growth.

The report looks at production to 2035 and capital investment to 2025. It warns that energy companies must avoid projects that would generate a total of 156 billion tons of carbon dioxide, in order to be consistent with the carbon budget in the International Energy Agency's 450 ppm demand scenario.

In my concluding remarks, I remind everyone that the IEA's 450 scenario sets out an energy pathway with *just* a 50% chance of meeting the UN 2°C climate change target. What we have all heard here this morning is a very conservative analysis.

Hastings, 26th November 2015

An evening talk that has an air of time travel about it. Hastings is the town where I was brought up. It is also Amber Rudd's constituency.

I walk to the venue along streets I remember well from a misspent boyhood. The event, organised by Transition Town Hastings, is in the basement of the building that used to house the local newspaper. I see from a phone message that that paper, the Hastings Observer, is keen to interview me on my differences of opinion on energy with Ms Rudd.

An audience of mostly elderly people sits in rows of deck chairs of the kind deployed on the beaches of Hastings in summer. There is no heating, and people have brought rugs to cover their legs.

My parents are in the audience. I can't remember the last time they heard me speaking in public. It could even have been when I was at school here, at what used to be the local grammar school, nearly half a century ago.

The organisers struggle with a microphone, which will be absolutely needed if most of the attendees are to hear me, not least my dad. After many minutes of fumbling, we work out that if I hold a frayed wire down with my thumb, the amplification works.

I give much the same talk as I did in the Church at Balcombe – an upbeat narrative of global energy transition unstoppably underway, with no mention of British politics and all the negativity that lies there. I am sure that the activists among the listeners are having difficulty with my failure to mention the UK government, much less to lambast them.

Here, unlike the church, I have had no requests to hold back for fear of offending local Conservatives. And when the question and answer session starts, I have a predictably immediate opportunity to offer my view of British energy policy. I can launch into it now in the context of a full account of the unfolding global clean-energy revolution, having already made an implicit case that the current UK energy situation is an aberration on the world stage.

And what an aberration it is shaping up to be.

Amber Rudd formalised all the government's low-carbon reversals in a speech slated as an "energy reset" for the UK last week. It was an exhortation to dash for gas and nuclear at all costs. Coal is to be phased out by 2025, she said, but the government will not do that unless gas is up and running on multiple fronts.

That, of course, will require a barrage of subsidy to the oil and gas industry.

She offered no solace for the anguished solar industry, and indeed barely referenced solar, other than to take a backhanded sideswipe at it. "We need to work towards a market where success is driven by your ability to compete in a market", she said, "not by your ability to lobby Government."

Meanwhile we learn that the UK has become the only G7 country to increase fossil fuel subsidies this year. A recent report from the Overseas Development Institute and Oil Change International reveals that the government is giving billions in ever increasing handouts to oil and gas majors. This applies at home and abroad. Another report, by Catholic aid agency Cafod, shows that the UK is spending £2.2bn supporting coal in developing countries, and less than half that – £1 bn – supporting clean energy.

As for nuclear, the latest setback for the government scarcely merits belief. EDF's employee shareholders have issued a collective plea to the company management to drop the Hinkley nuclear reactor project. They are saying it is so uneconomic that it puts the company's survival at risk.

The British government flounders on, notwithstanding.

The policy implications of their increasingly monstrous assault on low-carbon technologies become clearer all the time. In the first week of November, unexpected power plant outages required the National Grid to call on industrial users to reduce their power demand, for the first time ever. "UK's high-wire act on power supplies laid bare", the headline in the Financial Times reads.

More than ten percent of the 45 megawatts of electricity needed during the power crunch came from plants that will no longer be available next year.

What are Rudd's civil servants doing about this rapidly worsening high-wire act? Worrying about whether they will be joining the jobless solar thousands next year, in many cases. DECC, a small Whitehall department, is to lose 1 in 8 jobs as the government's drive for austerity in everything except fossil fuels and nuclear rolls on. Amber Rudd justifies this like a character from the novel Alice in Wonderland. "Subsidy should be temporary, not part of a permanent business model," she says.

Most of Britain's major cities seem to have given up on central government. The leaders of more than 50 Labour-run councils make pledges to eradicate carbon emissions in their areas, and to run on green energy by 2050.

Chapter 29

The absurd thing is we know exactly what needs to be done

Paris, 28th November–2nd December 2015

I hoped to join the biggest climate march in history today. The police have cancelled it, though, citing security danger in the wake of the 13th November terrorist attacks. Instead, thousands of shoes are laid out in the Place de la République, symbolising those who would have marched. Pope Francis and UN Secretary-General Ban Ki-moon are among those to donate their footwear. Elsewhere around the world, more than 700,000 people are marching in 175 countries.

With travel long since booked for yesterday, I have a day to fill. I sit in a Starbucks in Montmartre, listening to maddening Christmas songs one after the other, the penalty for access to high-speed wi-fi with which to tour news and weigh odds for the fortnight to come.

"*Have Yourself a Merry Little Christmas...*"

I won't if this climate summit comes off the rails.

I continue to find news to encourage me, though. 183 governments have now submitted their national emissions commitments to the UN. All the G20 countries have now filed, including Saudi Arabia. In a pre-summit announcement, the OECD has reined in export subsidies for coal power stations on a scale that that could put a stop to as many as 850 previously-planned coal plants. The new Canadian government is making progressive noises. The province of Alberta, home of the tar sands, has announced a ground breaking climate agreement aiming at a coal phase out by 2030, a renewable portfolio standard of 30%, a carbon tax and a cap on emissions from tar sands exploitation. In the corporate world, Unilever announces it will join the club of companies pledged to become 100% renewable powered: by 2030, in their case. Companies right across their vast supply chain will no doubt be wondering when they will be required to do the same.

As for the list of those offering potential for discouragement, going into the summit, India is near the top. The Indian government is saying it will oppose a deal in Paris to phase out fossil fuels by 2100. That would frustrate the G7, who – if they remain true to the commitment they made at their own summit in Germany this June – will be working hard to write a phase-out into the agreement. Not to agree with this would entail India telling the world that a global-warming ceiling of 2°C is unattainable. An Indian official justifies continuing their effort to protect coal with this thought: "The entire prosperity of the world has been built on cheap energy."

So why is China coming to such a different view? The air is as unbreathable in Delhi as it is in Beijing. And why assume the cheapest energy is going to coal, even in India? KPMG is forecasting Indian solar prices up to 10% lower than coal power prices these days. By 2025, they say, renewable energy could make up 20% of India's primary power, and be broadly in-line with the national share needed for a global 2°C target.

Climate scientists have announcements of their own, on the eve of the summit. We enter 2016 on the back of the hottest year ever in 2015, they say, with 400 parts per million of carbon dioxide already in the atmosphere, and one degree Celsius of global warming already in large measure the consequence. We will be in uncharted territory from now on, the UK Meteorological office warns. Two degrees looms just a few decades hence, unless we get on the road to deep cuts in global greenhouse-gas emissions soon.

On the eve of the summit, it looks to me as though there are three possible outcome scenarios, only one of them involving potential deep cuts in emissions. The worst would be "No Signal": a message telling the greenhouse-gas profligate organisations and institutions of the world that they can essentially maintain course with the ruinous status quo. This scenario would severely set back the rising tide of efforts by sub-national groups in the long run-up to Paris – states, cities, companies, communities, faith groups, and many others – to persuade the world to turn away from fossil fuels.

A "Contested Signal" scenario would not be so bad, but would hand plenty of opportunity to opponents of change to continue their current stalling.

The outcome that the world needs is "Clear Signal". This, not to be confused with the "Problem Solved" scenario that has never been on the table, will be most strongly felt in the energy sector, from where most of the emissions that threaten a liveable climate derive. Top of a long list of implications, "Clear

Signal" would tell energy incumbents that that the fossil-fuel era is over: that they are now in an era of transition, of rapid managed retreat, whether they like it or not. It would tell financial institutions that the hundreds of billions of dollars invested annually in clean energy today will become trillions.

For the Paris Agreement to send such a signal, it would need to possess a number of core components. It should set a low ceiling for global-warming, within this century. For years now 2°C has been the goal commonly assumed as a reasonable danger ceiling in policy discussions. But the low-lying island nations have long said this is too high to save them, and of late some richer governments – in the face of advice from their own scientists – have been coming to the view that it might be too high for them too. A global-warming ceiling of 1.5°C may be needed, they say. This is why G7 governments in their June 2015 statement adopted a milestone on the road to decarbonisation of "as much as 75% by 2050".

Given the realpolitik of these negotiations, wherein every party must decide it can live with all wording in the treaty adopted, I very much doubt that 1.5°C will make it to the final text. But some form of words for the target like "2°C or less, if necessary" would have to be part of a clear signal.

So too would the corollary: that fossil fuels have to be phased out, within the century. This can appear in coded form, as "net decarbonisation" or some such formulation that allows those who cling to carbon capture and storage to keep their hopes alive. But decarbonisation must be there, at minimum, as it is in the G7 statement.

Given that national emissions commitments made to date sum to global warming of 2.7°C at best, according to the UN, there must be what negotiators call a ratchet mechanism for tightening individual and collective targets. That must entail a transparent mechanism for measuring and reporting progress against existing targets, and a commitment to keep coming back to the table to toughen the treaty regime until the world is on course to stay below the agreed danger ceiling.

All this will be impossible without adequate provisions for finance, to state the blindingly obvious. There are some obvious bottom lines for a clear signal written in dollar signs. First, the rich nations must deliver on the Copenhagen commitment to $100 billion a year for poor nations from 2020. That sounds easy in principle, but there is a world of politics behind this issue, as will undoubtedly become clear in the fortnight ahead. Second, there should be enough treaty language to send a signal that whereas the world is currently investing a few hundred billion dollars a year on clean energy, soon it will be investing many hundreds of billions more, then a trillion a year, thereafter rising further.

There should be more, of course, in any treaty that sends a clear signal: extra action to preserve forests, help for the poorest nations with loss and damage from climate extremes, to name two. But I think of the core prescription for a clear signal can be reduced to a fairly simple 5-point list. Simple to write down, that is. Not simple to achieve.

2°C or less. Total decarbonisation. Effective ratchet. $100 billion to start. Trillions soon after.

Tomorrow I begin to see how that wish list stands up to reality.

Day One, Monday 30th November: I take a train from Gare du Nord to Le Bourget airport, where a vast site has been turned over to the UN to use 24/7 during the two weeks of the summit. 40,000 people are expected. 2,800 soldiers and police will guard them.

A constant stream of hybrid bendy buses ferries the delegates from the suburban train station to the site. Two towns made of structures somewhere between monstrous aircraft hangers and the biggest tents on Earth await us. One town, called the Blue Zone, will mostly host the government delegations and the press. The Climate Generations zone will host 360 civil society organisations and stage hundreds of events. Representatives of civil society organisations are allowed into the Blue Zone, but with caps on numbers per organisation.

I wonder how my Paris experience will compare with the logistical miseries I and almost every attendee I know endured at the Copenhagen climate summit in 2009. I soon find out.

I make my way to a bendy bus without queuing. After a ten-minute ride to the airport, I arrive outside the Blue Zone, and pick up my badge without queuing. I file through an x-ray security check without queuing. There are plenty of security officials in evidence, but all are polite, friendly even. Its as though someone has instructed them all so to be.

In Copenhagen, there were long queues at every stage, sometimes for hours, often extending outdoors in freezing cold, all supervised by police behaving as though they were itching not to help you but to beat you up.

I tell myself to be encouraged. Could this be an omen?

Inside the Blue Zone town, a kind of high street separates halls for negotiations and government offices to the left, and halls for media and non-governmental organisations (NGOs) to the right. I make my way to the Media

Centre. I am registered as a journalist, wearing a hat I can don when convenient by dint of my monthly column for Recharge magazine. My destination is a two-floor pavilion constructed, like most everything else in the halls, with sturdy wood-chip planks. Wires cross the floor, buried under plastic covers. The whole place has an air of well-thought-out impermanence.

The media centre houses 3,500 journalists, the limit set by the UN. Many more wished to attend. 150 heads of state and government are here for the first two days of the summit. Only a few of the journalists present will be allowed into the two vast plenary halls where the politicians and negotiators meet. Most will have to watch on the many big TV screens dangling around the media centre, and listen through headphones.

The action of the day has just started. Film crews are filming the TV screens. Other film crews film them doing so. I can't help but laugh.

Christiana Figueres is on the screen, giving a welcome speech. "Never before has a responsibility so great been in the hands of so few", she says.

The massed rows of diplomats don't look like a few to me. There are thousands of them. Therein lies part of the problem. They must all agree to sign off on every word of the Paris Agreement.

The UN has elected to bring the world leaders in at the beginning of this summit, rather than the end as was the case in Copenhagen. Their calculation is that the leaders can set a high bar, and put their negotiators under pressure to deliver during the remainder of the summit. In Copenhagen, the negotiators floundered around for the best part of two weeks, and relied on their leaders to fly in and sort things out on the last two days. They came ill-prepared, and failed.

Prince Charles has been asked to give the opening speech. I manage to find a seat in a far corner of the media centre, and settle in for the day.

"On an increasingly crowded planet, humanity faces many threats, but none is greater than climate change", the heir to the British throne says. "It magnifies every hazard and tension of our existence. It threatens our ability to feed ourselves, to remain healthy, and safe from extreme weather, to manage the natural resources that support our economies, and to avert the humanitarian disaster of mass migration and increasing conflict."

In November the Prince suggested in a TV interview that the Syria conflict was in part down to climate change. A long running and unusually severe drought had driven people off the land and into cities, helping to fan tensions. Much of the UK media immediately spun his caveated comment into an absolute statement, as is their wont.

Charles now urges negotiators to end fossil fuel subsidies and spend the money on sustainable energy instead.

Another of his current initiatives is a plan to expose climate polluters in the Commonwealth to big-money legal action should they fail to accurately disclose their impact on climate change.

The Prince is a force of nature himself, when it comes to the global environment. As he speaks, you can see clearly how deeply he cares.

"The whole of nature cries out at our mistreatment of her. If the planet were a patient, we would have treated her long ago. You, ladies and gentlemen, have the power to put her on life support, and you must surely start the emergency procedures without further procrastination.

"In damaging our climate we become the architects of our own destruction. While the planet can survive the scorching of the Earth and the rising of the waters, the human race cannot. The absurd thing is we know exactly what needs to be done."

Next up to address the thousands in the plenary hall, the tens of thousands listening on the Le Bourget site, and the millions tuning in around the world, is Ban Ki-moon. The master diplomat elects not to embroider the Prince's stirring rhetoric, and instead expresses his condolences to the hosts for their losses on 13th November. He invites delegations to hold a minute of silence for the victims.

Then President Hollande of France. Like Prince Charles, he links climate change and terrorism. Climate changes causes conflict like it causes storms, he says. What is at stake in this conference is peace. Life on the planet is also at stake. Billions are at watching us right now, and we have no choice but to deliver.

Laurent Fabius, the French foreign minister and former Prime Minister, follows. As president of the conference he must play a pivotal role in the days to come. The summit needs to be a turning point, he says.

"Today's generations are calling upon us to act, while tomorrow's generations will judge our action. We cannot hear them yet, but in a way they are already watching us."

They break for a family picture: 150 Heads of State and Government, standing together in rows as though for a school photograph. The last time such a thing happened, on such a scale, was the Rio Earth Summit in 1992. Then, in the days before modern terrorism, I was able to stand in the same room with the leaders. George Bush Senior, John Major, Fidel Castro, all the rest of them, lined up like schoolboys.

Back in the plenary hall it is the turn of President Obama. What better rejection of those who would tear down our world than marshalling our best efforts to save it, he suggests.

"What should give us hope that this is a turning point, that this is the moment we finally determined we would save our planet, is the fact that our nations share a sense of urgency about this challenge and a growing realization that it is within our power to do something about it."

Now that is key, I think to myself. It is the immense pressure built up by non-state actors of all types since 2013, as part of the UN's "groundswell", that allows you to say that.

Back in the USA, a solid majority of Americans now agrees that the US should join an international climate treaty. A new New York Times/CBS News poll shows 63% of Americans in support of domestic policy limiting carbon emissions from power plants. A slim majority of Republicans remains opposed.

Xi Jinping, President of China, follows Obama. The summit is not a finish line, he says, but a new starting point.

Back in his homeland, the capital is blanketed in choking smog. Beijing is on an "orange" pollution alert, the second-highest level. Xi does not mention this. Nor does he emphasise the solid start China has made on alternatives to coal. New data from Bloomberg shows that emerging nations are outspending rich countries for the first time. China alone is adding more renewables to its energy supply than the US, UK and France combined.

After Obama and Xi, representing between them 40% of the global emissions that need to be contained, heads of state and government one by one exhort each other to find common cause in the bargaining that lies ahead.

German Chancellor Angela Merkel observes that the target of 2°C is not sufficient for to save island states. More ambition is needed, she says.

The Climate Vulnerable Forum negotiating bloc, which includes the island nations, the Philippines, Bangladesh and Costa Rica, have a let it be known that they will break ranks with the G77 group, the traditional representatives of the developing countries, and sign a declaration urging the UN to adopt a more ambitious limit of 1.5°C. Hearing Merkel refer to increased ambition makes me think, for the first time, that a target tougher than two degrees might just be on the cards for the Paris Agreement.

I watch President Putin, fearing that he will take the opportunity for political grandstanding on the Ukraine, or Syria. But he avoids geopolitics. Another hopeful sign, I tell myself.

I have to head back into Paris to give an evening talk at the British Council. British business people resident in Paris have requested a briefing on the summit, and what it might mean for the business world. I take the train back into the city knowing I am missing some history in the making at Le Bourget.

In a separate hall, as leaders' speeches continue in plenary, India and France unveil a global alliance of 120 countries, mostly from the tropical countries, to accelerate the use of solar energy around the world. "The dream of universal access to clean energy is becoming more real", says Indian Prime Minister Narendra Modi. "This will be the foundation of the new economy of the new century."

This must be encouraging, as a clue to India's intentions for the endgame of this summit. Their approach is clearly not going to be a simple "coal or bust" defence.

President Modi has another incentive for the summit to work, a sad one. Every year at climate summit, there invariably seems to be a major climate-related disaster somewhere in the world. This year's seems to be appalling floods, after extreme downpours, in Chennai. Modi is publically ascribing them to global warming.

The global solar alliance is the first of several encouraging initiatives to be announced today. 21 governments announce what they call Mission Innovation: an effort to double the amount of public money going into clean energy innovation. Members include the US, the UK, Australia, Germany, China, South Africa and Brazil. The governments say that their collective efforts will double governmental assistance to renewables such as solar and wind energy to over $20 billion in the next five years.

An alliance of 25 mega-rich tech leaders from ten countries announces a clean energy investment coalition to operate alongside their governments. The Breakthrough Energy Coalition includes Microsoft founder Bill Gates, Facebook co-founder Mark Zuckerberg, Virgin founder Richard Branson and Amazon founder Jeff Bezos.

Gates offers only coded support for existing renewables technologies. "The renewable technologies we have today, like wind and solar, have made a lot of progress and could be one path to a zero-carbon energy future," he observes. "But given the scale of the challenge, we need to be exploring many different paths – and that means we also need to invent new approaches. Private companies will ultimately develop these energy breakthroughs, but their work will rely on the kind of basic research that only governments can fund. Both have a role to play."

I always find it difficult to understand why he is so lukewarm about existing renewables. More innovation is a certainly a great idea. But it will provide icing for what is already a thoroughly appetising cake, according to many practitioners in the renewables industries.

The Africa Renewable Energy Initiative is another newcomer to the list of new clean-energy coalitions. The African Group of Negotiators, the African Development Bank and the United Nations Environment Programme, among others, aim to facilitate the building at least 100 gigawatts of new and additional renewable energy generation capacity on the continent by 2020, and 300 gigawatts by 2030. Current electricity generation in the entirety of Africa is roughly 150 gigawatts.

My British Council commitment means I also miss David Cameron's speech. "Let's imagine for a moment what we would have to say to our grandchildren if we failed," the British Prime Minister says. "What we are looking at is not difficult. It is doable and therefore we should come together and do it."

Big British businesses do not seem to trust that his rhetoric will be matched by appropriate action at home. In an open letter they call on him to rethink his cutting of clean energy budgets. Signatories include Vodafone, Unilever, Tesco, Nestlé, Thames Water, BT, IKEA, Marks & Spencer, Kingfisher and Panasonic.

Flying home to Rome after a visit to Africa, Pope Francis is as clear as ever on the stakes in Paris. "Every year the problems are getting worse", he tells journalists. "We are at the limits. If I may use a strong word, I would say that we are at the limits of suicide."

Day Two, Tuesday 1st December: Laurent Fabius tells negotiators that they must iron out as many differences as they can by Saturday. Ministers will arrive at the weekend to resolve the most intractable remaining disputes. But they must be left with a manageable workload for the second week.

The negotiations split into many small-group sessions, known as "informal informals", to try and reduce the innumerable square brackets used to demark text not yet agreed. The UN has laid on 32 rooms for these bargaining sessions. They and the French Presidency calculate that splitting the negotiations into subsections will lubricate the process, and give voice to the minorities. One of the big mistakes the Danes made in Copenhagen was to agree a secret text with a limited number of countries. Once that became known, trust in their presidency was shot, and failure virtually assured.

At this early stage, one can only imagine the difficulties negotiators are facing in those small rooms, trying to turn their leaders' rhetoric into reality. The last time the co-chairs of these negotiations attempted to produce a slimmed-down draft text, in October, the G77 developing-country bloc rejected it as too biased towards the rich nations. A 20-page draft swelled to 50, riven with square brackets, as a consequence.

Coal will be at the heart of the bargaining. An organisation with a very similar name to Carbon Tracker, Climate Action Tracker, is all over the coal issue. Their quantification of the problem makes daunting reading. They report 2,440 coal plants planned around the world, totalling 1,428 gigawatts. 1,617 of them are in India and China. The rest are in Indonesia, Japan, South Africa, South Korea, the Philippines, Turkey, and the EU. If all of them were to be built, by 2030, emissions from coal power would be 400% higher than a trajectory consistent with 2°C. Yet it emerges now that more than 100 countries are calling for the Paris agreement to cite a warming limit of 1.5°C, at least as an aspiration.

It will be fascinating to see how that nexus of conflicting issues and interests unravels itself. If it does.

Forests are today's theme at Le Bourget. Prince Charles is again to the fore, assuring the summit that attitudes to protecting forests are beginning to change, that new initiatives are being introduced on multiple fronts. However, too many companies still turn a blind eye to the fact that their commercial activities destroy forests, he says. "It remains the case that many of the world's largest companies and their financial backers pay scant, by which I really mean no, attention to the deforestation footprint of their supply chains."

Marks & Spencer and other corporates, keen to show they are not in that camp, sign a new pledge at the summit, committing to prioritise the development of sustainable palm oil, beef, paper and other commodities.

I sit in the media centre with my old friend Tony Juniper, now an advisor to Prince Charles, soaking up everything that comes my way and trying to make sense of it. On computer terminals provided by the UN, I can surf between plenary halls, press conferences, and public events, without attending any of them. I head off periodically for meetings with informants in delegations, NGOs, and the many companies attending the summit. The coffee is good, the food is more than edible, the queues in the many cafes and restaurants are not too long. This, plus the many old friends and interesting acquaintances I keep running into in the corridors, means I would be in danger of enjoying myself if the stakes weren't so high.

One thing that is new about this summit is the amount of survey material that is instantly available on media coverage around the world. An hour scanning e-mail bulletins each morning can give a delegate a clear cross section of coverage and comment not just in his or her own country, but in many countries. This morning, scanning these surveys, I note there is a remarkably optimistic tone to the majority of coverage, in most countries, even India and China. The US and Chinese governments are earning particular plaudits for the games they have played so far.

The conservative press in America is a notable exception. A Wall Street Journal editorial assures American business that nothing significant will emerge in Paris. "The politicians want a deal so badly that they'll accept anything that can pass as one, but it won't amount to much," the Murdoch-owned paper opines. They are talking about my "No Signal" scenario.

And just in case anything does emerge, Republicans in Congressmen are planning to vote through legislation aiming to unravel President Obama's ability to enact policies consistent with text agreed in Paris. The President will veto it, of course, but the distasteful flat-earth political posturing of the Republican Party rages on, and it cannot help the American delegation here in Paris for the world to see what awaits if the they are elected.

Obama remains unflustered. "I'm anticipating a Democrat succeeding me", he says. "I'm confident in the wisdom of the American people on that front."

The Conservative press in the UK offers an interesting contrast. "We are going to move to a lower-carbon future whether it is needed or not", Telegraph columnist Philip Johnston opines, based on his reading of events in Paris. "The direction of travel is clear and irreversible."

Day Three, Wednesday 2nd December: My first event speaking for Carbon Tracker. We have a team of six at the summit. Anthony Hobley and Mark Campanale are doing most of the speaking engagements. I am filling in for them as required. This morning, I am on a platform organised by 350.org in the Climate Generations area, talking divestment.

Bill McKibben opens. More than 500 institutions with assets worth $3.4 trillion have now pledged to divest from fossil fuels, he announces. That's up from 400 commitments worth $2.6 trillion just 10 short weeks ago.

Astonishing, I think to myself. This is a snowball rolling downhill, picking up ever more snow and rolling ever faster. Might it evolve into an avalanche?

Bill recalls that the first institution to divest was a college in Maine, with an endowment of just $3 million. We were thrilled, he remembers. We never imagined we would spark a movement that would get to trillions. And every day the movement grows. Today it was the cities of Bordeaux and Dijon that joined.

He reprises the history: how he used Carbon Tracker's numbers to write in Rolling Stone magazine about the "terrifying new math" of the carbon bubble in July 2012, with a picture of Justin Beiber on the cover. How, just over three years later, when Mark Carney gave his speech in Lloyd's, the Governor of the Bank of England could have been reading from that Rolling Stone article. We didn't set out to be investment advisors, says Bill McKibben, but now it is easy for us to give investment advice. Just say no.

Kevin de Leon, President of the California State Senate, speaks next. CalPERS and CalSTRS, the State public employees' and teachers' pension funds, will now be heading out of coal investments by state dictat, thanks to California's politicians, he says proudly. We must continue to move public policy with seriousness of intent. We must export our policies around the world. We have put 500,000 into jobs in clean energy in California, and grown GDP. If we can decouple carbon and economic growth, we know the world can follow.

Stephen Heintz of the Rockefeller Brothers Fund next. Only 14 months ago, when we decided to divest, funds with only $50 billion under management had pledged to do so. Now the movement is valued at $3,400 billion and counting.

"It doesn't make any sense to invest in losers", he exhorts. "Let's invest in winners!"

Once again, I wonder what John D. Rockefeller would have made of his descendants and their collaborators.

Finally I say my piece, and then May Boeve, executive director of 350.org announces that while the panel has been speaking, one more French city has joined the divestment movement.

The attendees clap and cheer.

I return to the media centre, tour the headlines, and find that Sweden's AP2 state pension fund, the institution I spoke at in Gothenburg last month, has dropped yet more fossil-fuel investments, citing climate risks: 28 power utility stocks.

Todd Stern, US climate envoy, gives a press conference. He is guarded on likely outcomes for the negotiations. It is clear that most of the major issues remain unresolved, this early on.

Jim Hansen, the famous ex NASA climate modeller, now Columbia University professor, gives a solo press briefing. He was the first American scientist to testify in Congress that man-made greenhouse-gas emissions were almost certainly causing global warming. That was way back in 1988. Today, his contrast between the 2°C and 1.5°C targets is graphic. Two degrees would see the planet warmer than the Eemian, an interglacial period between 130,000 and 115,000 years ago, he says. Then, global sea level was between 6 and 9 metres higher than today. There is no question that 2°C is too high to avoid danger. 1.5 should be the target.

The UN High Commission for Refugees follows him. Climate change is already creating refugees, their representatives observe. They do not use caveats.

Refugees from Syria and other conflict-torn countries who have made it to Europe this year now number around a million. I wonder what a 2°C world would look like, refugee wise, with soaring levels of drought, land-based ice-sheet melting accelerating and global sea levels rising accordingly. I wonder what that would do to our world and its fragile social cohesion.

Back in the media centre, I try to take stock, pooling my conversations, the Todd Stern press conference, and press reports from journalists I trust and know to have good inside information of their own.

The big picture is that world leaders have been gone only a day or so, and their negotiators are already struggling. Dan Reifsnyder, one of the co-chairs responsible for editing down the text, has this to say: "We are not making anywhere near the progress we need to at this point."

The global-warming target is predictably contentious. The notion of a 1.5°C target is "beyond sensitive," a Saudi negotiator says. The island nations and their supporters respond that without it, they have no future.

Mention of decarbonisation is also therefore under dispute. The G7 countries are pushing their agreement that fossil fuels should be phased out this century, and they have many supporters, consistent with the requirements to keep global warming below 2 degrees. India, China and many oil-rich countries are resisting.

It is also not clear that there will be a ratchet in the agreement, whatever the temperature or carbon targets set, much less one with potential to be effective. Many countries are happy with the idea of a five-year review, regular reporting, transparency, and collective intent to tighten the ambition of national emissions commitments from 2020 on. But India, China and several countries in Latin America and the Middle East want to wait until up to 2030 to strengthen their targets. Too distant, say others: no sense of urgency.

Finance continues to be contentious. A newcomer to the negotiations would doubtless be pulling their hair out that the rich nations would hesitate for a moment on agreeing $100bn a year for the poor nations, to assist them with emissions cuts and adaptation to climate impacts. But the EU and US are wary about including specific sums in an agreement with an intended shelf-life measured in decades. Also, they point out that some countries classified as poor when the negotiations began in 1991 are now not so poor, and easily able to contribute finance for manifestly poor countries. Indeed, they already are. Witness Chinese investments in Africa.

The negotiators use the term "differentiation" to capture the evolving definition of who is a developing country and who isn't, and what that means for the provision of the trillions needed to fight climate change.

The long and fractious history of the talks suggests that observers should expect brinksmanship by both sides of the argument on finance, and differentiation more generally. For the Indians in particular this is a sore thumb of an issue. An article in the Times of India summarises the common view in Delhi. "There is a galling sense of disquiet, and some pique, that one of the world's poorest exploiters of resources may end up carrying the can for a clutch of effete, indulgent nations that have had a head start in industrialization, and are primarily responsible for punching a big hole in the ozone layer."

On top of all this, negotiators have to agree which elements of the agreement should be legally binding. Use of the word "shall" is crucial here: this is the label that signifies legal commitment. The US strategy is to avoid legally-binding language for emissions-reductions commitments, fearing to arm those who will insist that the Senate should ratify such a treaty. Most other countries are willing to accommodate them in that. But President Obama is equally clear that he wants the periodic review of emissions reductions targets to be legally binding.

On reparations for poor nations on loss and damage, the US and other wealthy countries have long opposed mention in the Paris accord, fearing it could expose them to open-ended claims for compensation. Now they appear to be saying they can accept reference to loss and damage, as long as it is abundantly clear they are not accepting responsibility for compensation. President Obama met with leaders of small-island nations yesterday to say that the USA would help them out with climate-risk insurance measures.

I watch the negotiators walk from one meeting room to another. My schedule of duties here is so much less stressful than theirs. Many of them are already sleep deprived. Sessions have been running hours past midnight.

I catch a bus back to the hotel. I chat to the person sitting beside me, as one tends to at these events. She is Egyptian, a translator on the Saudi delegation.

I don't want to embarrass her by fishing for clues as to how hard the Saudis are working behind closed doors to slow the process – I am told by many sources that they are – or whether she thinks they will ever support a Paris agreement that sends a strong signal to the world. But there is one question I can't resist asking. Back in London David Cameron is trying to persuade Parliament to give him a mandate to bomb Syria alongside the French. Pictures of British politicians braying, goading, and laughing are to be seen on news bulletins as they "debate" the issue. Nobody would imagine they are contemplating action that would inevitably involve the death of innocents as "collateral damage", even if they do succeed in killing existing terrorists. And how many surviving innocents, and Muslim onlookers, will they radicalise? On October 4th, after President Putin began bombing those who seek to oust Assad – ISIS and anti-ISIS alike – Cameron warned that the Kremlin's action "will lead to further radicalisation and increased terrorism." Now, after the November 13th terrorist outrage in Paris, different rules evidently apply to British bombs.

I ask the young Arab woman whether she agrees with me that aerial bombardment in Syria will only succeed in creating more terrorists.

Very much, she responds.

I reflect that if President Hollande is right that the Paris climate summit is really a peace conference, and if it does somehow succeed in sending a strong signal on fossil fuel phase-out this century, then a lot of quick thinking will have to be done on how the transition to clean energy can also become a transition to a more peaceful, or less violence-prone, world.

There are some obvious starting points, of course. If 100% of your energy is being created at home by renewables, you are much less likely to want to go to war for oil.

Chapter 30

A minimum of sleep

Paris, 3rd–6th December 2015

Day Four, Thursday 3rd December: The International Chamber of Commerce organises a one-hour invitation-only discussion in their Paris headquarters, to be recorded for a BBC Radio programme. Two dozen leaders from business and civil society sit round a boardroom table decked with microphones. One of the participants is Gerard Mestrallet, CEO of GDF Suez, now known as Engie. I listen to him carefully, fascinated by the U-turn his company has made from the 'gas-can-do-everything suppress-all-renewables' plan he was such a staunch advocate of until May this year. This was the man who I watched reveal, at the 2014 World Energy Congress, that his favourite target nation for all the fracked gas GDF Suez would be needing was the UK. How much of his new clean-energy business plan has been due to a genuine upheaval of belief system on his part? Or was he pushed into the move, with a degree of reluctance, by circumstances?

Another utility, RWE, has joined E.ON, GDF Suez and Enel in U-turning. On December 1st, exactly a year after E.ON announced it was splitting into two companies, one of them a new clean energy entity on which it will focus for growth, its main German competitor has done the same. RWE's share price had fallen 63% in the twelve months it delayed.

Mestrallet's interventions are cautious and bland. I don't see any evidence of genuine passion for clean energy shining through.

Leaving the meeting, I find myself in a lift with just him and an assistant. He doesn't know me from Adam. Unusually, I didn't find my hand going up in the BBC's discussion. I wonder whether to introduce myself, but decide against.

I have a few minutes to find somewhere with Wi-Fi so that I can tune into Carbon Tracker's main event at the summit. The only place possible is a rock music café. I tune in the UN's live relay web page, and put on my headphones just in time to catch Al Gore giving a speech to kick off the discussion.

"The TV news each evening is like a nature hiker's guide to the Book of Revelations", he says. I wonder if Pope Francis is also listening live.

The former US Vice President's rhetoric has an edge of triumphalism about it today, and the rock music backing it in my ears makes for an upbeat accompaniment. Energy additions from solar PV and onshore wind between 2015 and 2020 will be as large as all of additions of gas between 2010 and 2015, he says. This is because of plunging solar and wind costs, and how they have zero marginal cost once installed. The electricity is free. And don't forget the internet of things. If you have the choice between paying and free, you're going to go for free.

Meanwhile, there are trillions of dollars in carbon assets stranded, he continues. And there are illusions about their value. More and more fossil fuel companies are in danger of losing their licence to operate. I urge you to invest instead in the fantastic new opportunities that are now emerging in the low carbon economy, he concludes.

It is vintage stirring stuff from the tireless master climate campaigner.

I know how he is spending his mornings in Paris: in breakfast briefings with all the big-company CEOs in town. In small groups, he gives them his take on what is happening at the summit, and what it is going to mean for them and their businesses from 2016 onward. Beyond this close support for the UN Secretariat, there is no doubt he will be here in town for the finale, available for diplomacy and advice with other heavy hitters of the climate scene who are around, like Lord Nick Stern.

After Al's speech, Anthony Hobley and a panel of luminaries give their views on stranded assets. There is much talk of investors engaging with energy companies in order to pressure them to stress-test their business models against a world with a 2°C global-warming cap. Then they face questions. One comes from the World Council of Churches. We applaud the direction of travel, their representative asks, but you are not going fast enough. What's your 1.5°C stress test?

Mindy Lubber of Ceres replies. Let's begin with 2°C, she says. We may well need 1.5°C, but we believe that once companies have engaged and have processes, they will realise how much can be done, and actions previously deemed impossible will increasingly begin to seem doable.

I head off to a soiree at Bloomberg's Paris headquarters. Mark Carney is the guest of honour, and gives a short speech, giving us a preview of a major announcement he will make tomorrow. Climate change is a central issue for the financial markets now, the Governor of the Bank of England and Chairman of the International Financial Stability Board says. It is not an issue. It is a central issue. This manifests in three ways. The first is physical. Take insurance. The impacts of Hurricane Katrina were 30% worse because of sea level rise. The second is liability. Think of that risk from the perspective of a coal investor today. The third involves risks around the transition from traditional energy to clean energy. Can we make it smooth or will it be abrupt?

Minds should be focussed on the scale of stranded assets today and the scope for adding to them tomorrow, the Governor continues. The extractive industries have the biggest issues. It is very difficult to transition. For investors, that's the challenge. I don't care if you're a sceptic or a true believer, you can't have a market in transition yet. That's the reason for the initiative we are launching. It is a taskforce to get the switch going and the transparency and disclosures we need to get the markets moving. We launch it tomorrow. We look to have an interim result this time next year. It will be called the Task Force on Climate-related Financial Disclosures. It will be chaired by Michael Bloomberg.

I drift around the room when he has finished, judging the impact of this development. Al Gore, I find, is not the only luminary in danger of triumphalism today.

Dan Lashoff, an old friend who I worked opposite when he was with the Sierra Club, introduces me to Tom Steyer: American billionaire, former hedge funder, now philanthropist, major funder of Democrat contenders, and climate campaigner. Steyer, Bloomberg and Henry Poulsen, former US Treasury Secretary, published a highly influential report in June 2014 called "Risky Business." In it they argued that climate change needs to be treated by industry like any other business risk. Priced in a risk context, they reasoned, the imperative of greening energy comes into stark focus.

Please assure me that Donald Trump can't win the US Presidential election, I ask Tom Steyer.

I assure you, he replies. He fires poll statistics at me about Americans and their preferences, attitudes, and tolerance ceilings. It all sounds rather convincing.

I decide that another glass of champagne is justified, on that basis, hoping that I am not tempting providence.

Back at Le Bourget, champagne is definitely not in order. Negotiators and observers alike are pedalling disappointment now. The 54-page draft text has been cut down by just four pages, and all the substantive issues remain in square brackets.

Germany and France have joined the 1.5°C coalition, delighting the island nations. "This is historic", says the head of the Filipino delegation. "The call of the vulnerable has been answered by the presidency of the COP and the largest economy of the EU host region." But Saudi Arabia and India quash any reference in the draft to a UN "expert dialogue", published in June, which highlights the dangers of breaching 1.5C. They seem to be digging in.

On finance, there is no sign of movement yet. The US and EU are persisting with their argument that China and India should accept a role as donors themselves, as well as recipients. India pushes back hard. "The emphasis should be on the amount of money that is being raised and not on the number of countries in the donor list," says their head of delegation Ajay Mathur.

South African negotiator Nozipho Mxakato-Diseko warns that this is the "make or break" issue.

The amount of money is indeed an issue. Commitments by rich-nation governments to the Green Climate Fund – the organisation that would disburse the £100 billion – total just $2 billion, hardly consistent with the amount the rich nations have agreed to mobilise by 2020. Meanwhile, analysis circulating among delegations shows that the ratio of fossil fuel subsidies to Green Climate Fund pledges in 8 key countries is a shocking 40 to 1. Australia, Canada, France, Germany, Italy, Japan, the United Kingdom and the United States provide a total of around $80 billion per year to support fossil fuel production. This hardly sits comfortably with their rhetoric at this summit. Were I an Indian negotiator, I would find it hard to keep harsh words out of my narrative.

With the emergence of the 1.5°C coalition, the G77 seems to have essentially split into two groups. Rumours are circulating that members of the group opposing 1.5°C are trying to slow things down more generally. As one informant in the heart of the action puts it to me: this group would prefer that the agreement is weak, with as much business as possible left unfinished so that they can chip away and erode it further in 2016, once climate change is out of the spotlight again.

That said, of course, no country wants to be caught with their fingerprints on the wreckage.

Day Five, Friday 4th December: Mark Carney and Michael Bloomberg sit on the stage of the main press conference theatre. Those of us lucky enough to have had the preview last night know know that this duo is about to light a fire in the capital markets.

The G20 asked the Financial Stability Board, that I chair, to look at the issue of climate-change risk in the capital markets, Carney begins. We looked at the top thousand companies in the world and their disclosure of greenhouse-gas risk. Only around a third disclose anything. This has to change. We have to make sure companies have the right information for transition. Hence our announcement today. The Financial Stability Board is setting up a task force, led by the private sector to look at this issue, and make recommendations.

He hands over to the first chairman of the Task Force on Climate-related Financial Disclosures.

The business and finance communities are already playing a leading role on climate change, Michael Bloomberg says, through investments in technological innovation and clean energy. This task force I will be chairing will accelerate that activity by increasing transparency.

After short speeches, they move to a two-person panel discussion with a moderator.

Is business ahead of government, the moderator asks.

There are certainly a lot of companies looking at the opportunities arising from the risk, Carney replies guardedly.

Bloomberg is more forthcoming. No CEO would survive saying climate change isn't a risk, he suggests. If we give them information, they will act on it.

Both Carney and Bloomberg agree the target in the Paris Agreement should be zero carbon. Bloomberg says he thinks we can go below 2°C.

An end to fossil fuel burning just a few decades from now then, I think. Consistent with the G7's agreement, at their summit in June, to phase out fossil fuels this century. And not inconsistent with their acceptance that 70% decarbonisation by 2050 might prove necessary.

I am sitting with Tom Carnac, Christiana Figueres's Senior Strategy Advisor. We look at each other and smile.

Now questions from the floor. You're talking phase out of fossil fuels, Pilita Clark of the Financial Times observes. What does that mean for companies like ExxonMobil?

That is a question for Mr Bloomberg's task force, says Mark Carney. "In all of this I'm neutral", he continues. "I think there does need to be some element of what your strategy is, not just to manage downside risk, but to take advantage of opportunities in a world of transition."

A world of transition. That I like the sound of. Especially from the man responsible for the stability of the global capital markets.

Michael Bloomberg expands. "Philip Morris banned smoking in their offices," he observes laconically.

That I hadn't heard before. Fascinating. A tobacco company knowingly peddling death to the rest of the world, trying to protect the health of their own workers – or avoid being sued, more likely– by banning their own product under their own roof.

"I don't think you have to say anything else", he continues. "Oil companies are going to have to understand they've got a business to run, and they've got a world to live in, and they'll figure out some balance, some ways to do things."

We disband. I buy a coffee and sit to reflect.

I am fascinated by the implications for brands now that people like Michael Bloomberg are saying things like he just did. This is new.

Here in Paris a full propaganda battle is underway around brands. As it unfolds, I am seeing things I would not have believed credible just a short while ago – from both sides in the carbon war.

Shell has elected to take to YouTube to help make its case that gas is a vital part of the future, and renewables are not up to the job. Its PR department has dreamt up a short film, about a romance in a Paris-like city. The narrator speaks French. The subtitles are in English. The whole script, acted out by an attractive woman, and a stylish man, is as follows.

> *"Renewable Energy was lonely. Her love was intermittent. Only when the sun shone and the wind blew. She needed someone reliable. Predictable. Whose love was constant.*
>
> *Then she met Natural Gas. Like her, he was cleaner than most. And he was good value.*
>
> *He said: "What do you want?" She said: "I want someone consistent."*
>
> *He said: "I can love you when the sun goes behind the clouds, and the wind no longer blows."*
>
> *She asked: "Will you still be here in years to come?" He said: "Yes!"*
>
> *She asked: "How long will this last?" He said: "Much longer than you think."*

Renewable Energy was happy, and no longer alone. Together Renewable Energy and Natural Gas formed a Beautiful Relationship."

I had to watch it several times before I could persuade myself it wasn't some elaborate spoof.

I recall the letter Shell sent to its shareholders on May 16th about those of us who espouse stranded-asset risk, a community that has now expanded from Carbon Tracker analysts and their like to include Mark Carney and Michael Bloomberg. We promulgate a "fundamentally flawed" concept, that letter said. And "there is a danger that some interest groups use it to trivialize the important societal issue of rising levels of carbon dioxide in the atmosphere."

I try to imagine the scene where the script of Shell's silly film was pitched, agreed, and signed off. My imagination fails me. Never mind the inaccuracy of the premises, or the question of who is trivializing what, did anyone ask the basic question "might we risk alienating half the viewing population?" Or how about "that proportion of our staff that happens to be female?"

It tells us so much bout this company, the culture it has descended into, and the desperation of its case.

"This is what you get when you give men of a certain age a say in PR", I comment on Twitter.

The other side is no less direct in its outreach and messaging. Some 600 outdoor advertising sites in Paris carry posters bearing the names and logos of corporations, in their style, but purveying messages they will most certainly not be pleased with.

"We knew about the impact of fossil fuels but publically denied it," reads a fake ad for Mobil.

"We're sorry we got caught," one fake ad for Volkswagen reads.

You get the picture.

Brand value – how you are perceived by people – is very valuable to companies, of all kinds. The huge spending by oil companies on the arts makes that point. Any serious effort to malign a brand has to be taken very seriously by company management. But who could these companies sue? 80 renowned artists from 19 countries across the world were involved in designing the fake ads across Paris.

This is what a corporation can expect to face when civil society is turning against it, en masse, well-resourced and with seriousness of intent. Lawyering on their part can only make matters worse.

Both sides use film, of course, and the propaganda against Big Energy is vicious in that medium. A leading British pop musician has produced a video story of his own for the Paris summit. This film aims higher than YouTube. It can be viewed on the website of a national newspaper. In it, an undercover agent of some sort, a woman played by a well-known actress, shoots footage with a concealed camera at a drug-fuelled party overlooking the Eiffel tower. Using facial recognition software, her film identifies fictional Big Energy executives and spin doctors, plus their supporters in government departments, many of whom are committing gross acts as they feed their shared addiction. In one memorable scene, another well-known actress plays a woman the software reveals to be Emilia Knight, a fictional Vice President for Corporate Strategic Planning at ExxonMobil. She is licking a syringe we must assume is filled with heroin, before injecting it into the willing arm of Sam Johnson, a fictional official from the US Department of Energy.

The title of the film is "*La Fête est Finie*": The Party is Over.

Back in Le Bourget it is still not clear that fossil fuel addiction is threatened by the summiteers. The negotiators continue to struggle. There is opposition to a 2°C target for the global-warming ceiling, never mind the 1.5°C proposal. Behind the closed doors of their bargaining sessions, we can only imagine the scenes, and the roles the real-world Emilia Knights and Sam Johnsons are trying – minus syringes, no doubt – to play in them.

Meanwhile, the news from civil society participants continues to be uplifting. A thousand mayors announce that their cities are now intent on going 100% renewable by 2050. They hand a declaration to Ban Ki-moon supporting "ambitious long-term climate goals such as a transition to 100% renewables in our communities, or an 80% greenhouse gas emissions reduction by 2050."

The Mayor of Paris, Anne Hidalgo, led this charge. The Mayor of Sydney, Clover Moore, wants more. "We've developed a master plan to power the city of Sydney with 100% renewable energy by 2030", she tells the press. "We're building this transition from the ground up, showing negotiators here in Paris that they can and must commit to 100% clean energy and an end to fossil fuels as soon as possible."

Let's hope the negotiators are listening.

I ask Tom Carnac for a frank appraisal of the state of play. There will need to be a coming-together moment on finance, he says.

He is being diplomatic. French energy and environment minister Ségolène Royal spells it out more graphically for the press: "The fate of the United Nations global warming talks hinges on the willingness of richer countries to pay poorer ones more for climate-related projects."

China's President Xi Jinping now makes a move that shows up the rich nations: a $60 billion pledge for financing for development across Africa, nearly a doubling of the $30 billion that China promised over a three-year period from 2012.

If China can table that much alone, Beijing is inferring, how bad do the rich nations look holding back on a promise – made back in 2009 – to table $100 billion a year between them, from 2020?

The rich nations will delay until they have squeezed concessions they want from the developing world, of course. Most of those will be on the ratchet mechanism for the agreement: the key to the tightening of emissions commitments that will be needed to have a chance of 2°C or less.

But in all this negotiating brinksmanship, how much risk is there of chaotic meltdown in the closing days?

One development today flags worrying potential for that. The small island states negotiators are meeting in open spaces now, it turns out. They are worried that their designated offices are bugged.

In my tours of comment and commentators I detect the spread of pessimism about how the process is unfolding. I feel my own cautious optimism dip.

Christiana Figueres is as ever upbeat though. "We are where we thought we could be", she tells The Associated Press. Her greatest concern at this point is "that everyone remains focused, that everyone gets a least a minimum of sleep, that everyone remains healthy so that they can all do the work that needs to be done."

The ministers arrive at the weekend. The draft agreement is a catalogue of square brackets still. I don't think Christiana will see her wish about sleep materialise. The negotiators will surely be working all night tonight.

Day Six, Saturday 5th December: A networking breakfast with Michael Bloomberg, in a swank hotel. Once again I hear the great man offer his thoughts to luminaries at the summit. I love the plain way he speaks, and his track record both as a progressive mayor and businessman climate campaigner. But I don't always find myself agreeing with what he says.

Governments are out at Le Bourget shuffling bits of paper, he suggests this morning. Its all pretty meaningless stuff, designed to give journalists something to write about. Governments do what their populations force them to do. China is a great example today. You can't see across the street in their cities. They don't want to end up like Russia, where the Communist party didn't give the people what they wanted and now doesn't exist. Federal governments do basically nothing. Oh, they start wars. But cities are moving, and so are businesses.

Yes, I think to myself. Cities and many businesses are moving, making impressive commitments left, right and centre. But if federal governments were to act in any kind of congruence with that, would that not send the mother of all signals? I for one am not going to give up on federal governments so quickly.

Meaningless stuff or not, at least they seem to be trying. The negotiations did go all night. The final agreed draft text for ministers to review is presented to the UN minutes before Fabius's deadline of midday. It is down to 42 pages, with all major areas of disagreement unresolved.

Laurence Tubiana, French Ambassador for the climate negotiations, a key confederate of Christiana Figueres and Laurent Fabius, has this to say: "We could have been better, we could have been worse, important is that we have a text that we want an agreement next week and all parties want it. The job is not done, we need to apply all intelligence, energy, willingness to compromise and all efforts to come to agreement. Nothing is decided until everything is decided."

Today I am with the medics. The Global Climate and Health Alliance is staging a summit in the city. I have been invited to speak on a panel about what a clean energy future could mean for health. In preparing for it, I learn that 77% of countries have no comprehensive policies on climate change and health. Astonishing.

Speakers in the panel before mine show why. An epidemiologist recounts that ten years ago we didn't know that urban and indoor air pollution from fossil-fuel burning caused cardiovascular disease. Now that we do know, top of the action agenda is to stop to burning coal and kerosene.

The panel also talks of the high greenhouse gas emissions inherent in a meat diet and the much lower emissions for a vegetarian diet. And when you look at direct health benefits, the benefits are clear too, they say: much lower risk of type 2 diabetes and coronary heart disease in particular. Big statistical databases show a strong correlation with total mortality: the more you eat red meat, the earlier you tend to die. Processed meat is classified as a type 1 human carcinogen.

My panel begins with the thoughts of Gary Cohen, President of Healthcare Without Harm. He a man who takes no prisoners when it comes to the fossil fuel industries. "The captains of this public health holocaust are actually being subsidised!", he says. "This is an industry that is completely inimical to human healthcare." Medical organisations must divest, he says. "If we don't become campaigners on this, we are failing in our duty of care as medics."

The summit ends with half a dozen officials on stage representing health associations from around the world. They announce an unprecedented alliance of doctors, nurses, and other health professionals from every part of the health sector, all appealing for a strong agreement at the Paris climate summit. They announce the signatories of declarations representing over 1,700 health organizations, 8,200 hospitals and health facilities, and 13 million health professionals. The global medical consensus on climate change is now at a level never seen before, they say.

I had not realised that the international consensus and strength of feeling in the health sector was as strong as this.

We are all supposed to listen to our doctors, aren't we?

Day Seven, Sunday 6th December: Another venue within a city draped in climate posters, this time in a Palais. The International Renewable Energy Agency and other umbrella bodies convene a day with the European Commission and the government of Abu Dhabi entitled RE-Energising the Future. All too slowly the renewables industries are getting their act together on the climate scene. As things stand, any success at this summit will have had very little to do with the lobbying of the industries that have so much to gain.

National and regional solar associations from around the world do something today that they should have done years ago. Seventeen of them launch a united front, the Global Solar Council, a coalition that includes industry bodies from both established and emerging markets, including China and India. This belated initiative will find itself pushing on doors already ajar due to the earlier efforts of others.

"The levelized cost of solar electricity is 80% lower than it was during COP15 in Copenhagen in 2009", the publicity boasts. Solar, because of its plunging cost and versatility, is "the top candidate" to help countries tighten their national emissions commitments.

There is now many a delegate out at Le Bourget who wouldn't disagree with that.

At the event, I chat with Thierry Lepercq, CEO of Solairedirect, a French company somewhat similar to Solarcentury that has recently been acquired by GDF Suez / Engie as part of their business-model U-turn away from centralised power in Europe. I find him brimming with excitement about the capital that is going to be raised by the solar industry. The short term target should be a trillion dollars, he thinks. Just for solar.

After the lunchtime launch of the new solar council comes a fascinating afternoon session of the conference. A panel includes Statoil's Senior VP for Sustainability, Bjorn Otto Sverdrup; Facebook's Director of Sustainability, Bill Weihl; and Google's Vice President for Energy, John Woolard. I sit happily in the back row. Today is wholly a learning day for me. Increasingly I enjoy these more than the performing days.

The Statoil man surprises me. We are in energy transition, he says. When it comes to future energy, we are investing more and more into renewables. We want to be in the energy business, and for decades to come. We have seen some of the renewable investments outperform our oil and gas investments, for example in offshore wind.

He waxes lyrical about the company's new floating wind turbine project, with turbine blades as long as football pitches. This is real business, he says. It's taken very seriously. When it comes to existing energy, we need to replace coal with gas, and do carbon capture and storage. As for fossil fuel subsidies, of course we need to get rid of those.

Statoil are nearly there then, I think, if this is all company policy. As gas becomes more expensive than solar, and CCS becomes ever more clearly impossible at industrial scale, they will be completely there.

I wonder when that will be?

Facebook's Bill Weihl predictably outbids Statoil. We are trying to power our own datacentres as much as we can from renewables, he says. Our next target is 50% of our energy by 2018. Unless we have no choice, we are unlikely to go to a new area where we can't get clean energy.

Google's John Woolard trumps them both. We want to be 100% renewable powered in the next several years, well before 2020. Right now we're at 30%. We are currently the biggest corporate buyer of renewable electricity in the world.

The discussion is illuminating, as such events invariably are. I learn that Google and Facebook have been working together, notably on a green tariff with Duke Energy, a major US utility, in North Carolina. And both their representatives profess that innovation is key. We can't do what we need to do in the world only with what we have today, says Woolard.

Wiehl agrees. We need more early-stage investment to get companies across the valley of death. Bill Gates and Mark Zuckerberg announced something very important this week.

Back at Le Bourget, Richard Branson's B-team group of corporate leaders tables an appeal to governments to aim for 1.5°C. The bosses of Marks & Spencer, L'Oreal and Unilever are among those on the signature list. "We believe that net zero by 2050 would at least get us to two degrees, leaving the door open for further reduction to 1.5, which should be something we should be looking at in the future," says Jochen Zeitz, former CEO of Puma and co-founder of the B Team. "We believe the business case for net zero in 2050 is irrefutable."

"Carbon neutral by 2050: we will have 35 years to get there," Richard Branson adds. "It's actually just not that big a deal, but we need clear long-term goals set by governments this week. Give us that goal and we will make it happen."

I find all this predictably encouraging, the more so since the list of governments supporting 1.5°C is now well in excess of one hundred, I am told. To my amazement, both Canada and Australia have joined them. Though not New Zealand, to the bitter disappointment of the Pacific island nations.

Laurent Fabius convenes a team of 14 ministerial facilitators today to help him push the Paris agreement across the line. In order to leave time for the text to be translated into the UN's official languages, he says he wants the treaty complete by Wednesday.

CHAPTER 31

Red alert

Paris, 7th–10th December 2015

Day Eight, Monday 7th December: Ban Ki-moon gives a welcome speech to the ministers as they get down to work at Le Bourget. You are here this week to kick off a clean-energy revolution to rein in a climate catastrophe, he says. The world is expecting more than half measures. It is calling for a transformative agreement.

Paul Bodnar, President Obama's senior advisor on energy and climate, gives me his take on progress over a coffee. Differentiation is the key issue, he says, this is where the endgame will focus.

Paul has been a key player in the bilateral discussions the US has held with the Chinese over the last few years. I am struck by how sanguine he appears now, though he says nothing to suggest he is confident a strong-signal agreement can actually be delivered.

Finance is where the politics of differentiation play out most intensely. There are still no reports of any breakthroughs, and in that situation journalists and observers inevitably trawl for clues. I choose to find one in a statement by Shri Prakash Javadekar, India's Minister for Environment, Forests and Climate Change, who suggests that the real cost of the global energy transformation will be in the trillions, and that the $100 billion under dispute is "a symbolic but important gesture" en route to that.

Christiana Figueres elaborates on this issue in a press briefing. "I see a growing consensus that $100 billion will be the floor and not the ceiling", she says. "Are we there yet, at 100 billion? No. But we're certainly moving close."

We will find out in just a few days now.

John Kerry has arrived in town. "We need to have an agreement somewhat in shape by Thursday if we're going to meet the needs of Friday", he says, after a meeting with Laurent Fabius. "And I think everybody wants to try to get this done."

He would be managing expectations, not offering hostages to fortune, if he wasn't confident, I tell myself. I feel my optimism rise again after its dip at the weekend.

In the blizzard of press coverage from around the world, there is some useful fuel today for those who tend to glass-half-full analyses. Researchers publish data in Nature Climate Change suggesting that global carbon emissions have actually dropped in 2015, by 0.6%. This is extraordinary, given that the global economy grew. Though economic growth has decoupled from emissions in many countries, this cannot yet be said of the big emerging economies. Global emissions growth averaged 2.4% a year over the last decade, sometimes topping 3%. China's efforts to cut back its coal burning seem to be the dominant reason for the slight fall this year, alongside global growth of renewables and slower growth for oil and gas.

Does this mean that global emissions have peaked? Have they begun a structural decline, indeed? Or will they rise again in 2016? So much of the answer to that hinges on China and India, and their plans for their power sectors. And those plans in turn will hinge on the course of air pollution in both nations' cities.

Today Beijing announces its first ever "red alert" for air pollution: the highest level of public warning. Schools have been closed. Outdoor construction halted. Factories closed.

It is difficult to imagine that China can reverse its efforts to cut back on coal any time soon.

Renowned climate scientists stage an event at Le Bourget to explain why a 2°C global-warming ceiling does not avoid dangerous climate change, and what a 1.5°C target would entail. Their answer to the latter is a total phase-out of fossil fuel use worldwide by 2030, maybe even 2025, followed by large-scale use of negative emissions technology.

Glen Peters of research agency CICERO spells out the implications. "You would be shutting down coal power plants everywhere, you would be retiring oil everywhere, there would not be any place for gas", he says.

He doesn't view this scenario as likely. The 185 national emissions commitments now on the table in Paris imply a 1% annual growth in emissions, he points out. "Personally, I think if you look at progress in the negotiations, if you look at [national climate plans], you would have to say 1.5°C has an extremely slim chance."

I suppose he is probably correct, especially when one considers what the negative emissions technology entails: mass tree-planting to suck carbon dioxide from the air, and burning plant matter for energy then capturing and storing the emissions. But I wonder how many climate scientists have gone on field trips to Silicon Valley, and heard climate scientists there enthuse about the potential they see for explosive disruption of the incumbency by clean energy these days. There is no doubt this will embrace negative emissions technologies, as well as the rest of the clean-energy family.

Of course, unless there is clear seriousness of intent on decarbonisation in the Paris agreement, and a strong workable ratchet mechanism, all this is academic anyway.

It continues to be a scary thought for fossil-fuel producers though. Saudi Arabia and Venezuela win the NGOs' "Fossil of the Day" award today, for opposing any mention of decarbonisation in the agreement.

The professional purveyors of denial on behalf of fossil-fuel interests, those who style themselves as "sceptics", are much less in evidence at this climate summit than at any previous event. They have complained to the press of being "stifled". But the truth seems to be that there is simply much less interest in their message these days.

The ultra-conservative Heartland Institute, funded by oil and tobacco interests – a consistent promoter of climate disinformation encouraging scepticism – hosts an event in central Paris. This is the organisation that sent emissaries to Rome to try and persuade the Pope to amend or withdraw his encyclical on climate change. Reporters find only around 35 people attending their meeting, mostly old men. The event is advertised as a "day of examining the data", open to public and journalists. But UN-accredited journalists are barred from entering of it by the hosts.

What of the industry these old men seek to defend? CBC News accuses Big Oil today of "hiding" at the summit. "Energy companies prefer ad campaigns and sponsorships here in Paris, instead of setting up public pavilions and booths, as hundreds of other government organizations, NGOs and companies have done."

They, like the sceptics they often fund, certainly do not have the profile here that they normally do. The CEOs of BP, BG, and Statoil are exceptions, having braved the zeitgeist to announce a new initiative today. 45 governments, companies and organizations have committed to "Zero Routine Flaring by 2030". This entails a commitment to stop flaring gas at oil production sites, a process that emits more than 300 million tons of carbon dioxide annually, plus a lot of black carbon, a relaxed 14 years from now.

It is an initiative that highlights perfectly the vast gulf that would exist between the oil industry and the parties to a strong Paris agreement, if such is the outcome here. An agreement with seriousness of intent to decarbonise in time for a 1.5°C ceiling would mean no more oil and gas would be being used by 2030. The Zero Routine Flaring by 2030 target is aiming at operations that wouldn't exist by then.

None of which is to say that intense efforts are not needed by the oil and gas industry to stem the leakage of methane, deliberate and unintended, from their operations, starting yesterday. A new report is in circulation quantifying the amount of leakage from operations in the Barnett Shale of North Texas. A peer-reviewed study published in the Proceedings of the National Academy of Sciences today puts them at least 90% higher than government estimates. This study involves 20 co-authors from 13 institutions, including universities, government labs, the Environmental Defense Fund and private research firms.

Faced with a rising drumbeat of shocking revelations like this, increasingly directors at fossil fuel companies must be worrying about liability. If they aren't, they should be. Another paper published today, in Nature Geoscience, makes the case. In it, two experts in these matters from Client Earth, James Thornton and Howard Covington, note that concerned citizens may feel increasingly inclined to copy the Dutch group Urgenda by seeking to supplement international agreements on climate change in the courts. "We suggest that litigation could have an important role to play", they write. "Many avenues exist for bringing behaviour that drives climate change before the courts."

They go on to examine the options and prospects. Options include exploration of fiduciary duties of company officers and directors, as well as those of pension fund trustees, cases seeking to block state aid to carbon-intensive projects, cases to reform energy markets, and – the holy grail, as they put it – use of tort law: determination of liability that some entity, such as an energy corporation or a coal power plant, is responsible to a claimant for climate change damage. Prospects for success in the latter hinge on proving causality between someone's actions or negligence and a given damage. On this, the science is improving all the time. Thornton and Covington are bullish, on this basis: "We argue that, if science and the law join forces, there is a good chance of achieving court orders for the restriction of greenhouse gas emissions in the not-too-distant future."

In the evening, I head off to a reception in Paris for one of the nations that will be at the top of the list of potential claimants in the event that use of tort law materialises: the Republic of Kiribati, a low-lying Pacific island nation.

President Anote Tong speaks for a few minutes. He describes the impacts of rising sea-level on his coral-atoll homeland: the seawater invasion of the freshwater lens that sustains life on the islands, the increasing ferocity of the cyclones.

Yet Australia intends to open up 50 new coal mines, as things stand, knowing what this will do to us, he says quietly. I'm not an emotional man, but I can't understand the mindset. I ask you to pray for us when you go to bed tonight.

Greenpeace Executive Director Kumi Naidoo also gives a short speech. I spent five days on Kiribati, he says. It was the most emotional time of my years with Greenpeace. He explains why.

I can picture the scenes as he speaks. I visited Kiribati in my own time with Greenpeace, right back at the beginning of my long experience of the carbon war, in October 1990. I remember the happy, trusting, laid back people. A meeting with a president wearing flip flops. Talk of a danger well understood even then, six months after the publication of the first Intergovernmental Panel on Climate Change report.

What must the people of Kiribati think, a quarter of a century on, about the rich nations and how we have collectively ignored their voice for so long? What must they think of people prepared, even now, to open new coal mines, introduce new leaky drilling techniques for extracting oil and gas in ever more extreme ways, mine the tar sands ever deeper?

"I can't understand the mindset", says Anote Tong.

If I were from Kiribati, I'd have stronger words for it than that.

Day Nine, Tuesday 8th December: A Sustainable Innovation Forum opens at the French national stadium, the Stade de France, a venue more accustomed to staging rugby and soccer matches. Another intense but efficient security barrier to get through. Flowers scattered in the street outside the stadium to remind us why it is necessary.

The Carbon Tracker team has hired a box for an elite discussion. Outside in the drizzle the pitch is being re-laid. In this slightly surreal setting, twenty people sit around the room in a circle. Half them them represent serious money, many billions of dollars: asset owners and asset managers. I'm not able to report what is said, beyond what is now probably obvious: a lot of companies can expect to be asked by their shareholders to produce stress-tests of their business plans against a world heading for a 2°C global-warming ceiling in 2016. They will not find it easy.

The occasion adds to my rising sense of hope for the rest of this week, and beyond.

I chat afterwards to Bob Litterman, a former Goldman Sachs executive, now with a hedge fund. He has been famously shorting coal, in a fund he set up wherein significant proceeds go to bankrolling environmental group WWF. The fund is up 40% since its creation in June last year, and a further 7% just in the last few days.

Fossil fuels are in meltdown, Litterman tells me. Analysts are saying that it's because Saudi Arabia is still pumping oil. I think it's at least in part because of the rain of news coming out of Paris. People see writing on walls.

Back at Le Bourget, ministers from the so-called BASIC countries – Brazil, South Africa, India and China – give a press conference. Any agreement must contain clearly differentiated responsibilities for developed and developing nations, they say. It is imperative that rich nations take a visible lead in cutting emissions. They call for more ambition from developed countries. They profess themselves ready to agree a provision in the agreement that would ask them to offer climate finance if they were "in a position to do so" or "willing to do so". But they would agree to such a provision only if it is totally voluntary and if the developed countries themselves explicitly pledge to raise at least US$100 billion every year after 2020, as they promised to do in Copenhagen.

As for the 1.5°C target, that is still on the table as far as they are concerned.

The continuing survival of this 1.5 initiative, widely unforeseen entering the summit, is amazing many people I talk to. Will it be traded away in the endgame? I pray not, especially for the people of Kiribati, and all the other island nations.

Then, late in the day, a very surprising development. A unique group of nations announces they have formed a new coalition. They are calling themselves the "high ambition coalition". Members include the US, the EU, African, Pacific and Caribbean states, and a few Latin American countries.

It turns out that they had agreed their initiative in secret, at the suggestion of the Federated States of Micronesia, six months ahead of the talks.

The group seeks a legally binding global and ambitious deal on climate change, one that sets a clear long-term goal on global warming, in line with scientific advice; that introduces a mechanism for reviewing countries' emissions commitments every five years; that creates a unified system for tracking countries' progress on meeting their carbon goals. "Scientific advice" means the 1.5°C target is an integral part of their high ambition.

Now the game is really on. This is a group that cross-cuts all established coalitions in the climate negotiations. It is unprecedented in the history of the negotiations, and the members in favour of an agreement of the strong-signal type. They will clearly now be endeavouring to recruit others fast.

It emerges that President Obama has been phoning world leaders, pushing hard for an ambitious agreement, encouraging them to intervene constructively with the delegations in Paris.

The parallel efforts of the corporate world remain impressive. Today 114 companies sign up for carbon targets in line with a two-degrees global target. They call themselves the Science Based Targets Initiative. Members include Coca-Cola, Ikea, Kellogg, Proctor and Gamble, Sony, Walmart and many other household brand names. Collectively, the 114 are responsible for emissions the size of South Africa and make nearly $1 trillion a year in profits.

The International Energy Agency joins the Global Solar Council in suggesting solar energy may be the biggest winner in Paris. The agency releases an analysis of the energy-related impacts of the national emissions pledges. "The most common energy-related measures are those that target increased renewables deployment (40% of submissions), or improved efficiency in energy use (one-third of submissions)", the IEA concludes. It also calculates that the national pledges collectively would required $14 trillion additional investment in renewable energy and energy efficiency through 2030.

Barclays Capital analyses who the losers will be, and by how much. If the current pledges are enacted, the fossil fuel industry will lose around $34 trillion in revenues cumulatively from 2014–2040.

I celebrate all this intended business decarbonisation at yet another evening reception, this time hosted by the New York Times. Their celebrated columnist Thomas Friedman interviews Carlos Goshn, CEO of Renault Nissan Alliance, on the future of the electric vehicle.

EVs are 1% of the global car market, Friedman observes. Why should we be excited?

Cars cannot be fuelled by oil going forward, Goshn replies. It's not sustainable. Emissions regulation will drive the growth that is coming. And innovation. The improvements in battery technology are so very encouraging.

So what the hell happened at VW, Friedman asks.

It would be unethical of me to respond to that, says Goshn. Except to say that now everyone in the car industry is under suspicion. I think all of us will now become increasingly attentive to the new technologies, because emissions restrictions will become tougher and tougher.

Friedman waxes lyrical about driving in the Google autonomous vehicle. I felt safer within a few minutes, he says. Where are driverless cars going, he asks.

If the regulations allow, Goshn replies – a big if – in 5 years you will have autonomous cars ready to go, in cities as well. We are working with the Japanese, French and US governments on this. Safety will be a huge factor – 95% of auto accidents are a result of human error.

Dinner is served after their discussion. I talk to neighbours: a man from Microsoft to my left. A lady from Siemens to my right. I ask them what they make of it all, and where their companies' heads are.

We went carbon neutral at Rio+20 four years ago, Microsoft man says. Even if the governments agree nothing here, the energy transition is underway. We'll be signing 30-year power purchase agreements off into the future.

Siemens lady agrees. We pledged to go climate neutral in Climate Week this year, she tells me.

I find myself tempted to think that it might not matter too much if the governments mess up the endgame at the summit. So much of the road to transition seems already seemingly signposted in the markets.

But then I think to myself: so just imagine what the impact will be, on top of all this, if a strong signal does get sent in the Paris agreement.

Day Ten, Wednesday 9th December: In the media centre it is now becoming very difficult to find a seat along any of the dozens of fifty yard desks. The sofa room, a space laid out for relaxing, has turned into a dormitory full of sleeping journalists. Many media representatives are burning as much midnight oil as the negotiators.

John Kerry gives a speech: forty-five minutes of reason delivered with controlled passion. "'We didn't come to Paris to build a ceiling that contains all that we ever hope to do", he says. "We came to Paris to build a floor on which we can and must altogether continue to build." All nations need to play their part. Rich nations have a moral responsibility to help poorer ones, he he says, but developing nations make up 65% of greenhouse gas emission these days and so need to make bold steps to reduce emissions as well. "The situation demands and this moment demands that we do not leave Paris without an ambitious and durable agreement."

The new draft text comes out mid afternoon.

Negotiations overnight have been in what the UN calls the indaba format, wherein only ministers and a few senior advisors are in the room seeking the necessary compromises. As a result, the thousand-plus square brackets of the 3rd December text are down to 226, the 141 options down to 32, 100-plus pages down to 29, 34,000 words down to 19,000.

British pundits are soon ready give their take to the UK press, and do so in one of the three press conference theatres. Michael Jacobs, former senior advisor to former Prime Minister Gordon Brown, says that he and others familiar with climate summits are encouraged by progress, and explains why. Adaptation, capacity building, technology and compliance are pretty clean now, he thinks. Differentiation remains to be fixed: mitigation, finance, and transparency still have different options. Two very different visions can be read in the brackets, and they remain very much open. Only one of these options sends the clear signal that is needed.

The first question is from the Sunday Times. So 1.5°C is still in the draft, albeit in brackets. That means zero carbon by 2050, does it not. How realistic is that?

The treaty needs to send a market signal capable of accelerating the processes we would need in multiple sectors of industry to have a chance of hitting that target, Jacobs responds: in finance, in cities and indeed right across society. We have to remember that solar prices have come down 80% since Copenhagen. This gives a sense of what is possible going forward.

Michael Jacobs was in the thick of it in Copenhagen, right alongside Gordon Brown. Someone asks him what the difference feels like.

At this point in Copenhagen, he responds, we had stopped talking. The summit had broken down.

The journalists disperse. They and others begin their forensics on who played a good game and who were the laggards in the indaba. Informants from inside the room, all speaking unattributably, offer their views. It seems that Saudi Arabia and Malaysia had strong reservations about even this text of 200-plus brackets. The EU, once leader of most charges at climate summits, seems to have almost marginalised itself in these negotiations. The Japanese take a particular caning for their lack of leadership, and willingness even now to cling to coal.

Meanwhile, there is no respite in pressure on negotiators from the business world. Big initiatives are seemingly being announced every day, and today is no exception. A coalition of twenty investors worth £352 billion launch a new push for clean energy. They include the investment arm of one of the UK's biggest insurers, Aviva Investors, the French public sector fund ERAPF, and Norway's largest private pension fund KLP. They will be calling on the biggest companies on the London Stock Exchange to switch to 100% renewable electricity use, and to make public pledges to that effect by joining the existing fifty-plus companies committed to 100%.

Here, then, negotiators will be able to sense a sizeable microcosm of what can be expected if they succeed in sending a big signal in the Paris agreement: capital abandoning fossil fuels actively to favour renewables.

The denier organisations continue to suffer setbacks. Today the Global Warming Policy Foundation has fallen for a Greenpeace sting operation that lays bare its true colours. One of its leading advisors has agreed to accept cash for writing a paper for a sham oil company spinning climate change as scientifically doubtful. The Charity Commission is now investigating whether the Foundation, founded by former Conservative Chancellor Lord Lawson, should have its charitable status revoked.

Exhausted negotiators return to Le Bourget for another long session, seeking the multilateral magic that will close all the brackets.

Fatigue must increasingly be a risk factor as it begins to look like the negotiations will over-run. Laurent Fabius wanted to finish today. He fell more than 200 brackets short of that. John Kerry targets tomorrow. Many experienced players are now betting on Saturday.

Christiana Figueres answers an obvious question from the media. "Of course I get tense. My daughters keep me fed, and they make sure that I'm sleeping. I wouldn't say it's the most restful sleep, but that's not what we are here for. We are here to get a task done."

Day Eleven, Thursday 10th December: My day begins with a panel on fossil-fuel subsidies in the EU Pavilion. The experts assembled on the stage tell a grim story. The OECD counts almost 800 ways that governments subsidise fossil fuels, a representative says. Pre-tax subsidies for fossil fuels add up to between 600 billion and a trillion dollars. Consumer subsidies such as subsidized fuels and company cars add up to $548 billion annually, four times the subsidies going to renewables globally. Producer subsidies that support companies to develop fossil fuels or fossil fuel infrastructure amount to up to $542 billion by G20 governments alone. All these lock in fossil fuel dependency. Many governments spend more on fossil-fuel subsidies than they do on healthcare and education.

Things are changing as pressure builds on climate change and wasted capital, but not fast enough. The International Energy Agency, the OECD, and a growing number of governments are calling for a complete end to fossil-fuel subsidies. Yet subsidies will not feature in the Paris agreement. Few countries mention subsidies in their national commitments, including the EU.

When my turn comes I suggest that the present swelling stream of investor capital fleeing fossil fuels is set to become a mighty river after Paris, hoping I am not tempting providence. The challenge for renewables companies and investors with foresight is to work together on the best ways to divert as much of the river as they can to clean energy. The current hundreds of billions have to become trillions, and soon. Cancelled subsidies will add significantly to that "trillionisation", I suggest. I mean, what is the point of subsidising fossil-fuels if investors are increasingly pulling capital out of them? So, to all the campaigning organisations present, to all those working on this particular battle within the carbon war, more strength to your arm, I say. Let us hope the Paris agreement provides you with a powerful lever. There can't be any fossil fuel subsidies in a 2°C world, much less a 1.5°C world.

There is much milling around today. A "final" text is due at 3 p.m., but 3 comes and goes. It is clear now that the summit will extend into Saturday.

Briefing sessions and huddles are to be observed throughout the halls and corridors of the Blue Zone. Conflicting stories emerge, inevitably. Rumours spread. Only one thing is clear. Any outcome of this summit is still possible, from the good to the bad to the ugly.

Laurent Fabius reminds negotiators of their duties again. "Compromise requires us to forget the ideal solution for everybody so we can get something desirable for everybody."

The Economic Times of India has an observation that rings of truth. "Experts say that the core issue is one of trust", it tells its readers. "Developing countries need to feel that the burden of effort will not fall on them, while developed countries need to feel that as countries move up the development ladder they too will do their share."

The Energy Desk website captures the mood well. "It is as if Le Bourget is full of anxious relatives waiting for a new birth. The pregnancy has gone more smoothly than expected, but tension is growing as the due date approaches. The French hosts have shown great skill as midwives. They are busy trying to work out the words of encouragement – or perhaps the medical intervention – that can bring an agreement into the world."

I smile as I read this. They should have added that the relatives come from a very large and extremely fractious family more accustomed, in their mutual dealings, to the rattling of swords, putting up of fences, and worse, than to endeavours at universal co-operation.

The new text is delivered at 9:20 p.m.

We are down from 29 pages to 27.

The text seeks to keep temperature rises to "well below" 2°C and to "pursue efforts" to limit the temperature increase to 1.5.

Excellent. So long as there is a tight ratchet, and the $100 billion a year that lights the way to the trillion then the trillions.

There is.

In shooting for well below 2°C, and trying for 1.5°C, countries would peak climate change-causing emissions "as soon as possible", and "undertake rapid reductions thereafter towards reaching greenhouse gas emissions neutrality in the second half of the century." They would review and potentially raise the level of pledges countries make, on a five-year cycle. Developed nations would mobilize finance beyond a floor of $100 billion per year.

It is potentially a strong-signal text, if the brackets are closed the right way.

But will they be? While there are square brackets, scope exists for derailments.

There is also some confusion over what "greenhouse gas emissions neutrality" means. Pundits are quick to assert it is simply another way of saying "zero net emissions": greenhouse gases put into the atmosphere are offset to zero, measurably, by the creation of "sinks" like new biomass that take them out.

The Marshall Islands minister who has been the driving force of the 1.5°C campaign, Tony de Brum, immediately lets it be known that he can live with the new draft. He urges all countries to support it.

The next indaba is due to start at 11:30.

I head for my bed, feeling sorry for the negotiators.

Or more exactly, sorry for the majority of negotiators: the ones listening to Laurent Fabius.

Can the few that aren't yet sink the ship?

Chapter 32

Act de Triumph

Paris, 11th & 12th December 2015

Day Twelve, Friday 11th December: The final draft will not be ready today, but tomorrow, the UN announces. Veteran journalists are quick to persuade inexperienced colleagues not to read too much into this. No climate summit ever finishes on time, they say.

I start the day with another panel for 350.org, this time talking to a constituency from cities. Bill McKibben and I perform together once again. I always learn something new when I hear him speak. Today it is where Nelson Mandela went first upon visiting America after his release from jail in South Africa. It wasn't the White House. He went to the University of California to thank the students for their divestment campaign. Apartheid could not have fallen without you, he told them.

When it comes to fossil-fuel divestment today, the cities are right there with the campuses, Bill says. Most recently, Portland has signed up for no fossil fuel infrastructure, and a number of cities up and down the west coast have joined them.

More than 640 cities have now divested, May Boeve, 350.org Executive Director adds.

Its not just about divestment, says Divest-Invest Executive Director Clara Vondrich. Its about where you reinvest the capital you have withdrawn from fossil-fuels: ideally in clean energy, wherever feasible. Well, 130 foundations have shown the way by committing to divest-invest, she says. And their bottom lines are doing just fine.

While this is going on, a group of eminent climate scientists is giving a press conference. They focus on the potential 1.5°C target. John Schellnhuber, Director of the Potsdam Institute for Climate Impact Research – arguably

Germany's most famous climate scientist – says that aspiring to 1.5°C is one thing, but governments need to be clear in the agreement what that target entails.

"What I feel is insufficient in the current treaty is that if you say 1.5°C then you need [to be] phasing out carbon dioxide by the middle of this century", he argues. "You need zero carbon emissions by 2050. If that would also appear in the text than I would be more than happy, and entitled to open a bottle of champagne at Champs-Élysées."

It is another day of much milling around, for press and observers alike. Not so negotiators. They are spotted from time to time scurrying from one side meeting to the next.

News emerges in dribbles. Most significantly, Brazil joins the "high ambition coalition", breaking their longstanding alliance with China and India.

I head into Paris again, this time for an appearance at an arts festival. It proves to be a wonderful experience, just what I need to escape the kicking of heels and general fretting at Le Bourget. The Summit of Creatives is all about using art, in its many forms, to spread the word on the dangers of climate change and to stimulate action in society. I am on a panel of five, sitting in armchairs in the atrium of a grand old Parisian theatre, in conversation, being recorded for live streaming. The audience sits at tables, cafe style, either watching us close to, or on screens if they are further away. This format is a first for me.

The five include Beka Economopoulos, co-founder of a group campaigning to persuade museums to ditch all oil-industry sponsorship, on the grounds that it comes from companies wilfully trashing the beauty of the planet that the museums seek to depict. She tells a compelling story of campaigns around the world, including the one I know about in the UK: the innovative mini-performance based student activism against BP and Shell sponsorship of the Tate, the British Museum and the Science Museum. She shows a short and moving video shot clandestinely on a mobile phone by two young women in police custody here in Paris. Their crime has been to spill molasses posing as oil on the marble floor in the Louvre, protesting about oil industry sponsorship of the arts in France.

Another panellist is Allison Akootchook Warden, described in the programme as activist representative of Alaska Natives. She is an extremely multitalented extravert, whose art ranges across disciplines from paintings to rap music performances. She talks passionately of her efforts to help persuade Shell to quit the Arctic, and leave her beloved animals in peace.

I rap in the voice of the polar bear, and the whale, she explains. I try to explain how it might look from their perspectives. Maybe I can show you all later.

I talk about my book – this book: why I am writing it, my hopes for it. I revel in my single opportunity to talk about the story of the winning of the carbon war in the language of a writer, not that of a creature of science, energy, and capital markets.

After we have all had our say, a moderator leads us in conversation as a group and asks for questions from the audience.

Allison finds her moment to give us a taste of her rap music. She dons a white woolly hat and hits a button on her iPhone – which is already plugged into big a set of the theatre's speakers, it turns out.

Chica boom, chica boom – the music fills the theatre, and Allison warms up for her song with a dance improvisation, in her armchair, of what looks like a swimming polar bear. An exhausted swimming polar bear.

I contemplate shrinking into my armchair. I am English. I used to teach at the Royal School of Mines. I don't share platforms with dancing rap artists. Even if they are seated.

But Allison turns out to be good at this, and has a light touch. She is very a compelling polar bear. Her lyrics are precisely what I would have to say, were I a polar bear.

She has a chorus in her song. "Ooooooooh, where has all the icccccccccce gone?"

The third time she gets to this, she motions to her fellow panellists to join her in song.

Oh no, I think. I am English. Perleease.

But sometimes in life it's easier to have no choice.

"Ooooooooh", I sing. Toneless is fine in rap, I guess. "Where has all the icccccccccce gone?"

Around me, the whole theatre, panel and audience, are polar bears in song now.

Back at Le Bourget, press releases are being prepared. I hear that many organisations are drafting two, one each for a positive and negative outcome. The biggest global public relations firms – the likes of Edelman and Fleishman Hillard – have been hired by governments to handle their messaging. Everybody will want a share of the limelight if history is made. Nobody will want any fingerprints if there is a corpse.

Still the brinksmanship continues.

Todd Stern, US climate envoy, tells Associated Press: "I think there's more of a sense that something is going to get done. But we're not there yet."

Indian environment minister Prakash Javadekar suggests the talks could roll over into Sunday. "Rich nations will have to show a spirit of accommodation and flexibility."

Nigerian environment minister Amina Mohammed spells out what accommodation and flexibility entail. "The $100bn we are asking for is a signal from the international community that they are serious about the financing challenge for climate change", she says. "We need trillions, not billions. The first one hundred billion is the signal that trillions will be attainable."

This is the core of the endgame perfectly summarised. It is all about money. And not much of it at that, in order to start the ball rolling.

Surely they can't let the patient die for want of the capex bill for two giant oilfields, I tell myself. (I speak of Kashagan, the one known in Shell as Cash-all-gone.)

Then a vital piece of information begins to circulate. President Obama has been on the phone to President Xi Jinping, and American and Chinese negotiators have been seen spending suspiciously long in each others' company.

Just before midnight, a remarkable headline appears on the Reuters wire:

"In final push for landmark climate deal, end of fossil fuel era nears."

"At the tail end of the hottest year on record, climate negotiators in Paris will aim on Saturday to seal a landmark accord that will transform the world's fossil-fuel-driven economy within decades and turn the tide on global warming."

Day Thirteen, Saturday 12th December: I make my way to Le Bourget once again, hopefully for the last time. I have a Eurostar back to London booked for this evening.

Outside my hotel, students are gathering for a demonstration 350.org has planned for today. Similar groups are gathering all over Paris. This one speaks Dutch and English with Australian accents. They wear red headbands. The demo is all about red lines, the term climate negotiators use for the issues they supposedly can't compromise on. The point is that the demonstrators hope that red lines have been evaporating at Le Bourget overnight.

So do many people out at Le Bourget.

I find a far-flung spot in the media centre, stock up with food and water, fire up my screens, and wait. I am surrounded by strangers, sitting amid more than 3,000 journalists. But this is where I am going to see the most action I possibly can today.

The plenary hall fills up around 11:30. I watch delegates chatting, many taking selfies. Ban Ki-moon and Francois Hollande walk in together. Delegates stand and clap.

Laurent Fabius calls the thousands to order and begins to speak. Francois Hollande sits to his right, Ban Ki-moon to his left. Christiana Figueres sits to the left of Ban Ki-moon.

Al Gore sits in the front row, Nick Stern next to him. The UN has been putting their experience of the high politics of climate change to good use in the elite discussions for the last few days.

"Today we are close to the end of the process", Fabius says. "For I firmly believe that we have reached an ambitious and balanced draft agreement which reflects the positions of the Parties. It will be distributed to you in a few minutes."

They haven't seen it yet then, beyond what must have been a very small final drafting group.

"This text, which is necessarily a balanced text, contains the main steps forward which many of us thought it would be impossible to achieve."

Fabius and the last-draft editors, whoever they were beyond the obvious guess of the USA and China, are going for a strong-signal text then.

"The proposed draft agreement is differentiated, fair, sustainable, dynamic, balanced and legally binding."

Excellent. I picture international lawyers high-fiving all over the world.

"It acknowledges the notion of "climate justice" and takes into account, for each issue, the countries' differentiated responsibilities and their respective capabilities in the light of different national circumstances. It confirms our central, even vital objective of holding the increase in average temperature to well below 2°C and pursuing efforts to limit this increase to 1.5°C."

Many delegates now clap and cheer.

The high-ambition coalition has won the day, provided the draft isn't torpedoed in plenary.

But who would do that, after all the consultation the French have facilitated? Who would want to stand exposed on the international stage now as the wrecker of hope on global warming?

"It sets an ambitious but necessary long-term target", Fabius continues. "It makes reducing greenhouse gas emissions everyone's responsibility, through the submission or updating, every five years, of national contributions, which in this case can only become more ambitious. It focuses heavily on adaptation to the effects of climate change. It recognizes the permanent, preeminent need for cooperation on loss and damage. It provides for the resources needed to enable universal access to sustainable development, by mobilizing appropriate means of implementation; in addition, the draft decision of our Conference stipulates that the $100 billion per year planned for 2020 will need to be a floor for post-2020 and that a new quantified target will need to be set by 2025 at the latest."

Every one of the boxes in the strong-signal scenario is ticked, then.

"It provides for a global stocktake of our progress every five years, which will enable to us to take collective action if our efforts fall short of the targets set. If it is adopted, this text will therefore mark a historical turning point. And more generally, this COP21 is a real turning point, both in terms of non-governmental action – that of local governments, businesses and many organizations – and in terms of the establishment of a universal legal agreement."

He appeals to delegations to come together in compromise now. The text won't please everyone, he says, but everyone needs to ask themselves whether they can realistically hope for more than the overall balance that is being offered.

"And the answer, as I firmly believe and I hope you do too, is clearly that this text, which we have built together, our text, is the best balance possible, a balance which is both powerful and fragile, which will enable each delegation, each group of countries, to return home with their heads held high, having gained a lot. Today is therefore the moment of truth for all of us. Before you examine the text and we are able, I hope, to approve it a little later in the day, I would like to end by saying this.

"This agreement is necessary both for the world as a whole and for each of our countries.

"It will help island States, for example in the Pacific and the Caribbean, to protect themselves from the rising sea levels which are beginning to submerge their coasts. It will speed up the process of giving Africa access to the financial and technological means that are indispensable for the continent's sustainable development. It will support Latin American countries, in particular to preserve their forests. It will support the countries that produce fossil fuels in their efforts towards technological and economic diversification. It will help us all to make the transition to resilient, low-carbon development based on sustainable ways of life. For above and beyond the climate issues per se, this agreement will support major causes such as food production and security, public health, poverty reduction, essential rights and, lastly, peace."

A peace treaty for our times, then. Just as President Hollande said on Day One.

Fabius reminds the assembled thousands of the failure in Copenhagen. "There were failings and mistakes, and the stars were not aligned; today, they are. At the time, some still hoped that the failure of the moment would be overcome. But if, today, we were so misfortunate as to fail, how could we rebuild hope? Confidence in the very ability of the concert of nations to make progress on climate issues would be forever shaken. Beyond that, it is the credibility of multilateralism and the international community as an entity capable of addressing universal challenges that is at stake. None of us can or will neglect that aspect. The citizens of the world – our own citizens – and our children would not understand it. Nor, I believe, would they forgive us.

"I therefore call upon everyone to bear in mind what our Heads of State and Government said loudly and clearly during the opening of this Conference. What did they say? What mandate did they give us? 'Conclude this universal climate agreement.' At this moment, as we hold the fate of the agreement in our hands, we cannot allow any doubt about the sincerity of the commitments of those very senior leaders, or our own ability to honour the commitments they made.

"Let me conclude. One of you mentioned the other day a famous quote by Nelson Mandela, most suited to the occasion. 'It always seems impossible until it's done.' I would like to add a few more words, by the same hero. 'None of us acting alone can achieve success.' Success is within reach of all our hands working together. Together, in this room, you are going to decide on a historic agreement. The world is waiting with bated breath and is counting on us all."

An eruption of applause.

Delegates standing everywhere.

I brush the tears from my eyes, and check Twitter. Pictures from the demonstration on the streets are streaming in. 15,000 people are evidently marching.

"This is what thousands of people who want real solutions look like", a 350.org tweet says. A sea of people along the Champs-Élysées, red ribbons, scarves, and umbrellas all the way to the Arc de Triomphe on the horizon.

Yes, I think. And I'm looking at another sea of people, on the screen of my UN computer, who also evidently want real solutions today.

Another Twitter picture shows ten square inflated balloons, maybe three metres on a side, labelled as carbon bubbles, bouncing above the heads of the demonstrators.

Ban Ki-Moon follows Fabius, warning delegates not to risk ending up on the wrong side of history, trying to make it politically impossible for any government to wreck a deal now.

"The end is in sight. Let us now finish the job. The whole world is watching."

Then Francois Hollande.

This man and his team have played an absolute blinder, as we English like to say, though rarely of the French.

"You choose for your country, your continent, and for the world. History is here. France calls on you to change the world, to agree the first universal climate agreement in history, that our planet may live a long time."

This gets a standing ovation.

And what happens next, do you think, dear reader?

I invite you to look away, have a think, and guess.

They break for lunch.

This is France, after all.

Reconvening time is set at 3:45.

A French lunch, evidently.

Greenpeace, WWF, Oxfam and Christian Aid give a press briefing. "The climate wheels turn slowly", says Kumi Naidoo for Greenpeace, "but in Paris they have turned. The deal falls short of what we need but puts the fossil fuel industry on the wrong side of history."

Mohammed Adow of Christian Aid: "For the first time in history, the whole world has made a public commitment to reduce greenhouse gas emissions and deal with the impacts of climate change. Although different countries will move at different speeds, the transition to a low carbon world is now inevitable. Governments, investors and businesses must ride this wave or be swept away by it."

Responses to Fabius's speech are flooding into my inbox now, in the form of hurried press releases and simple e-mail quotes to lists of journalists.

Bill McKibben from the march. "This didn't save the planet but it may have saved the chance of saving the planet."

From London, Anthony Hobley gives Carbon Tracker's perspective. "A 1.5°C carbon budget means the fossil fuel era is well and truly over. There is absolutely no room for error. Fossil fuel companies must accept that they are an ex growth stock and urgently re-assess their business plans. New energy technologies have leapt down the cost curve in recent years. The effect of the momentum created in Paris means this is only going to accelerate. The need for the financial markets to fund the clean energy transition creates unparalleled opportunity for growth on a scale not seen since the industrial revolution."

John Schellnhuber, eminent climate scientist: "If agreed and implemented, this means bringing down greenhouse-gas emissions to net zero within a few decades. It is in line with the scientific evidence we (the Intergovernmental Panel on Climate Change) presented of what would have to be done to limit climate risks such as weather extremes and sea-level rise. To stabilize our climate, carbon dioxide emissions have to peak well before 2030 and should be eliminated as soon as possible after 2050. Technologies such as bio-energy and carbon capture and storage as well as afforestation can play a role to compensate for residual emissions, but cutting carbon dioxide is key. Governments can indeed write history today, so future generations will remember the Paris summit for centuries to come."

The corporate world begins to respond, only slightly slower than the NGOs. Paul Polman of Unilever hits the nail hard on the head: "The consequences of this agreement go far beyond the actions of governments. They will be felt in banks, stock exchanges, board rooms and research centres as the world absorbs the fact that we are embarking on an unprecedented project to decarbonise the global economy. This realisation will unlock trillions of dollars and the immense creativity and innovation of the private sector who will rise to the challenge in a way that will avert the worst effects of climate change."

Rob Bernard, Chief Environmental Strategist, Microsoft: (this) "will provide the certainty required for corporations around the world to accelerate their low-carbon investments and foster the creation of a true low-carbon global economy."

I like the certainty bit.

Stephanie Pfeifer, CEO of IIGCC, a network of 120 institutional investors with over €13 trillion in assets under management: "The Paris Agreement is a historic turning point for investors. If adopted, 195 countries will together opt to signal their intention to decouple their prosperity and development from fossil fuel use – sending a very strong signal to business and investors that there is only one future direction of travel to reduce emissions in line with a 1.5°C pathway. Investors across Europe will now have the confidence to do much more to address the risks arising from high carbon assets and to seek opportunities linked to the low carbon transition already transforming the world's energy system and infrastructure."

Reading the cascade of comment and analysis from around the world, and chipping in some of my own on Twitter, time flies.

I suddenly notice that 3:45 has come and gone.

News comes that the restart has been delayed to 5:30.

The rumour mill immediately kicks in.

This is where Twitter comes into its own. Many journalists are glued to it.

Can this delay signify something bad? Twitter information-sharing suggests not.

One journalist tweets that the Indian environment minister has pronounced the text good.

A picture slowly emerges in multiple tweets from trustworthy sources. China, India, the Alliance of Small Island States and Norway have all backed the text.

The US is fine with it apart from what they are saying is "a couple of legal tweaks".

What can that mean, I think. They must surely have been at the table drafting the final form of words. They want to re-edit their own text?

Then, wonderful news. Saudi Arabia is fine with the text, apparently.

Isabel Hilton, a former FT journalist, now has a tweet for the UK government. "Who is going to explain to @GeorgeOsborne that the world just changed and he's looking In wrong direction."

I retweet it, adding: "Can I have a crack at that one please @David_Cameron?"

An American friend drops by my lonely post in the melee of media strangers: Alden Meyer, from the Union of Concerned Scientists. He has a song for the incumbents, he says. "It's the end of their world as they knew it."

And where are the incumbents? Nowhere to be seen. Silent in the media so far.

Bill McKibben sticks it to them further. "Leaders adopt 1.5 C goal", he tweets, "and we're damn well going to hold them to it. Every pipeline, every mine: 'You said 1.5!'"

Finally, the delegates start drifting back into the plenary hall.

I watch anxiously, trying to sense the mood.

Then I see Ban Ki-moon hugging Al Gore and Ségolène Royal, beaming.

The hall fills up.

The minutes pass, with much milling around, hand shaking, and taking of photos, selfies and otherwise.

But they don't come to order.

Huddles of negotiators can be seen, talking to each other seemingly in urgent voices.

Surely there can't be a problem, not now.

A rumour spreads that the hold up is over the words "shall" and "should" in a single sentence of the draft.

An hour has passed.

I have missed my Eurostar.

An hour and a half.

The media centre is full of incredulous faces and shrugs.

At 19:16 Laurent Fabius finally calls the plenary hall to order. He explains he is going to cover legal issues, typos and mistakes in the draft, formally adopt the agreement, then hear speeches from delegates.

He must be extremely confident of unanimity, or something very close to it, to be able to take that approach. It means dissenting governments would have to register any complaint with the wording of the accord after it has been adopted. Fabius must know almost everyone, if not everyone, is on board.

A representative of the secretariat apologises for typos and mistakes made in the document by sleep-deprived officials. Among them, "shall" becomes "should" in Paragraph 4.4.

I now invite the COP to adopt the document the Paris Agreement, says Fabius.

I see no objections, he says, barely looking up. The Paris agreement is adopted.

The cheering is ecstatic.

Clapping delegates stand, tears in many eyes now.

In the media room, rows of journalists join them. Fists pump the air.

So much for dispassion in reporting.

Christiana Figueres hugs Laurence Tubiana. Francois Hollande hugs Christiana.

Christiana Figueres reminds Laurent Fabius that he has forgotten something rather important.

In the excitement of the moment, he forgot to use his gavel.

Fabius lifts the gavel, and raps it on the table, smiling broadly.

The whole row of UN bosses and French hosts on the platform joins hands and lifts arms.

Act de Triumph, I think.

Epilogue

Kent, 31st December 2015

No need for a log fire this New Year's Eve. It's shirtsleeve weather outside. The daffodils are coming out.

Weird weather is in the news on most continents as the year turns. Floods threaten cities in northern England. Tornadoes hit America. Texas struggles with snowstorms yet temperatures reach above freezing at the North Pole. Scientists are quick to find the fingerprints of climate change. Yes, there is a record El Niño event in the Pacific. But such El Niños may well themselves be enhanced by the warming world.

I am reminded of a question I first heard an insurer, actuarial expert Andrew Dlugolecki, ask in the early 1990s: at what point does a climate in meltdown begin destroying wealth faster than it can be created? His guesstimate then, for a world making no effort to curb emissions, was around 2065. Babies born today would be 50 then. An interesting world, we would be bequeathing them.

I am trying not to let the reminders of our race against time dampen my bullish post-Paris mood. I consider that I witnessed a unique milestone in human history at the Paris climate summit, one that holds significant positive transformative power for humankind. I hold this view for three reasons: stakes, scale, and scope.

There has clearly never been a gathering of world leaders to discuss stakes of the kind under discussion in Paris: a threat to global food and water supply, to the global economy and indeed – as many heads of state and government saw it – to life on the planet.

Never in human history has there been a summit to negotiate a treaty on such a scale: one that all nations would have obligations under. 195 governments took part. That is every independent nation on the planet, as listed by worldatlas.com. These governments elected to set aside all the many other areas where they are in dispute to face down a shared global threat, and with seriousness of intent. It may have taken them a quarter of a century to get there, but it was a display of global co-operation without precedent in history.

The scope of the Paris Agreement involves a total system change to the lifeblood of the global economy: decarbonisation of energy. Governments wrote into the treaty the common understanding that they probably need to accomplish this incredible U-turn within the lifetimes of most schoolchildren and indeed college students alive today.

It must seem unbelievable, for people who haven't been following the long-running negotiations. But it is true.

How was it possible? Crucially, the story of the Paris Agreement is about both national actors and non-national actors. As for the national actors, the federal governments, I know more of what went on behind the closed doors, from talking to top-table participants since 12th December, than I have written in my eye-witness account. But definitive accounts of the summit for policy experts will soon be written by people who were actually in those rooms. I am writing my book for the policy experts' relatives, as it were. I do not think my narrative will need correcting materially once detailed forensics of the negotiations are published by practitioners.

In the brew that made for ultimate success in Paris, an unprecedented legion of key groups in civil society descended on the summit with the express purpose of egging governments on. A thousand cities committed to go 100% renewable, some as early as 2030. So did more than fifty of the world's biggest companies, one as early as 2020. Investors with funds worth more than 3 trillion dollars pledged to divest from fossil fuels and/or put shareholder pressure on traditional energy companies to decarbonise. The list of such inducements to national governments is long, as chapters 29 to 32 show.

I believe this double act of governments and civil society is going to change the world in unprecedented and mostly positive ways in 2016 and beyond. My early contacts with the corporate world after the summit suggest an encouraging number of other business people share this view.

Others disagree, of course. They point to the inability of the emissions-reductions commitments made in the Paris Agreement to contain global warming near 2°C, and the lack of legally-binding commitments even for those.

True, the targets themselves are not binding, but a global process capable of achieving them and much more is written in legally-binding language. The ratchet, as negotiators refer to it informally, involves obligations to prepare, communicate and maintain targets and pursue domestic measures to achieve them. It also includes provisions on how to measure, report and verify emissions reduction commitments in a unified way, with the same same frequency and format for all governments, and have them verified through an independent technical process. The word "shall" appears 143 times.

Legally-binding agreements under international law have been likened to credit rating agencies. There are no physical penalties involved, yet companies pull out all stops to stop their ratings falling. The penalties entail loss of credibility, standing, trust among peers – and of course they usually involve decreased ability to raise capital.

Only the passage of more time will show whether the bulls or the cynics are correct in their predictions. But here is mine. The post-Paris era will see the rapid emergence of a major new theme in the worlds of business, policy and politics: transition management.

A chronological selection of extracts from the journey the reader has taken in this book, viewed through the rear view mirror, paints a picture of what the great global energy transition might look and feel like in its early years.

In May 2013 we began the journey by visiting a vast "gigawatt fab" factory making solar panels in China. We can expect many more of these to proliferate across the world now, plus their equivalents in other clean energy industries, and not just in China. Brazil, India and the USA are shaping up to be the biggest national solar markets, as things stand. As the cost of solar energy falls below gas and then coal we can expect large national solar markets to rely increasingly less on imported solar products.

In June 2013 we visited the UK boardroom of BHP Billiton, in the aftermath of the May 2013 Carbon Tracker report warning of capital being wasted in developing new fossil fuel assets, especially in coal. We found complacent, combative, men there. Where is the BHP Billiton share price, and those of other coal miners, today? Low as they are, we can expect further falls, and bankruptcies as a result. These companies have probably left it too late to change course.

In early August 2013 we saw the Bank of England, like BHP, essentially rejecting the idea of stranded assets, and professing that they did not consider that climate risk was particularly relevant to the capital markets. Two years later, Governor Mark Carney led a U-turn in thinking in the run up to Paris, and at the summit. We can now expect more central bankers to follow suit. It would be difficult enough for investors to ignore the concerns of one regulator.

But multiple regulators? The river of capital that has flowed almost by default to fossil fuels hitherto can expect a growing amount of diversion.

In mid August 2013, we visited the first intended shale-fracking site in southern England. Two years later, there has been no fracking in the south of the country yet, and very little in the north. Public protest has seen to that. After Paris, we can expect even more delay, in the face of a protest movement with a tail wind. I predict that the UK Prime Minister and Chancellor will ultimately be forced to retreat from their illusion of plentiful domestic British oil and gas extracted from shale. In the interim they will be reminded of their Paris Agreement obligations time and again. As Bill McKibben put it, speaking of all leaders: "we're damn well going to hold them to it. Every pipeline, every mine: 'You said 1.5!'"

In late August 2013, we sat in the World Economic Forum's HQ in Switzerland, and saw for the first time behind the scenes as the UN laid its careful plan for Paris before a willing vanguard of the global business world. How well that plan played out, we now know. We can expect an emboldened UN in 2016 and beyond. The Paris Agreement is the biggest feather in their cap to date.

In early October 2013 we stood outside the Russian embassy in London and worried about the excesses of President Putin's state in dealing with a peaceful Greenpeace protest against oil and gas drilling in the Arctic. Two years later in Paris, despite much sabre rattling and worse in the Ukraine and Syria in the interim, Putin kept his counsel and went with the flow. We can probably expect major efforts to play green-tech catch-up in Russia in 2016 and beyond. The Russians will not want to be left behind as a green industrial revolution sweeps the world.

In mid October 2013, at the World Energy Congress in Korea, we saw how very far the energy incumbency then had to travel, were it to embrace the Paris agenda. Yet by then the top 20 European utilities, worth €1 trillion in 2008, had already by then lost half their value because of the rise of renewables. The writing was on the wall, for those looking. By the time the Paris climate summit was over, four of those utilities were in the process of executing U-turns with their business models, looking to clean energy not fossil fuels for growth. We can certainly expect more to do so in 2016. Big energy will be morphing before our eyes.

In early November 2013, in Oslo, we sensed for the first time how even the biggest of pension funds was becoming vulnerable to pressure in seeking to make or maintain fossil fuel investments. Little did we imagine then that divestment would snowball as fast as it did subsequently, especially in the month before the Paris summit. We can expect more accretion to the snowball in 2016. We shouldn't be surprised if the flight of capital from fossil fuels turns into an avalanche.

In mid November 2013, in Berlin, we went on our first street demonstration. In the two years to follow, as the activists from Avaaz, 350.org and other organisations wove their bottom-up magic, these protests became increasingly potent in projecting public anguish about global warming. Christiana Figueres has said that she knew we would win when she witnessed the massive street protests in New York in September 2014. The success in Paris will have emboldened and energised a global movement already brimming with youthful vigour. We can expect all would-be foot-dragger governments and corporations to face higher levels of peaceful public protest, in all its varied and increasingly innovative forms, in the transition.

In December 2013, we joined military and other experts in energy security for our first discussion of oil-supply concerns. I laid out my fears for a global supply shortfall as I held them at the time. As I write, global oil production is descending from its highest-ever level at nearly 97 million barrels. Is that the global peak? We will only know in the rear-view mirror, after the passage of more time.

The UK Industry Taskforce on Peak Oil and Energy Security that I convened in 2006 concluded in 2010 that the global peak of oil production would happen in 2015, at around 92 million barrels a day. Maybe we got the year right, but we underestimated the amplitude: like so many others, including in the oil industry, we did not foresee the extent to which the US industry would be able to keep lifting shale oil production by feasting on junk debt. We also underestimated erosion of demand around the world.

There is still cause for concern about supply though. The protracted low oil price since late 2014 has cut the number of active drill rigs deeply. And if you are drilling less and less, at some point production has to drop. Improvements in the efficiency of production processes, such as those achieved by American shale drillers, can only go so far. If that 97 million barrels a day in August 2015 was the global peak, the fall since can still slide into the the accelerating collapse that I and others have worried about so much – not least the International Energy Agency, until their about-face in 2012. Then the price would go right back up to levels that threaten economies with recession and worse. This volatility-potential of oil is another huge incentive to accelerate clean energy.

What to expect with oil supply, in the transition? I have to admit I am unsure now, beyond predicting more bankruptcies among US oil and gas companies. Currently 40 have gone under. The oil price stands below $40 as I write, with many pundits forecasting further descent and a long stint in the doldrums.

In January 2014, in the UN, we took a close look at the gathering wave of investor concern about climate change: 500 investors convened by Ceres in one hall, representing $22 trillion worth of funds under management, all there to discuss the risks inherent in fossil fuel dependency. Even then, with all that money in one room, there was little to indicate the acceleration of concern that was to come, especially in the months immediately before Paris. Neither was there at the 2014 World Economic Forum meeting in Davos shortly thereafter. Climate change and stranded assets figured very low on the agenda. Few talked of a great global energy transition. We can expect a very different agenda at the UN and in Davos in January 2016. Many companies, I predict, will leave Davos desperate to draw up new plans based on the premise that radical disruption is now inevitably descending on the energy sector, with all the knock-on implications that will have for every other sector.

In February 2014, in rural Kenya, we toured the front lines of oil eradication in African lighting, with a SolarAid field team. We saw that it is not difficult at all to replace a whole sector of fossil fuel use at the bottom end of the energy ladder. As significant additional capital flows into Africa for clean-energy development, consistent with announcements in Paris, we can expect rising appreciation of this fact in 2016 and beyond, and acceleration of kerosene phase-out.

The giant consultancy companies will no doubt be at the forefront of transition management, and in April 2014 we had dinner with Accenture's energy team as they endeavoured to peer into the crystal ball. We found them more than ready to contemplate changes to their business model, in tune with transitioning client companies. All those present worked on oil, gas and utilities at that time. By the end of 2016 we can expect a very different picture.

In May 2014, in Stockholm, we encountered for the first time European utility bosses accepting of the need for transition, and not denying that some of their number will be bankrupted by the rise of renewables. We pondered the cultural barriers they will face, making such changes. We can expect expediency increasingly to overcome culture in 2016.

We witnessed the dysfunctional culture in the oil industry coming into the open that month, as ExxonMobil and Shell presented their shareholders with the company's arguments against Carbon Tracker's analysis of wasted capex potential. Their case for business as usual, and their assumption of zero risk of stranded assets, did not play well with investors, even then, a year and half before Paris. We can expect to see the oil giants facing a wall of investor pressure in the post-Paris world.

In September 2014, it became clear that some oil companies were already mulling the process of transition themselves. Contingency planning, of course, is a long way from action. But at this point in our journey another big change was brewing. Having been above $100 for years, the oil price was starting to fall.

It was an interesting time for a first visit to the Kingdom of Saudi Arabia, where we discovered a solar industry-in-waiting even in the nation with the world's biggest oil reserves. If the Saudi government were to press the green light on the use of solar to replace oil burning in power plants – which it has said it will – many companies stand ready to execute, both domestic and foreign. We can surely expect the government finally to press that green light in 2016.

In late September 2014, back in New York again, we attended the launch of Carbon Tracker's coal report. Coal could well be a sinking ship, the authors concluded. We can expect their conclusion to become crystal clear, on a global basis, by the end of 2016.

In October 2014, we visited Germany's Fraunhofer Institute, a centre of excellence on the front lines of the solar revolution. We saw how Germany's best scientists have fed the technology needs of the national energy transition, or "energiewende", for years now. We can expect labs like this, and the US National Renewable Energy Lab, to come into their own as R&D funding becomes available in volume off the back of the various governmental and industry solar innovation initiatives announced in Paris. This will be a vital theme in transition. We could provide 100% of the energy for modern economies with existing technologies if we wanted to, so many renewables advocates believe. But we won't have to use only existing technology. There will be further innovation. And when it comes to the negative emissions needed for a 1.5°C target cap on global-warming, there will certainly need to be.

In late October 2014, we visited beleaguered Occupy Democracy students protesting about austerity outside the British parliament. For them climate change is about more than regulation of emissions. It is about the broken form of capitalism that drives fossil fuel profligacy. Little did they suspect that the leader of the Catholic church would come out on their side of this argument less than eight months later. We can expect more speaking out by religious leaders in 2016. We can expect their views to change the minds of others, in a world where believers are measured in billions. We can expect more students and young people camped out opposite parliaments, in greater numbers even than in 2015. Companies, even if they themselves are changing in the post-Paris zeitgeist, will need to keep in mind that they will be drawing their customers and workforces from this demographic pool. These are the people who are going to have to live with the some of the worst impacts of climate change, and execute the endgame of the global energy transition. Increasingly, they know it.

In December 2014, in the City of London, we attended a conference of conservatives mulling climate change on the day the Bank of England let it be known it was investigating whether or not fossil fuels pose a risk to the global capital markets, and E.ON announced a U-turn in its business model, becoming the first Big Energy company to officially begin a total retreat from fossil fuels. We saw how the conservative mind tends to be so closed on climate change that even seismic developments like these two do not rock mental boats. This conservative tendency to denial will surely continue to be a major barrier to progress on climate change, especially in the US and UK. But we can probably expect a rising drumbeat of developments as big as the two on 1st December 2014 to begin eroding it in 2016. The US Presidential election in November will be a major litmus test. Climate change is sure to figure strongly.

A year before Paris, in December 2014, we attended the annual climate summit in Lima. There we saw the US and China working together for the first time, having engaged in secret bilateral talks for months in advance. What might be possible, in 2016 and beyond, now that the two giants have seen how effective bilateral co-operation can be in bringing the world with them on an issue of global security? It is an open question, given how intractable some of their differences seem to be on other issues. But Paris has done nothing if not raise hopes for the future in this regard

In January 2015, we visited Total, the French oil and gas giant, in its Paris headquarters. We asked the question of how long it will take for the first oil major to follow the utility giants and effect a business-plan U-turn. Total and Statoil would appear to be the most likely candidates, then and now. I speculated at the time that it would take three years for the first oil giant to turn. I think the outcome of the Paris climate summit has probably shortened that timeframe. The increased difficult of lining up shareholder support for new big-capex projects is likely to be the deciding factor.

We also attended the World Future Energy Summit in Abu Dhabi, and witnessed how serious that government is about renewables. The scale of their support, and the tenor of their PR at that summit, seemed counterintuitive for a state built on oil and gas wealth. But in that spectacle, with the benefit of hindsight, we can sense how it would subsequently prove possible for Gulf nations to go along with the Paris Agreement, in the endgame. We met the Saudi Arabian climate envoy, and we saw how keen he was for the Kingdom not to be seen as obstructive. He, like the executives we met in Total, genuinely does not believe 100% renewables by 2050 is feasible. Let us see how the march of events in 2016 influences opinions like his. I predict a lot of mind-changing.

February 2015 provided a perfect example of why this might be the case. We learned then that Apple is intent on mass producing solar-charged electric vehicles by 2020. How many announcements like that would it take before disbelievers in a renewable-powered future come to re-examine their premises?

In early March 2015, we met with the BBC's environment correspondent and learned that the energy transition is not covered on TV as much as he would like because of a dearth of moving pictures. Today, the success in Paris should have painted a label of "unavoidable" on the issue. Climate change coverage in the media can be expected to rise. Increased coverage of both the energy transition and climate change will make it more difficult for the oil-and-gas establishment to keep pumping out the messages it needs to convince the world of, if it is to keep a business-as-usual licence to operate.

In mid March 2015, at a conference in Bloomberg's HQ, we heard incumbency captains in industry and government on the offensive, professing that those who oppose the fracking of shale are "misguided" and "naïve". They seemed blind to, or negligent of, the growing problems with junk debt, pollution, and methane leakage in American shale fracking: the model they seek to import to the UK. But then so many people seem to be unaware of these problems with fracking. First, they rarely see the emerging problems in the news. Second, they are often simply too busy with their day jobs to follow play. In April 2015, at a conference in PWC, we saw how even the solar industry's practitioners are ill-informed about the what the oil and gas industry is up to. We can expect this to change, too, in the post-Paris world. One reason will be the increasing emergence of confident advocates on the side of clean energy, on the offensive themselves.

In early May 2015, we watched one such, Elon Musk, launch Tesla Energy online. He was not slow to point out the downsides of fossil fuels to his whooping supporters. A week later, with $800 million in indicative orders for Tesla's batteries – for buildings and industrial sites, not electric vehicles – he can have had few regrets about picking an enemy so visibly. The energy incumbency can expect more of this kind of confrontational marketing in 2016.

In late May 2015, we sat in an expert panel at work in Rome, one of many plugging perspectives into the Vatican as the Pontifical Council for Justice and

Peace drafted the Papal encyclical on climate change. We were watching two very different constituencies at work on climate change in these May scenes. The Pope and his advisors went to great lengths to bullet-proof the advice they drew on in the encyclical, before it went out to the world's billion-plus Catholics. Elon Musk and his innovative colleagues would have prepared no less hard before unveiling their batteries at the launch of Tesla Energy. The devotees of tech around the world number no fewer than Catholics, I imagine. Each community holds the potential to be a powerful voice for climate action. The two voices together are so much bigger than the sum of the parts. And of course, these are just two of the constituencies from civil society that were so vocal in Paris. We can expect many more synergies of this kind, planned or inadvertent, in 2016 and beyond.

In June 2015, we saw in Bonn that synergy among sub-national constituencies was already at impressive levels, in what the UN calls the "groundswell". We also went to New Orleans to attend a solar conference as big as an oil and gas junket, with a thousand presenters. There we could sense that the solar revolution has a human-resource engine room, one that is priming the industry for the accelerated growth it will need to achieve now that Paris has succeeded. Match the global groundswell with the solar talent pool, add the battery link to motive power and the ethical arguments of leaders of world religions and you would have had a potent recipe for transition even if Paris had failed. And it didn't fail.

In late June 2015 we went to a Formula E grand prix race in Battersea Park. We heard engineers enthusing about the pace of innovation compared to Formula 1, especially in battery technology, with expectations of doubled battery power within five years. We heard Formula E boss Alejandro Agag express hope that in 5 years the world will see Formula E teams from Apple and Tesla alongside those from Virgin, Renault and the rest racing that day. But now, after Paris, why would they wait five years?

In July 2015, we visited Solarcentury in London and SolarAid in Nairobi to take a dive into the world of solar innovation. We met people who are driving the kind of innovation that will make the electricity flowing into the batteries of Tesla, Formula E, and the rest ever cheaper, and ever closer to zero emissions. We also encountered some of the problems faced by organisations endeavouring to be innovative, on the solar front lines. Fast growth and clever innovation does not equal certainty of commercial success, especially in a world where goalposts of all kinds are being moved by those working against transition. SolarAid and our retail arm SunnyMoney were in dire trouble with cash in July. I was successful in raising most of the hybrid grant / loan capital I needed to plug the gap for a while, and other supporters kindly donated to help out. As I write, we live to regroup and fight on in 2016. But in a war, it is important to remember that a victorious side can lose platoons and even regiments and still win. Over the long course of the carbon war many solar companies have gone under, but the global solar market has expanded apace. Let us hope SolarAid, with its unique charitable model, can find a way to continue helping out materially at the bottom of the pyramid, where more conventional organisations might fear to tread.

In August 2015, we had tea in London with a conscience-stricken executive from a fossil fuel company. We found that he is contemplating flight from his role stoking global warming, on the wrong side of a rising tide of anguish in civil society. I made the point that I have been surprised, for a quarter of a century now, that I don't meet more people like him. I am going to stick my neck out and predict that this too will change in 2016. Too many people in the incumbency will be encountering too much grief at their breakfast tables, especially those with sons and daughters old enough to understand what the greenhouse effect is. Neither will many of them have the stomach, on their own account, to stay on the wrong side of history. The great global energy transition will consist of many rising tides, and one of them will be converts.

In September 2015, in Oslo, Bucharest and London, we witnessed different kinds of companies wrestling with the notion of profound energy transition, even ahead of a result from Paris. The entire executive team of utility Statkraft needed no persuading. Partners of an un-named accountancy-and-advisor firm reviewed opportunities and options at a strategy retreat. Boutique investment bank GMP Securities enthused about opportunities in Africa. Now that we do have a result from Paris, we can expect an explosion of such companies acting in anticipation of transition in 2016.

In late September 2015, we listened to Pope Francis deliver his historic speech to the combined houses of the US Congress. We marvelled at his diplomatic yet clear messaging. We noted the traction he is achieving with those messages, even in conservative quarters. Along the way in 2014, we also noted leadership initiatives from other religions, including Islamic leaders. In 2016, we can expect the ethical arguments of religious leaders to add considerable momentum to the direction of travel.

In October 2015, we went to California in search of clues for the energy transition as Silicon Valley is pursuing it. On this trip we saw that the astonishing advances of the information revolution have barely begun to overlay on the energy transition. If someone were to ask me to pick just one leg of the journey between May 2013 and November 2015 to justify why I am cautiously optimistic about the post-Paris world, it is this one.

And so, at the end of November 2015, we arrived in Paris. And the rest is history.

Now, with 2015 coming to a close, I contemplate 2016 beside my fireless hearth, trying not to be too discombobulated by the weird weather.

My first two invitations for the year ahead are from Saudi Arabia and China. The Saudi national business daily Al-Eqtissadiah wants me to write an op-ed on the kingdom's solar future post-Paris. A Chinese solar manufacturer wants me to come and present their board with my bullish view of what the post-Paris world involves for their company. Two better signposts to a world potentially shifting on its axis it would be hard for someone like me to be handed. And I am confident that very many people like me, in many countries, will be receiving invitations akin to these in the opening weeks of 2016.

I have called this book *The Winning of the Carbon War*. So has it indeed been won then?

I think not. My first two invitations are no more than potential signs, at the level of one foot soldier, of a wider winning trajectory. I chose the wording of the book title carefully when I started publishing monthly instalments of my chronicles from the front lines in March 2015. My argument is that we turned a corner in the first half of 2013, and started the process of winning. The success in Paris does not change this. It marks, to coin a much-used phrase of Winston Churchill's, "the end of the beginning". It is not a definite "beginning of the end". It makes an ultimate win more likely of course, but the run of geopolitical events could yet derail and reverse the process of winning. Anyone who reads a newspaper regularly can come up with a derailment scenario or six. Their list might begin with America electing Donald Trump in November 2016.

But let us set the potential for derailment aside for a moment. 195 governments agreed in Paris to bat on the same side, to a meaningful degree, at least on this one issue, and very many sectors of civil society elected to join forces with them. As I have argued, this is a milestone without precedent in human history. Many would say, on this basis, that the notion of a carbon war is not even applicable any more: that we are all, at some level, on the same side now. That might be prove to be a happy possibility. But much depends on whether key players in the incumbency elect to switch sides, to retire from the field, or to dig in and "fight ugly", as that charming fossilista lobbyist put it in October 2014 (Chapter 13). Some of the dirtiest fighting in wars seems to happen towards the end. Think of Berlin and Okinawa in 1945. Sadly, my instinct tells me this is the way some of the high priests of the incumbency will elect to go. There is just too much machismo enculturated in their belief system. I hope I am wrong.

But even if ExxonMobil and the rest elect to go down fighting that way, their cause is likely to be forlorn now, barring external events of the derailment kind. That is how I choose to interpret the history I describe in this book. The harder they fight now, the more their executives are likely to face jail time in their dotage, imposed by a society looking back in anger.

And if the incumbency's case is indeed as forlorn as I suggest, absent derailment scenarios? Then what?

Well, we might find that peace breaks out in interesting ways. Seven decades after Berlin and Okinawa, some of my closest collaborators are German and my wife is Japanese. I bet my dad would never have imagined that in a million years, flying his planes in 1944. He is still alive to wonder at the spectacle.

Could we realistically dream of a happy ending, then, once and if the carbon war is finally won?

I am consumed by this question now, as I endeavour to peer into the future. Over the Christmas break I have been thinking about the road beyond the carbon war in the context of human history, or more exactly the history of the species that has come to dominate the planet – ours, *Homo sapiens* – since all other species of genus *Homo* became consigned to the fossil record.

Our ability to use fictive language, starting around 70,000 years ago, had much to do with our success. Our sole dominion began around 13,000 years ago. We were hunter-gatherers then. Around 12,000 years ago we discovered agriculture and began to settle down in communities. Since then, to simplify the narrative, we have tended to come together more and more: to become ever more aligned and unified as the millenia have passed, notwithstanding all the gruesome violence along the way. There seems to be a direction to our history.

In his book *Sapiens*, Yuval Noah Harari argues that the single biggest factor in the success of our species has been an ability to dream up intangible belief systems and organise around them. Hence kingdoms, the first of which appeared around 5,000 years ago; empires, the first of which appeared around 4,250 years ago; and other such constructs including religions, limited liability companies, and so on. He posits three main reasons, involving these belief systems, for the unification tendency in our history: the creation of empires, the use of money, and the emergence of monotheistic religions. Most *sapiens* to have lived so far have done so under the rule of empires, and empires have tended to merge peoples under a common belief system. Money has facilitated trade, and the spread of trading has tended to merge peoples under a common belief system too: most recently it has been called capitalism. As for religions, the emergence of monotheistic faiths took belief in deities global. The polytheistic religions of earlier times tended to focus on deities more pertinent to the local than the global.

Let me take the overall coming together megatrend as a premise, then, and overlay that with the recent discoveries of neuroscientists. I precis those in Chapter Seven. The discoveries of theirs that intrigue me most are our inate tendency to favour community, and (believe it or not) our tendency to abhor violence (most of us, that is), notwithstanding what the neuroscientists refer to as our "predictable irrationality" in individual and collective thought.

This pooling of observations about human history and neuroscience might incline a futurologist to cautious optimism, I suggest, and the optimism might be augmented if we add to our recipe the notion that the digital age has tended to foster empathy on a global basis. The idea here, simplifying matters again, is as follows. Consider the real-world example of footage recorded on a mobile phone, but screened on American television, showing a young Iranian woman, dressed in jeans, being shot and killed while protesting in her homeland. Is this more or less inclined to make young Americans swallow Fox News-type propaganda suggesting that Iran is a nation composed entirely of fanatically religious mullahs and their equally fanatical acolytes? If the answer is "less inclined", might that make it more or less difficult to persuade young Americans to go to war against Iran? The answer is yes, because modern communications have encouraged young Americans to feel a degree of empathy for young Iranians.

This is but one example of an impressive body of evidence for a general trend of rising empathy in the world over time. Jeremy Rifkin, in his book on the subject, goes so far as to envision "The Empathic Civilisation".

Let us now view Paris through these rosy spectacles. The agreement achieved there shows what we can do, when we pull together: all 195 of our governments in harness with so many civil-society actors, of such great variety. There can be little doubt that the collective success will encourage governments to attempt other acts of multilateral gymnastics. It will certainly send many sectors of civil society into 2016 with the wind of social good strong in their sails.

Can all this amount to a direction of travel for *sapiens*, in the post-Paris world, wherein we might be able to organise ourselves well enough – drawing on our tendency to empathy and community, and our general abhorrence of violence – to take steps back from the brink of many of our worst problems?

Might it even amount to a road to renaissance in the state of humankind?

Perhaps I should end my book with that question. But at one level the answer to it is obvious: this outcome is very far from guaranteed. So stopping without a few more paragraphs is not an option.

A world well on the road to being run 100% on renewable energy, and all the other social benefits a global action plan consistent with the Paris Agreement would bring, can still find itself sliding into darkness scenarios, all too easily. The list of reasons why might begin with the possibility that global warming will prove to be worse than the best estimates of scientists today. We are in a race against time, in this respect. Simply stated, we must decarbonise fast enough to stop the impacts of global warming running away from us. Many climate scientists tell us there is a point of no return, beyond which the greenhouse-gas emissions man-made warming has stimulated from the natural environment – from melting permafrost, destabilising methane hydrates, warming oceans, burning forests and many other potential natural amplifiers of warming – mean it no longer matters that we are on our way to zero anthropogenic emissions. Global warming would keep right on going.

Even short of this apocalyptic scenario, the emerging impacts of global warming might destroy our social cohesion, over-running any survival instinct. Searing heat across the Middle East, for example, might drive the already-considerable flow of refugees ever higher, translating into a rise of support for authoritative regimes in Europe and North America.

At which point another potential darkness scenario could kick in. The same sophisticated use of the internet-of-things that holds such potential to amplify the power of clean energy also potentially creates the perfect apparatus for police states. Think of drones, sensors, facial-recognition software, mega-data processing power, and let your imagination do the rest, should things go wrong at the ballot box. We should never forget that Adolf Hitler was elected. In this scenario, democracy itself comes under threat, with the real prospect of no return ticket.

I could continue, but I think I have made my point. The carbon war is not won, and even if it is, it may simply be replaced by the next great struggle of human belief systems.

But one thing is clear to me. What happened in Paris in December 2015, and the run-up to it from 2013, keeps the dream of human renaissance alive. I do not think it would still be breathing, had Paris failed to send a strong signal.

Perhaps that signal can be a trigger for a wider survival instinct in our species. I hope my book has done justice to that candle for hope in our world, as I think I have seen it flicker into life between 2013 and 2015.

Index

B

C

D

E

F

I

J

K

L

M

N

O

P

Q

R

S

T

U

V

Lightning Source UK Ltd.
Milton Keynes UK
UKOW06f0227080416

271802UK00002B/100/P

9 781364 429492